The Superalloys

Superalloys are unique high temperature materials used in gas turbine engines, which display excellent resistance to mechanical and chemical degradation. This book presents the underlying metallurgical principles which have guided their development and practical aspects of component design and fabrication from an engineering standpoint. The topics of alloy design, process development, component engineering, lifetime estimation and materials behaviour are described, with emphasis on critical components such as turbine blades and discs.

The first introductory text on this class of materials, it will provide a strong grounding for those studying physical metallurgy at the advanced level, as well as practising engineers. Included at the end of each chapter are exercises designed to test the reader's understanding of the underlying principles presented. Additional resources for this title are available at www.cambridge.org/9780521859042.

ROGER C. REED is Professor and Chair in Materials Science and Engineering at Imperial College London. From 1994 to 2002, he was Assistant Director of Research in the Rolls-Royce University Technology Centre at the University of Cambridge. From 2002 to 2005, he held a Canada Research Chair in the Department of Materials Engineering at the University of British Columbia, Vancouver. He is widely known in the gas turbine community for his work on the physical metallurgy of the superalloys, and has taught extensively in this field.

The Superalloys Fundamentals and Applications

Roger C. Reed

CAMBRIDGE
UNIVERSITY PRESS

CAMBRIDGE UNIVERSITY PRESS
Cambridge, New York, Melbourne, Madrid, Cape Town, Singapore, São Paulo

Cambridge University Press
The Edinburgh Building, Cambridge CB2 8RU, UK

Published in the United States of America by Cambridge University Press, New York

www.cambridge.org
Information on this title: www.cambridge.org/9780521859042

First published 2006
This digitally printed version 2008

A catalogue record for this publication is available from the British Library

ISBN 978-0-521-85904-2 hardback
ISBN 978-0-521-07011-9 paperback

But if I were to say, my fellow citizens, that we shall send a rocket to the moon, 240 000 miles away from the control station in Houston, a giant rocket more than three hundred feet tall, the length of this football field, made of new metal alloys, some of which have not yet been invented, capable of standing heat and stresses several times more than have ever been experienced, fitted together with a precision better than the finest watch, carrying all the equipment needed for propulsion, guidance, control, communications, food and survival, on an untried mission, to an unknown celestial body, and then return it safely to Earth, re-entering the atmosphere at speeds of over 25 000 miles per hour, causing heat about half that of the temperature of the sun, almost as hot as it is here today, and do all this, and do it right, and do it first before this decade is out, then we must be bold . . .

From John F. Kennedy's speech at Rice University, Houston, Texas, 12 September 1962

Contents

Contents

Foreword

I am grateful to Cambridge University Press for allowing me to make some comments about Roger Reed's new textbook *The Superalloys: Fundamentals and Applications*. Nickel-based superalloys represent a very important class of engineering material, finding widespread application for example in critical components within the gas turbine engines used for jet propulsion and electricity generation. This is due to their superior mechanical properties that are maintained to elevated temperatures. Indeed, new classes of superalloy are continually being sought by gas turbine manufacturers around for the world for applications in the hottest parts of the engine. This is because higher temperatures result in improvements to the efficiency of the engine and therefore lower fuel burn. Engine performance is a major factor in any power plant competition, which helps to explain why all the engine manufacturers spend so much money developing future generations of superalloys.

The author has provided us with a textbook covering both the fundamentals and applications of superalloy technology. This is a significant and unique achievement, especially given the broad range of subject matter dealt with. In Chapter 1, the requirement for materials capable of operating at elevated temperatures is introduced along with the historical development of the nickel-based superalloys and their emergence as materials for high-temperature applications. Chapter 2 concerns the physical metallurgy of the superalloys, with an emphasis on the details which distinguish them from other classes of engineering alloys, for example the gamma prime strengthening phase, the role of defects such as the anti-phase boundary, the unique particle-strengthening mechanisms, the anomalous yield effect and the creep deformation behaviour. Chapters 3 and 4 deal with superalloy technology as applied to two major gas turbine components of critical importance: the turbine blade and discs. Of note is the balanced coverage given to the processing, alloy design and microstructure/property relationships relevant to superalloys used for these distinct applications. Chapter 5 deals with surface coatings technologies, which are becoming increasingly critical as operating temperatures continue to rise. In Chapter 6, projections are made about the future of superalloy technology.

In view of the author's exemplary treatment of the subject, this tectbook deserves to become the definitive textbook in the field for the foreseeable future. I have found myself referring to it on many occasions already! It is recommended to all those with an interest in the field of high-temperature materials, particularly those involved with gas turbine technology or those embarking on a higher degree in the subject. Also, since the superalloys represent a considerable success story in the field of materials science and engineering, it is recommended for use within the materials-related curricula at universities.

Dr Mike Hicks
Chief Technologist – Materials
Rolls-Royce plc

Preface

Based upon nickel, but containing significant amounts of at least ten other elements including chromium and aluminium, the superalloys are high-temperature materials which display excellent resistance to mechanical and chemical degradation at temperatures close to their melting points. Since they first emerged in the 1950s, these alloys have had a unique impact. Consider the aeroengines which power the modern civil aircraft. The superalloys are employed in the very hottest sections of the turbines, under the heaviest of loads, with the utmost importance placed on assuring the integrity of the components fabricated from them. Indeed, the development of the superalloys has been intrinsically linked to the history of the jet engine for which they were designed; quite simply, a modern jet aeroplane could not fly without them. Further improvements in temperature capability are now being actively sought, for example for the engines to power the two-decked Airbus A380 and the Boeing 787 Dreamliner. Superalloys are being employed increasingly in the land-based turbine systems used for generating electricity, since fuel economy is improved and carbon emissions are reduced by the higher operating conditions so afforded. But new developments in superalloy metallurgy are required for the next generation of ultra-efficient power generation systems. Over the next 25 years, the world's installed power generation capacity is expected to double, due to the rapidly growing economies and populations of the developing countries, and because most of the current plant in the developed countries will need to be replaced. Thus the superalloys have never been more important to the world's prosperity.

The remarkable performance of today's superalloys is not merely fortuitous. Numerous researchers and technologists have worked to develop the basic understanding of their physical behaviour and the more practical aspects required to put these alloys to best use. With this book, the reader is presented with an introduction to the metallurgical principles which have guided their development. It turns out that the topics of alloy design, process development, component engineering, lifetime estimation and materials behaviour are very closely inter-related. The book is aimed at those pursuing degrees in Materials Science and Engineering, but those studying Mechanical Engineering, Aerospace Engineering, Physics and Chemistry will also find it useful. It has been developed with two audiences in mind: first, final-year undergraduate students taking an elective course in high-temperature materials technology, and, secondly, students embarking on a higher degree in this field who seek an introduction to the subject. Included at the end of the first five chapters are questions designed to test the reader. Many of these require numerical working using a calculator, spreadsheet or computer programming tool. Most of these exercises have been tested on students at Imperial College, the University of Cambridge and the University of British Columbia, where I have held teaching positions. I would like to receive comments from those who have attempted the problems – I will attempt to answer all queries as promptly as I can. My email address can be found at www.cambridge.org/9780521859042.

In preparing the book, the reader should be aware that it has been necessary to be selective about the content included in it. I have included sufficient material to satisfy the requirements of a one-course semester – about 30 hour-long periods, some of which should be used for classwork, problem sets and design exercises. In my own teaching I have used the book's contents to emphasise the relationships between the structure/composition of these materials, their mechanical/chemical behaviour, the processing of them and the design of the components which have provided the technological impetus for their development. Thus, a proper balance between these different topics has been sought; the need for this makes the superalloys an excellent case study in the field of materials.

Acknowledgements

It is a great pleasure to acknowledge the considerable number of friends and colleagues who have generously provided information during the writing of this book. The following deserve special mention: Dr Cathie Rae, Dr Sammy Tin, Dr David Knowles, Professor Harry Bhadeshia of the University of Cambridge; Professor Malcolm McLean, Dr Barbara Shollock, Professor Peter Lee and Dr David Dye of Imperial College, London; Professor John Knott of the University of Birmingham; Dr Nirundorn Matan of Walaikak University; Dr Philippa Reed of the University of Southampton; Professor Tresa Pollock of the University of Michigan; Professor Alan Ardell of UCLA; Professor Kevin Hemker of Johns Hopkins University; Professor Hael Mughrabi of the University of Erlangen; Professor Brian Gleeson of Iowa State University; Professor Mike Mills of The Ohio State University; Dr Vladimir Tolpygo of the University of California at Santa Barbara; Professor Alec Mitchell, Dr Steve Cockcroft and Dr Ainul Akhtar from the University of British Columbia; Dr Mike Winstone of DSTL Laboratories, Farnborough, UK; Dr Hiroshi Harada, Dr Hiro Hono, Dr Seiji Kuroda, Dr Hongbo Guo and Mr Atsushi Sato of the National Institute for Materials Science, Japan; Dr Bruce Pint, Dr Suresh Babu and Dr Mike Miller of Oak Ridge National Laboratory, USA; Dr Jim Smialek and Dr Tim Gabb of NASA; Dr Pierre Caron from ONERA, France; Dr Jeff Brooks of Qinetiq, UK; and Dr Dan Miracle of the Wright-Patterson Air Force Laboratory, USA. I would like to thank Dr Bob Broomfield of Rolls-Royce plc, who was particularly helpful during the period 1994 to 2002 whilst I was working at the University of Cambridge. Neil Jones, Jim Wickerson, Dr Andrew Manning, Dr Henri Winand, Colin Small, Dr Duncan MacLachlan, Paul Spilling, Dr Steve Mckenzie, Steve Williams and Dr Ken Green of Rolls-Royce also provided information very graciously. More recently, very helpful advice has been received from Dr Scott Walston of GE Aircraft Engines, Dr Ken Harris and Dr Jackie Wahl of Cannon-Muskegon Corporation, Dr Allister James of Siemens Westinghouse, and Dr Gern Maurer and Dr Tony Banik of Special Metals Corporation. Last but not least, thanks to Joanna Copeman of University College London for her love and support.

1 Introduction

1.1 Background: materials for high-temperature applications

1.1.1 Characteristics of high-temperature materials

Certain classes of material possess a remarkable ability to maintain their properties at elevated temperatures. These are the *high-temperature materials*. Their uses are many and varied, but good examples include the components for turbines, rockets and heat exchangers. For these applications, the performance characteristics are limited by the operating conditions which can be tolerated by the materials used. For example, the thrust and fuel economy displayed by the modern aeroengine is strongly dependent upon, and limited by, the high-temperature strength of the nickel-based superalloys used for its hottest sections.

What are the desirable characteristics of a high-temperature material? The first is *an ability to withstand loading at an operating temperature close to its melting point*. If the operating temperature is denoted T_{oper} and the melting point T_m, a criterion based upon the homologous temperature τ defined as T_{oper}/T_m is sensible; this should be greater than about 0.6. Thus, a superalloy operating at 1000 °C in the vicinity of the melting temperature of nickel, 1455 °C, working at a τ of $(1000 + 273)/(1455 + 273) \sim 0.75$, is classified as a high-temperature material. But so is ice moving in a glacier field at -10 °C, since τ is $263/273 \sim 0.96$, although its temperature is substantially lower. A second characteristic is *a substantial resistance to mechanical degradation over extended periods of time*. For high-temperature applications, a time-dependent, inelastic and irrecoverable deformation known as *creep* must be considered – due to the promotion of thermally activated processes at high τ. Thus, as time increases, creep strain, ϵ_{creep}, is accumulated; for most applications, materials with low rates of creep accumulation, $\dot{\epsilon}_{creep}$, are desirable. In common with other structural materials, the *static* properties of yield stress, ultimate tensile strength and fracture toughness are also important – and these must be maintained over time. A final characteristic is *tolerance of severe operating environments*. For example, the hot gases generated in a coal-fired electricity-generating turbine are highly corrosive due to the high sulphur levels in the charge. Kerosene used for aeroengine fuel tends to be cleaner, but corrosion due to impurities such as potassium salts and the ingestion of sea-water can occur during operation. In these cases, the high operating temperatures enhance the possibility of oxidation. Under such conditions, any surface degradation reduces component life.

1

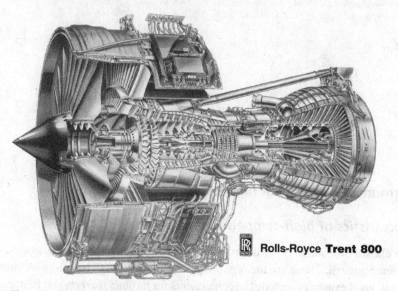

Fig. 1.1. Artist's impression of the turbomachinery in Rolls-Royce's Trent 800 engine, which powers the Boeing 777 aircraft. (Courtesy of Rolls-Royce.)

1.1.2 The superalloys as high-temperature materials

When significant resistance to loading under static, fatigue and creep conditions is required, the nickel-base superalloys [1–3] have emerged as the materials of choice for high-temperature applications. This is particularly true when operating temperatures are beyond about 800 °C. This is the case for gas turbines used for jet propulsion, for example, the 100 000 lb thrust engines used for the Rolls-Royce Trent 800 and General Electric GE90 which power the Boeing 777 [4] (see Figure 1.1), but also the smaller 1000 lb engines used for helicopter applications. Gas turbines are used also for electricity generation, for example, the 250 MW gas-fired industrial plant which can generate enough power to satisfy a large city of a million people (see Figure 1.2), or the smaller 3 MW gas-fired generators suitable for back-up facilities [5]. When weight is a consideration, titanium alloys are used, but their very poor oxidation resistance restricts their application to temperatures below about 700 °C [6]. For some electricity-generating power plant applications which rely upon super-heated steam at 565 °C, high-strength creep-resistant ferritic steels are preferred on account of their lower cost. However, the latest generation of ultra-supercritical steam-generating coal-fired power stations requires boiler tubing that can last up to 200 000 hours at 750 °C and 100 MPa – new types of superalloy are being developed for these applications [7], since ferritic steels cannot be designed to meet these property requirements. Generally speaking, ceramics such as silicon carbide and nitride are not used for these applications, despite their excellent oxidation and creep resistance, due to their poor toughness and ductility. Zirconia-based ceramics do, however, find applications in thermal barrier coatings which are used in association with the superalloys for high-temperature applications.

Since the technological development of the superalloys is linked inextricably to the gas turbine engine, it is instructive to consider the functions of its various components [8].

Fig. 1.2. Artist's impression of the turbomachinery in Siemens Westinghouse's W501F gas turbine engine, used for electricity generation. (Courtesy of Siemens Westinghouse.)

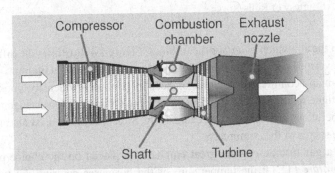

Fig. 1.3. Diagram illustrating the basic features of a very basic gas turbine engine: the turbojet. (Courtesy of Rolls-Royce.)

Consider the turbojet engine, Figure 1.3. The role of the *compressor*, consisting of compressor blades and discs, is to squeeze the incoming air, thus increasing its pressure. The compressed air enters the *combustor*, where it is mixed with fuel and ignited. The hot gases are allowed to expand through a *turbine*, which extracts the mechanical work required to drive the compressor; this necessitates a *shaft*, which transmits the torque required for this to happen. In the case of the turbojet, thrust arises from the momentum change associated with the incoming air being accelerated and its emergence as exhaust gas at a significantly higher velocity. Variants of this basic design are possible. In a turbofan engine, an additional low-pressure compressor, or *fan*, is added to the front of the engine. Although this necessitates an additional shaft and an associated turbine to drive it, the weight penalty associated with the extra turbomachinery is offset by the greater fuel economy – this arises from the

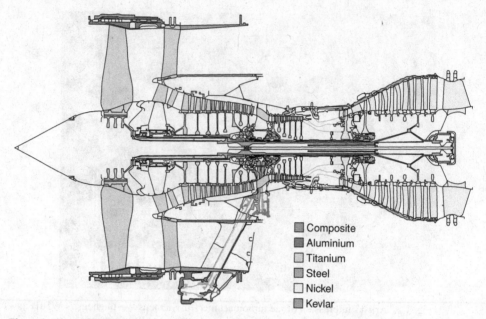

Composite
Aluminium
Titanium
Steel
Nickel
Kevlar

Fig. 1.4. Illustration of material usage in the Trent 800 engine. Note the extensive use of nickel-based superalloys in the combustor and turbine sections (see the back cover for a colour version of this figure). (Courtesy of Rolls-Royce.)

thrust provided from the air, which *bypasses* the turbines. Figure 1.4 illustrates the different materials used in the various parts of the Trent 800 aeroengine. One sees that titanium alloys are chosen for the fan and compressor sections, on account of their low density, good specific strength and fatigue resistance. However, in the combustor and turbine arrangements, the nickel-based superalloys are used almost exclusively. Superalloys are used also in the final (high-pressure) stages of the compressor.

When designing a gas turbine engine, great emphasis is placed on the choice of the *turbine entry temperature* (TET): the temperature of the hot gases entering the turbine arrangement [9]. There, the temperature falls as mechanical work is extracted from the gas stream; therefore the conditions at turbine entry can be considered to be the most demanding on the turbomachinery and the nickel-based superalloys from which are they made. As will be seen in Section 1.2, the performance of the engine is greatly improved if the TET can be raised – and over the 50 years since their conception, this has provided the incentive and technological impetus to enhance the temperature capability of the superalloys, and to improve their processing and the design of components fabricated from them. The success of these enterprises can be judged from the way in which the TET of the large civil aeroengine has increased since Whittle's first engine of 1940 (Figure 1.5); a 700 °C improvement in a 60-year period has been achieved [10]. Of course, the TET varies greatly during a typical flight cycle (see Figure 1.6), being largest during the take-off and climb to cruising altitude. Turbines for power-generating applications experience fewer start-up/power-down cycles but very much longer periods of operation, during which the TET tends to be rather constant.

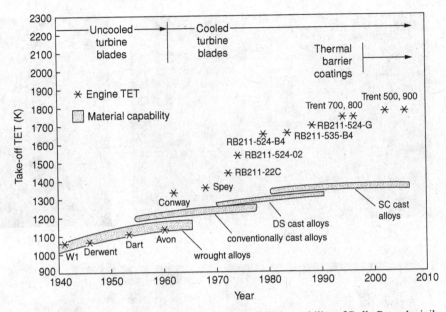

Fig. 1.5. Evolution of the turbine entry temperature (TET) capability of Rolls-Royce's civil aeroengines, from 1940 to the present day. Adapted from ref. [10].

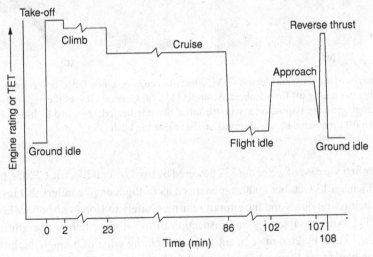

Fig. 1.6. Variation of the turbine entry temperature (TET) during a typical flight cycle of a civil aircraft.

1.1.3 Instances of superalloy component failures

The failure of components fabricated from the superalloys is a rare occurrence, since the applications demand very conservative designs and significant safety factors. Nevertheless, failures have occurred and undoubtedly there will be further ones in the future. Examination of some examples emphasises a number of important points.

Fig. 1.7. Images of the General Electric CFM56 turbofan engine, which failed in service in October 2000, shortly after taking off from Hobart, Australia [11]. (a) General view of the high-pressure turbine; (b) high-pressure turbine rotor with the failed blade indicated; (c) failed high-pressure turbine blade; (d) tip notch damage observed on the remaining blades.

Consider first the case of a Boeing 737 powered by two General Electric CFM56 turbofan engines, which, on 13 October 2000, experienced an in-flight engine failure that resulted in the engine being shut down and the aircraft returning safely to Hobart airport in Tasmania, Australia [11]. After disassembly and inspection of the engine, failure was attributed to the loss of a 15 mm by 20 mm segment from the trailing edge of a single high-pressure (HP) turbine blade (see Figure 1.7) fabricated from the Rene 125 alloy. This passed into the low-pressure turbine stages, where it caused overloading and collapse of the entire blade array. Metallographic examination of the HP blades indicated that radial cracks near the blade tips were common, as a result of severe thermal cycling, high thermal gradients and thermal fatigue. Many of the tips had been weld-repaired using the Rene 80 alloy. In the failed blade, tip cracks had grown into a v-shaped notch because of oxidation and corrosion effects, and had intercepted a deep underlying repair weld fabricated from the lower strength superalloy Inconel 625. On reaching the base of the repair weld, the resultant fatigue stresses were sufficient to propagate the cracking to the point of final failure. The failed

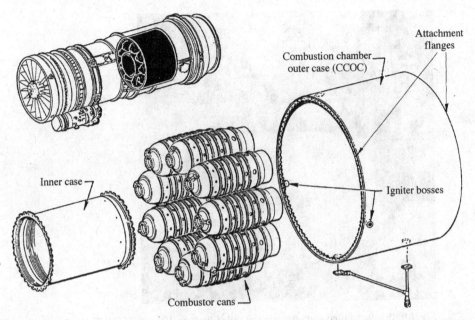

Fig. 1.8. Schematic diagram of the arrangement of the combustor in the Pratt & Whitney JT8D-15 engine [12].

blade had completed 17 928 flight cycles, and had flown 5332 cycles since its repair and overhaul.

A second example had very much more tragic consequences. On 22 August 1985, at Manchester Airport in the United Kingdom, a British Airways Boeing 737 carrying 131 passengers and 6 crew suffered an uncontained engine failure during take-off, which was therefore aborted by the crew [12]. Unfortunately, pieces from the port Pratt & Whitney JT8D-15 engine were ejected from the engine – and these punctured a fuel tank causing a catastrophic fire, in which 55 persons on board lost their lives. The aeroplane had not left the ground. The origin of the failure was a 360° separation of the No 9 combustor can (see Figure 1.8), which consisted of 11 pieces of the superalloy Hastelloy X in sheet form, welded together; this allowed hot gases to escape from it and impinge upon the inner surface of the combustion chamber outer case, which ruptured catastropically during take-off due to localised overheating, see Figure 1.9. In November 1983, after 3371 cycles, the No 9 can had been inspected and circumferential cracking of 180 mm combined length had been repaired, by fusion welding. Solutioning and weld stress-relief heat treatments had not been applied. It lasted a further 2036 cycles before failure.

In these instances, lessons should be learned from these tragic circumstances. The following should be clear to the reader. First, the modern jet engine relies very heavily upon the superalloys to withstand the significant loads and temperatures developed during operation. Secondly, where defects exist due to manufacturing or else due to damage accumulated during service, these can lead to catastrophic failure; hence, for these applications it is absolutely critical to ensure the components fabricated from the superalloys are of the highest

Fig. 1.9. View of failed combustor in the failed Pratt & Whitney JT8D-15 engine, August 1985, Manchester Airport, UK [12].

integrity and quality. Thirdly, procedures for inspection and repair are critical if ongoing operation is to remain safe. Finally, a close integration of procedures for component design, manufacturing and lifetime estimation is required in order to ensure the safe working conditions of components fabricated from the superalloys, and gas turbines in general.

1.2   The requirement: the gas turbine engine

The gas turbine used for jet propulsion and electricity generation is an example of a heat engine. The power output therefore depends upon the temperature increment through which the working fluid is raised – it is improved if this can be increased.

In order to reinforce this concept, consider Carnot's reversible heat engine, which consists of an ideal gas that cycles between hot and cold reservoirs of temperatures T_1 and T_2, respectively; see Figure 1.10(a). Once each cycle, an amount of heat, Q_{AB}, is absorbed from the hot reservoir and a quantity, Q_{CD}, is transferred to the cold one; a fraction, η, of the heat, Q_{AB}, absorbed from the hot reservoir is converted into useful work, W, such that the efficiency, η, equals W/Q_{AB} with conservation of energy requiring that $W = Q_{AB} - Q_{CD}$.

It is relatively easy to determine expressions for η and W. Writing the ideal gas equation as $pV = nRT$, where p is pressure, V is volume, T is temperature, R is the gas constant and n is the number of moles, one must consider the following four steps [13].

(1) The gas, in perfect thermal contact with the hot reservoir, undergoes an infinitely slow expansion from volume V_A to V_B. During this process, the work done by the gas, W_{AB},

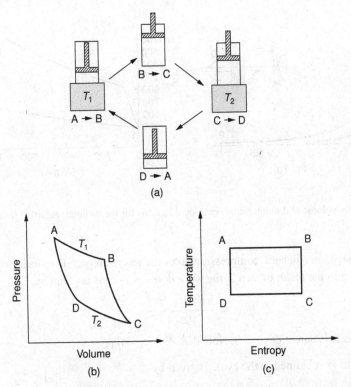

Fig. 1.10. The Carnot cycle for an ideal gas. (a) The four steps of the cycle: two isothermal and two adiabatic. (b) Corresponding pressure–volume diagram. (c) The associated temperature–entropy diagram. Adapted from ref. [13].

is given by

$$W_{AB} = \int_{V_A}^{V_B} p \, dV = \int_{V_A}^{V_B} \frac{nRT_1}{V} \, dV = nRT_1 \ln \left\{ \frac{V_B}{V_A} \right\} = Q_{AB} \qquad (1.1)$$

(2) A second step, in which thermal contact to the reservoir is broken and the gas is insulated from its surroundings: the gas undergoes an adiabatic expansion from state B to state C along the adiabat BC, such that $p_B V_B^\gamma = p_C V_C^\gamma$, where γ is the ratio of specific heats at constant pressure and volume. Since the gas is isolated, $Q_{BC} = 0$. The work done *by* the gas, W_{BC}, is given by

$$W_{BC} = \int_{V_B}^{V_C} p \, dV = \int_{V_B}^{V_C} \frac{p_B V_B^\gamma}{V^\gamma} \, dV = \frac{p_C V_C - p_B V_B}{1 - \gamma} = \frac{nR(T_1 - T_2)}{\gamma - 1} \qquad (1.2)$$

where T_1 and T_2 are the initial and final temperatures, respectively.

(3) A third step: the gas, in contact with the cold reservoir at temperature T_2, undergoes an isothermal compression from V_C to V_D. Work is done *on* the gas such that

$$W_{CD} = \int_{V_C}^{V_D} p \, dV = \int_{V_C}^{V_D} \frac{nRT_2}{V} \, dV = nRT_2 \ln \left\{ \frac{V_D}{V_C} \right\} = Q_{CD} \qquad (1.3)$$

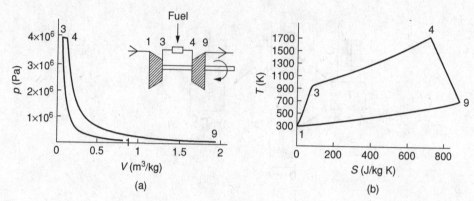

Fig. 1.11. Pressure–volume and temperature–entropy diagrams for the turbojet. Adapted from ref. [10].

(4) In the final step, an adiabatic compression takes the gas from state D to its initial state A. Appealing to the result of step 2, the work done *on* the gas is given by

$$W_{DA} = \frac{nR(T_2 - T_1)}{\gamma - 1} \tag{1.4}$$

Since the gas is isolated, $Q_{DA} = 0$. Note that $W_{DA} = -W_{BC}$.

The total work, W, obtained in the cycle is given by

$$W = W_{AB} + W_{BC} + W_{CD} + W_{DA} = Q_{AB} + Q_{CD} = nRT_1 \ln\left\{\frac{V_B}{V_A}\right\} - nRT_2 \ln\left\{\frac{V_C}{V_D}\right\} \tag{1.5}$$

and is equal to the area enclosed by the loop ABCD in Figure 1.10(b); this follows from a consideration of the four integrals, $\int p \, dV$, defining W. The efficiency, η, defined as the ratio of the useful work obtained, W, to the heat absorbed from the heat source, Q_{AB}, is given by

$$\eta = \frac{W}{Q_{AB}} = 1 - \frac{T_2 \ln\{V_C/V_D\}}{T_1 \ln\{V_B/V_A\}} = 1 - \frac{T_2}{T_1} \tag{1.6}$$

where use has been made of relationships such as $p_A V_A = p_B V_B$ and $p_B V_B^\gamma = p_C V_C^\gamma$, etc. It follows that if T_1 is raised, or T_2 is reduced, the work extracted from the cycle is increased and the efficiency improved. In practice, raising T_1 is the more practical option – with the limit for T_1 being the capability of the turbomachinery to withstand the high-temperatures and stresses involved. Figure 1.10(c) shows the temperature–entropy plot for the reversible Carnot cycle.

The assumptions made in the Carnot cycle make it only a first approximation for the gas turbine engine, particularly as additions of mass and heat are made to the working fluid via the burning of fuel. However, the fuel-to-air ratio is typically only 0.02–0.03 at maximum power and 0.01 under cruise conditions. Therefore, the pressure–volume cycle for a simple turbojet (see Figure 1.11(a)) exhibits many of the features expected from the Carnot cycle; note in particular the 40-fold increase in pressure. Figure 1.11(b) shows the corresponding temperature–entropy plot (the numbers 1, 3, 4 and 9 have been chosen for

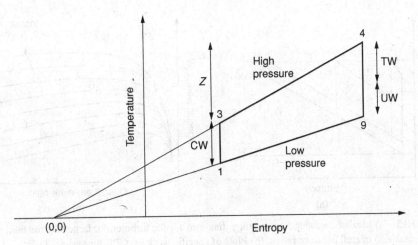

Fig. 1.12. Simplified temperature–entropy diagram for the turbojet, with compression and turbine stages assumed to be isentropic. CW = compression work; TW = turbine work; UW = useful work; Z = ideal work of combustion.

consistency with commonly used terminology in gas-turbine design). Note that neither the work of compression nor the work of expansion in the compressor are isentropic, implying irreversibility, and that the chemical energy introduced in the combustion chamber leads to a significant increase in temperature of about 800 K. These effects are not accounted for in the Carnot analysis.

To make approximate but quantitative calculations, one can assume that the compressor and turbine stages are isentropic, so that the lines 1–3 and 4–9 in Figure 1.11(b) are vertical. Furthermore, the lines 3–4 and 1–9 can be assumed to be straight but diverging; they must of course pass through the point $(0, 0)$ since at zero kelvin gases have no entropy. The situation is then as depicted in Figure 1.12. The distance 1–3 is proportional to the compression work (CW), and the distance 4–9 is proportional to the sum of the turbine work (TW) and the useful work (UW): the work of expansion (EW). The vertical distance between 3 and 4 measured along the temperature axis is proportional to Z, the rate of addition of chemical energy from the fuel, or the ideal work of combustion. Because the lines 3–4 and 1–9 diverge, and since for an ideal cycle CW = TW, the quantity UW is finite and positive; it can be used for propulsion or else to drive an external shaft for electricity generation.

Example question

A turbojet is operating with a TET of 1400 K, an overall pressure ratio (OPR) of 25 and an air-mass intake, $\dot{M}$, of 20 kg/s. Assuming an ideal cycle such that the compression work equals the turbine work, determine the rate of useful working (or specific power) available to provide a propulsive force. Hence estimate the thermal efficiency, η_{th}. Assume adiabatic, reversible processes where necessary. As a further exercise, estimate (i) the thrust, F_N, developed and (ii) the propulsive efficiency, η_{prop}, if the velocity of the intake air is 250 m/s. (Assume $c_p = 1000$ J/(kg K), $\gamma = 1.4$ and the intake air to have a temperature of 288 K and velocity V_0. Take atmospheric pressure to be 101 325 Pa.)

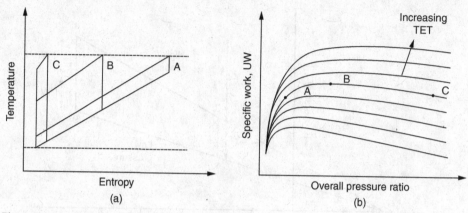

Fig. 1.13. (a) Idealised temperature–entropy diagrams for the turbojet, for large (C), medium (B) and small (A) overall pressure ratios. (b) Plots of specific work vs. OPR for various TETs.

Solution For an adiabatic, reversible process one has

$$\frac{T_3}{T_1} = \left(\frac{p_3}{p_1}\right)^{\frac{\gamma-1}{\gamma}}$$

and since the pressure ratio $p_3/p_1 = 25$, this yields $T_3 = 723\,\mathrm{K}$ if $T_1 = 288\,\mathrm{K}$. In a similar way, $T_9 = 558\,\mathrm{K}$ if $T_4 = 1400\,\mathrm{K}$. Then $Z = \dot{M}c_p(T_4 - T_3) = 13.6\,\mathrm{MW}$, CW = TW$=$ $\dot{M}c_p(T_3 - T_1) = 8.7\,\mathrm{MW}$ and EW $= \dot{M}c_p(T_4 - T_9) = 16.8\,\mathrm{MW}$. It follows that the useful work, UW, is $16.8 - 8.7 = 8.1\,\mathrm{MW}$.

The thermal efficiency is defined as the ratio of useful work to the work of combustion, i.e.

$$\eta_{\mathrm{th}} = \mathrm{UW}/Z = 8.1/13.6 = 0.60$$

These considerations demonstrate that the turbine entry temperature, T_4, needs to be as high as possible, if the useful work is to be maximised. Since the constant pressure lines diverge, UW is greatest when T_4 is large. There does, however, need to be a balance between a high TET and a large enough OPR; see Figure 1.13. This is because at very low or very high OPRs, the *difference* between CW and the EW is minimal, and, since CW = TW, then negligible work is available for propulsion.

The useful work, UW, is spent increasing the kinetic energy of the gas stream passing through the engine; hence,

$$\mathrm{UW} = \frac{1}{2}\dot{M}\left[V_{\mathrm{jet}}^2 - V_0^2\right]$$

so that $V_{\mathrm{jet}} = 934\,\mathrm{m/s}$. The thrust, F_N, is then given by (from momentum considerations)

$$F_\mathrm{N} = \dot{M}[V_{\mathrm{jet}} - V_0]$$

or $20 \times (934 - 250) = 13\,680\,\mathrm{N}$.

The propulsive efficiency is a measure of the proportion of jet kinetic energy (relative to the engine) which is converted to rate of working (by the thrust) on the aeroplane, defined

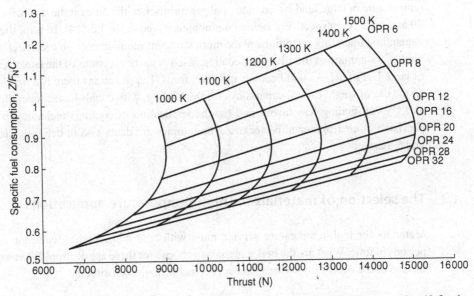

Fig. 1.14. Performance diagram illustrating the variation of the specific fuel consumption (defined as normalised fuel burn/thrust developed) against thrust, for various values of the TET and OPR.

according to

$$\eta_{prop} = \frac{F_N V_0}{\frac{M}{2}\left(V_{jet}^2 - V_0^2\right)}$$

This yields a value of 0.42.

Such calculations, although simplistic, can be used to justify the use of the bypass design for a modern civil aeroengine; see Figure 1.1. Thrust is required for propulsion, and this depends upon both η_{th} and η_{prop}. A high η_{th} demands large TET and reasonable OPR. Hence, if the available useful work, UW, is used primarily to accelerate the air passing through the engine, then it is inevitable that the jet velocity will be very high, and the thrust, F_N, and specific fuel consumption (SFC), too large. This is demonstrated in Figure 1.14. The propulsive efficiency, η_{prop}, is correspondingly very low. The only way around this dilemma is to use a bypass design. Most of the useful work developed by the core of the engine is used to accelerate a much greater mass of air through a fan to a modest exit velocity. The remaining useful work is used to accelerate the core air to a rather modest V_{jet}.

These considerations explain why a modern civil turbofan aeroengine has a bypass ratio between 6:1 and perhaps 9:1 – to develop reasonable η_{prop} and SFC. For a military aeroengine, however, a low bypass ratio of 1.2:1 is more appropriate since a high thrust-to-weight ratio is required; furthermore, the fan diameter must be small to prevent the engine protruding from the airframe so that speed and acceleration is maximised. For further reading, refer to ref. [10] for a thorough and lucid introduction to the aerodynamic design of modern engines for civil and military aircraft. The analysis also explains why there has been a tremendous drive to increase both the pressure ratio and the turbine entry temperature of large land-based industrial gas turbines, with values at the time of writing of 20 and 1425 °C, respectively, being commonplace; see Table 1.1 [14]. In turn, these more demanding operating conditions place more stringent requirements on the materials used. Table 1.2 summarises the alloys and coatings used in the hot sections of the latest generation of large IGTs [14]. As will become apparent, for IGT applications there is a trend towards (i) the use of single-crystal superalloys for the blading, with cobalt-based superalloys such as ECY-768 finding less favour and (ii) the application of coatings technologies for the provision of environmental protection. These topics are dealt with in detail in Chapters 3 and 5, respectively.

1.3   The selection of materials for high-temperature applications

Materials for high-temperature service must withstand considerable loads for extended periods of time. What are the best materials to choose for these applications? Can we justify the use of the superalloys which have nickel as the major constituent?

1.3.1   Larson–Miller approach for the ranking of creep performance

Resistance to creep deformation is a major consideration. For many materials and under loading conditions which are invariant with time, the creep strain rate, $\dot{\epsilon}_{ss}$, is constant; i.e. it approaches a steady-state [14]. This implies a balance of creep hardening, for example, due to dislocation multiplication and interaction with obstacles, and creep softening, for example, due to dislocation annihilation and recovery. Very often, it is found that

$$\dot{\epsilon}_{ss} = A\sigma^n \exp\left\{-\frac{Q}{RT}\right\} \tag{1.7}$$

where σ is the applied stress, n is the stress exponent, A is a constant and Q is an activation energy. When a value for Q is deduced from the experimental creep data, one often finds that it correlates with the *activation energy for self-diffusion*, see Figure 1.15 [15]. This implies that some form of mass transport on the scale of the microstructure is rate-controlling.

Design against creep usually necessitates a consideration of the *time to rupture*, t_r, which usually satisfies the so-called Monkman–Grant relationship

$$t_r \times \dot{\epsilon}_{ss} = B \tag{1.8}$$

where B is a constant which is numerically equal to the creep ductility, i.e. the creep strain to failure. Then at constant σ one has

$$t_r \exp\left\{-\frac{Q}{RT}\right\} = C \tag{1.9}$$

or

$$\log_{10} t_r - 0.4343\frac{Q}{RT} = D \tag{1.10}$$

Table 1.1. *Evolution of the features of large land-based industrial gas turbines [5]*

Year of introduction	1967	1972	1979	1990[a]	1998[b]
Turbine inlet temperature (°C)	900	1010	1120	1260	1425
Pressure ratio	10.5	11	14	14.5	19–23
Exhaust temperature (°C)	427	482.	530	582	593
Cooled turbine rows	R1 vane	R1, R2 vane, R1 blade	R1, R2 vane, R1, R2 blade	R1, R2, R3 vane, R1, R2, R3 blade	R1, R2, R3 vane, R1, R2, R3 blade
Power rating, MW	50–60	60–80	70–105	165–240	165–280
Efficiency, simple cycle (%)	29	31	34	36	39
Efficiency, combined cycle (%)	43	46	49	53	58

Note: [a] Corresponds approximately to GE 7F, 7A/9F, Westinghouse 501F/701F and Siemens V84.3/94.3 turbines.
[b] Corresponds approximately to GE 7H/9H, Westinghouse 501G/701G, Siemens V84.3A/94.3A and ABB GT24/26 turbines.

Table 1.2. *Alloys and coatings used in the hot sections of various industrial gas turbines, as at the end of 1997* [5]

Manufacturer	Model	Vanes	Blades	Coatings
ABB	11N2	IN939	IN738LC	NiCrAlY+Si
	GT24/26	DS CM247LC (R1)/ MarM247LC (R2,3)/ IN738 (R4,5)	DS CM247LC (R1–3)/ MarM247LC (R4,5)	TBC (R1V)/NiCrAlY+Si (R2–4 B,V) Uncoated R5V, Chromised (R5B)
GE	7/9EA	FSX-414 (all stages)	GTD-111/IN738/ Udimet 500	RT22/GT29-In+ (R1B)
	7/9FA	FSX-414 (R1)/ GTD-222 (R2,3)	DS GTD-111 (R1)/ GTD-111 (R2,3)	GT33-In/GT-29-In+/ Chromise (R3)
	7H	SC Rene N5 (R1)/ FSX-414/GTD-222	SC Rene N5 (R1)/ DS GTD-111 (R2,3)	TBC (R1,2 B,V)/ All others GT33
Siemens	V84/94.2	IN939	IN738LC/IN792 (R4)	CoNiCrAlY+Si
	V84/94.3A	SC PWA1483 (R1,2)/ IN939	SC PWA1483 (R1,2)	TBC (EB-PVD R1B)/MCrAlY+Re
Westinghouse/ Mitsubishi	501D5/701D	ECY-768/X-45	Udimet 520	MCrAlY
	501/701F	ECY-768/X-45	IN738LC	TBC (R1 B,V)/ MCrAlY/Sermalloy J
	501/701G	IN939	DS MarM002 (R1,2)/ CM247	TBC (R1,2 B,V)/ EB-PVD/MCrAlY

Note: R1, R2, etc. refer to the first, second rotor sections, etc.; B and V refer to blade and vane, respectively.

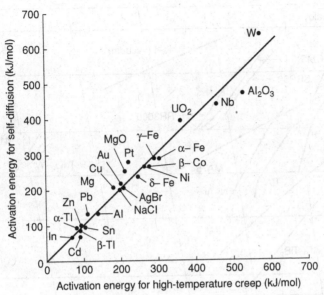

Fig. 1.15. Correlation between the creep activation energy and the energy for diffusion for several materials. Data taken from ref. [15].

where C and D are constants. Equation (1.10) can be written in the form

$$T[E + \log_{10} t_r] = P \tag{1.11}$$

where P is known as the Larson–Miller parameter and E is the Larson–Miller constant; this is found to vary between 15 and 25 $\log$ h and is taken to be 20 $\log$ h. When creep tests for various combinations of (σ, T) are carried out, one usually finds a strong correlation between P and $\log \sigma$, consistent with the assumptions made. These plots are known as Larson–Miller diagrams.

In Figure 1.16, the performance of various materials are compared. Note that the alloys SRR99, CMSX-4 and RR3000 are first-, second- and third-generation single-crystal superalloys, which are used for the turbine blades of modern aeroengines. Waspaloy is a polycrystalline superalloy used for the turbine discs required to house the blading. MA754 is an oxide-dispersion-strengthened superalloy made by powder metallurgy. Ti-6242 is a titanium alloy used widely for the compressor sections of modern aeroengines; it has the highest creep resistance of any of the titanium alloys. Also shown are various intermetallic compounds such as NiAl, Ni_3Al, TiAl, Ti_3Al and FeAl, and other exotic alloys such as Nb-I, Mo-41Re and thoria-doped tungsten; these show promise as high-temperature materials, but they have not yet found application in the gas turbine. A number of points emerge from a consideration of Figure 1.16. First, for any given material the Larson–Miller line is approximately straight, although there is a tendency for some non-linearity, which results in the lines curving gently downwards. Second, in comparison with the other materials, it is clear that the single crystal superalloys behave excellently; the shift towards the right as one moves from SRR99, CMSX-4 to RR3000 has arisen due to the metallurgists who have been

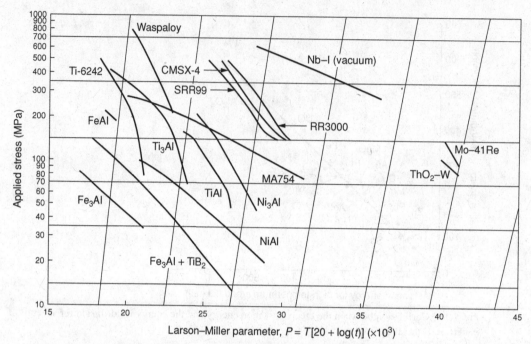

Fig. 1.16. Values of the Larson–Miller parameter, P, for a number of high-temperature materials. The horizontal and near-vertical lines have spacings equivalent to a factor of 2 change in creep life and 200 °C temperature capability, respectively; the materials which display the best high-temperature performance lie towards the top right of the diagram. Adapted from a graph provided by Dan Miracle.

able to produce alloys of increasing creep resistance; these improvements turn out to be very substantial, although clearly if a step change could be made to a new alloy system, for example, one based upon niobium, then this would be very advantageous. Polycrystalline superalloys such as Waspaloy are less good in creep, but their behaviour exceeds that of many other alloy systems. The intermetallic compounds are not bad, particularly if one makes the comparison on a density-corrected basis; it turns out, however, that their toughness is no match for that displayed by the superalloys. Finally, titanium alloys behave well, although it has been found that their oxidation resistance beyond about ~700 °C is very poor, and this limits their range of application – it has been found necessary, for example, to use superalloys for the last stages of the compressor sections of many aeroengines, even though titanium alloys are used for the first stages. Alternative materials based upon W or Mo show promise, although their density tends to be significant. Niobium alloys seem attractive, but their resistance to oxidation is very poor.

1.3.2   Historical development of the superalloys

Over the latter part of the twentieth century, a concerted period of alloy and process development enabled the performance of the superalloys to be improved dramatically. Although further improvements are being actively pursued, it is important to recognise this historical

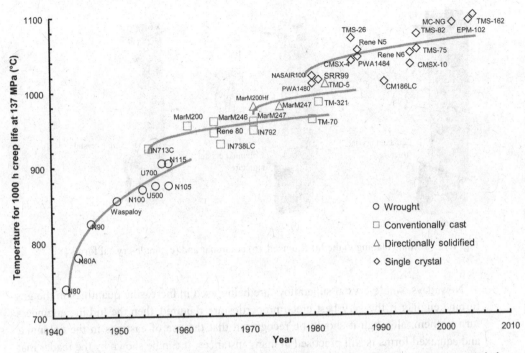

Fig. 1.17. Evolution of the high-temperature capability of the superalloys over a 60 year period since their emergence in the 1940s.

context, since much can be learned from it. Figure 1.17 provides a perspective for the alloy and process development which has occurred since the first superalloys began to appear in the 1940s; the data relate to the materials and processes for turbine blading, so that the creep performance (here taken as the highest temperature at which rupture occurs in not less than 1000 h, at 137 MPa) is a suitable measure for the progress which has been made. Various points emerge from a study of the figure. First, one can see that, *for the blading application,* cast rather than wrought materials are now preferred since the very best creep performance is then conferred. However, the first aerofoils were produced in wrought form. During this time, alloy development work – which saw the development of the first Nimonic alloys [1] – enabled the performance of the blading to be improved considerably; the vacuum induction casting technologies which were introduced in the 1950s helped with this since the quality and cleanliness of the alloys were improved dramatically. Second, the introduction of improved casting methods, and later the introduction of processing by directional solidification, enabled significant improvements to be made; this was due to the columnar microstructures that were produced in which the transverse grain boundaries were absent (see Figure 1.18). Once this development occurred, it was quite natural to remove the grain boundaries completely such that monocrystalline (single-crystal) super-alloys were produced. This allowed, in turn, the removal of grain-boundary strengthening elements such as boron and carbon which had traditionally been added; this enabled better heat treatments to reduce microsegregation and eutectic content induced by casting, whilst avoiding incipient melting during heat treatment. The fatigue life is then improved.

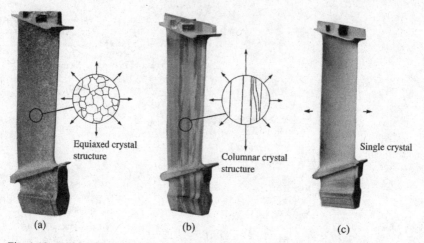

Fig. 1.18. Turbine blading in the (a) equiaxed, (b) columnar and (c) single-crystal forms [8].

Nowadays, single-crystal superalloys are being used in increasing quantities in the gas turbine engine; if the very best creep properties are required, then the turbine engineers turn to them, although it should be recognised that the use of castings in the columnar and equiaxed forms is still practised in many instances. It can be shown by the reader that the data in Figure 1.16 indicate that creep rupture lives of the single-crystal superalloys have been lengthened from about 250 h at 850 °C/500 MPa for a typical first-generation alloy such as SRR99, to about 2500 h for the third-generation alloy RR3000. Under more demanding conditions, for example, 1050 °C/150 MPa, rupture life has improved four-fold from 250 h to 1000 h. These improvements were won in a period of approximately 15 years between 1980 and 1995, primarily as a consequence of a better appreciation of the physical factors which confer high-temperature strength in these materials. It should be appreciated that each of the different engine manufacturers and materials suppliers have searched for their own proprietary compositions; Table 1.3 lists the compositions of common cast superalloys, including those used in single-crystal form. One can see that as many as 14 different alloying additions are added. The so-called first-generation single-crystal superalloys, such as PWA 1480, Rene N4 and SRR99, contain appreciable quantities of the γ' hardening elements Al, Ti and Ta; the grain-boundary-strengthening elements C and B, which were added routinely to the earlier directionally solidified alloys, are no longer present. Second-generation alloys, such as PWA 1484, CMSX-4 and Rene N5, are characterised by a 3 wt% concentration of Re, which is increased to about 6 wt% for the third-generation alloys such as CMSX-10 and Rene N6. Generally speaking, the modern alloys are characterised by significantly lower concentrations of Cr and higher concentrations of Al and Re. Concentrations of Ti and Mo are now at very modest levels. The period since 2000 has seen the emergence of the fourth-generation single-crystal superalloys, such as MC-NG, EPM-102 and TMS-162, which are characterised by additions of ruthenium.

Table 1.4 lists the chemical compositions of common wrought superalloys, which are employed therefore in polycrystalline form. When their compositions are compared with those of the cast superalloys, various points emerge. The concentrations of the γ' hardening

Table 1.3. *The compositions (in weight%) of some common cast superalloys*

Alloy	Cr	Co	Mo	W	Al	Ti	Ta	Nb	Re	Ru	Hf	C	B	Zr	Ni
AM1	7.0	8.0	2.0	5.0	5.0	1.8	8.0	1.0	—	—	—	—	—	—	Bal
AM3	8.0	5.5	2.25	5.0	6.0	2.0	3.5	—	—	—	—	—	—	—	Bal
CM186LC	6.0	9.3	0.5	8.4	5.7	0.7	3.4	—	3.0	—	1.4	0.07	0.015	0.005	Bal
CM247LC	8.0	9.3	0.5	9.5	5.6	0.7	3.2	—	—	—	1.4	0.07	0.015	0.010	Bal
CMSX-2	8.0	5.0	0.6	8.0	5.6	1.0	6.0	—	—	—	—	—	—	—	Bal
CMSX-3	8.0	4.8	0.6	8.0	5.6	1.0	6.3	—	—	—	0.1	—	—	—	Bal
CMSX-4	6.5	9.6	0.6	6.4	5.6	1.0	6.5	—	3.0	—	0.1	—	—	—	Bal
CMSX-6	10.0	5.0	3.0	—	4.8	4.7	6.0	—	—	—	0.1	—	—	—	Bal
CMSX-10	2.0	3.0	0.4	5.0	5.7	0.2	8.0	—	6.0	—	0.03	0.03	—	—	Bal
EPM-102	2.0	16.5	2.0	6.0	5.55	—	8.25	0.07	5.95	3.0	0.15	0.10	0.014	0.007	Bal
GTD-111	14.0	9.5	1.5	3.8	3.0	5.0	3.15	0.8	—	—	—	0.08	0.004	0.02	Bal
GTD-222	22.5	19.1	—	2.0	1.2	2.3	0.94	—	—	—	—	0.18	0.014	0.06	Bal
IN100	10.0	15.0	3.0	—	5.5	4.7	—	—	—	—	—	0.05	0.01	0.10	Bal
IN-713LC	12.0	—	4.5	—	5.9	0.6	—	2.0	—	—	—	0.11	0.01	0.04	Bal
IN-738LC	16.0	8.5	1.75	2.6	3.4	3.4	1.75	0.9	—	—	—	0.07	0.016	0.018	Bal
IN-792	12.4	9.2	1.9	3.9	3.5	3.9	4.2	—	—	—	—	—	0.009	0.10	Bal
IN-939	22.4	19.0	—	2.0	1.9	3.7	—	1.0	—	—	—	0.15	0.015	0.03	Bal
Mar-M002	8.0	10.0	—	10.0	5.5	1.5	2.6	—	—	—	1.5	0.15	0.015	0.05	Bal
Mar-M246	9.0	10.0	2.5	10.0	5.5	1.5	1.5	—	—	—	1.5	0.15	0.015	0.03	Bal
Mar-M247	8.0	10.0	0.6	10.0	5.5	1.0	3.0	—	—	—	1.5	0.15	0.015	0.03	Bal
Mar-M200Hf	8.0	9.0	—	12.0	5.0	1.9	—	1.0	—	—	2.0	0.13	0.019	0.04	Bal
Mar-M421	15.0	10.8	1.8	3.3	4.5	1.6	—	2.3	—	—	—	0.18	—	—	Bal
MC2	8.0	5.0	2.0	8.0	5.0	1.5	6.0	—	—	—	0.1	—	—	—	Bal
MC-NG	4.0	—	1.0	5.0	6.0	0.5	5.0	—	4.0	4.0	0.1	—	—	—	Bal
MX4	2.0	16.5	2.0	6.0	5.55	—	8.25	—	5.95	3.0	0.15	0.03	—	—	Bal

Table 1.3. (cont.)

Alloy	Cr	Co	Mo	W	Al	Ti	Ta	Nb	Re	Ru	Hf	C	B	Zr	Ni
Nasair 100	9.0	—	1.0	10.5	5.75	1.2	3.3	—	—	—	—	—	—	—	Bal
PWA1422	9.0	10.0	—	12.0	5.0	2.0	—	1.0	—	—	1.5	0.14	0.015	0.1	Bal
PWA1426	6.5	10.0	1.7	6.5	6.0	—	4.0	—	3.0	—	1.5	0.10	0.015	0.1	Bal
PWA1480	10.0	5.0	—	4.0	5.0	1.5	12.0	—	—	—	—	—	—	—	Bal
PWA1483	12.2	9.2	1.9	3.8	3.6	4.2	5.0	—	—	—	—	0.07	—	—	Bal
PWA1484	5.0	10.0	2.0	6.0	5.6	—	9.0	—	3.0	—	0.1	—	—	—	Bal
PWA1487	5.0	10.0	1.9	5.9	5.6	—	8.4	—	3.0	—	0.25	—	—	—	Bal
PWA1497	2.0	16.5	2.0	6.0	5.55	—	8.25	—	5.95	3.0	0.15	—	—	—	Bal
Rene 80	14.0	9.0	4.0	4.0	3.0	4.7	—	—	—	—	0.8	0.03	0.015	0.01	Bal
Rene 125	9.0	10.0	2.0	7.0	1.4	2.5	3.8	—	—	—	0.05	0.16	0.017	0.05	Bal
Rene 142	6.8	12.0	1.5	4.9	6.15	—	6.35	—	2.8	—	1.5	0.11	0.015	0.02	Bal
Rene 220	18.0	12.0	3.0	—	0.5	1.0	3.0	—	—	—	—	0.12	0.010	—	Bal
Rene N4	9.0	8.0	2.0	6.0	3.7	4.2	4.0	5.0	—	—	—	0.02	—	—	Bal
Rene N5	7.0	8.0	2.0	5.0	6.2	—	7.0	0.5	3.0	—	0.2	—	—	—	Bal
Rene N6	4.2	12.5	1.4	6.0	5.75	—	7.2	—	5.4	—	0.15	0.05	0.004	—	Bal
RR2000	10.0	15.0	3.0	—	5.5	4.0	—	—	—	—	—	—	—	—	Bal
SRR99	8.0	5.0	—	10.0	5.5	2.2	12.0	—	—	—	—	—	—	—	Bal
TMS-75	3.0	12.0	2.0	6.0	6.0	—	6.0	—	5.0	—	0.1	—	—	—	Bal
TMS-138	2.9	5.9	2.9	5.9	5.9	—	5.6	—	4.9	2.0	0.1	—	—	—	Bal
TMS-162	2.9	5.8	3.9	5.8	5.8	—	5.6	—	4.9	6.0	0.09	—	—	—	Bal

Table 1.4. *The compositions (in weight%) of some common wrought superalloys*

Alloy	Cr	Co	Mo	W	Nb	Al	Ti	Ta	Fe	Hf	C	B	Zr	Ni
Alloy 10	11.5	15	2.3	5.9	1.7	3.8	3.9	0.75	—	—	0.030	0.020	0.05	Bal
Astroloy	15.0	17.0	5.3	—	—	4.0	3.5	—	—	—	0.06	0.030	—	Bal
C-263	16	15	3	1.25	—	2.50	5.0	—	—	—	0.025	0.018	—	Bal
Hastelloy S	15.5	—	14.5	—	—	0.3	—	—	1.0	—	—	0.009	—	Bal
Hastelloy X	22.0	1.5	9.0	0.6	—	0.25	—	—	18.5	—	0.10	—	—	Bal
Haynes 230	22.0	—	2.0	14.0	—	0.3	—	—	—	—	0.10	—	—	Bal
Haynes 242	8.0	2.5	25.0	—	—	0.25	—	—	2.0	—	0.15	0.003	—	Bal
Haynes R-41	19.0	11.0	10.0	—	—	1.5	3.1	—	5.0	—	0.09	0.006	—	Bal
Incoloy 800	21.0	—	—	—	—	0.38	0.38	—	45.7	—	0.05	—	—	Bal
Incoloy 801	20.5	—	—	—	—	—	1.13	—	46.3	—	0.05	—	—	Bal
Incoloy 802	21.0	—	—	—	—	0.58	0.75	—	44.8	—	0.35	—	—	Bal
Incoloy 909	—	13.0	—	—	4.7	0.03	1.5	—	42.0	—	0.01	—	—	Bal
Incoloy 925	20.5	—	—	—	—	0.20	2.1	—	29.0	—	0.01	—	—	Bal
Inconel 600	15.5	—	—	—	—	—	—	—	8.0	—	0.08	—	—	Bal
Inconel 601	23.0	—	—	—	—	1.4	—	—	14.1	—	0.05	—	—	Bal
Inconel 617	22.0	12.5	9.0	—	—	1.0	0.3	—	—	—	0.07	—	—	Bal
Inconel 625	21.5	—	9.0	—	3.6	0.2	0.2	—	2.5	—	0.05	—	—	Bal
Inconel 690	29.0	—	—	—	—	—	—	—	9.0	—	0.025	—	—	Bal
Inconel 706	16.0	—	—	—	2.9	0.2	1.8	—	40.0	—	0.03	—	—	Bal
Inconel 718	19.0	—	3.0	—	5.1	0.5	0.9	—	18.5	—	0.04	—	—	Bal
Inconel 738	16.0	8.5	1.75	2.6	0.9	3.4	3.4	1.7	—	—	0.11	0.01	0.05	Bal
Inconel 740	25.0	20.0	0.5	—	2.0	0.9	1.8	—	0.7	—	0.03	—	—	Bal
Inconel X750	15.5	—	—	—	1.0	0.7	2.5	—	7.0	—	0.04	—	—	Bal
LSHR	13	21	2.7	4.3	1.5	3.5	3.5	1.6	—	—	0.030	0.030	0.050	Bal
ME3	13.1	18.2	3.8	1.9	1.4	3.5	3.5	2.7	—	—	0.030	0.030	0.050	Bal
MERL-76	12.4	18.6	3.3	—	1.4	5.0	4.3	—	—	0.35	0.050	0.03	0.06	Bal
Nimonic 75	19.5	—	—	—	—	—	0.4	—	3.0	—	0.10	—	0.06	Bal
Nimonic 80A	19.5	—	—	—	—	1.4	2.4	—	—	—	0.06	0.003	0.06	Bal
Nimonic 90	19.5	16.5	—	—	—	1.5	2.5	—	—	—	0.07	0.003	0.10	Bal
Nimonic 105	15.0	20.0	5.0	—	—	4.7	1.2	—	—	—	0.13	0.005	0.04	Bal
Nimonic 115	14.3	13.2	—	—	—	4.9	3.7	—	—	—	0.15	0.160	0.02	Bal
Nimonic 263	20.0	20.0	5.9	—	—	0.5	2.1	—	—	—	0.06	0.001	—	Bal
Nimonic 901	12.5	—	5.75	—	—	0.35	2.9	—	—	—	0.05	—	—	Bal

Table 1.4. (cont.)

Alloy	Cr	Co	Mo	W	Nb	Al	Ti	Ta	Fe	Hf	C	B	Zr	Ni
Nimonic PE16	16.5	1.0	1.1	—	—	1.2	1.2	—	33.0	—	0.05	0.020	—	Bal
Nimonic PK33	18.5	14.0	7.0	—	—	2.0	2.0	—	0.3	—	0.05	0.030	—	Bal
N18	11.5	15.7	6.5	0.6	—	4.35	4.35	—	—	0.45	0.015	0.015	0.03	Bal
Pyromet 860	13.0	4.0	6.0	—	0.9	1.0	3.0	—	28.9	—	0.05	0.01	—	Bal
Pyromet 31	22.7	—	2.0	—	1.1	1.5	2.5	—	14.5	—	0.04	0.005	—	Bal
Rene 41	19.0	11.0	1.0	—	—	1.5	3.1	—	—	—	0.09	0.005	—	Bal
Rene 88DT	16.0	13.0	4.0	4.0	0.7	2.1	3.7	—	—	—	0.03	0.015	0.03	Bal
Rene 95	14.0	8.0	3.5	3.5	3.5	3.5	2.5	—	—	—	0.15	0.010	0.05	Bal
Rene 104	13.1	18.2	3.8	1.9	1.4	3.5	3.5	2.7	—	—	0.030	0.030	0.050	Bal
RR1000	15.0	18.5	5.0	—	1.1	3.0	3.6	2.0	—	0.5	0.027	0.015	0.06	Bal
Udimet 500	18.0	18.5	4.0	—	—	2.9	2.9	—	—	—	0.08	0.006	0.05	Bal
Udimet 520	19.0	12.0	6.0	1.0	—	2.0	3.0	—	—	—	0.05	0.005	—	Bal
Udimet 630	18.0	—	3.0	3.0	6.5	0.5	1.0	—	18.0	—	0.03	—	—	Bal
Udimet 700	15.0	17.0	5.0	—	—	4.0	3.5	—	—	—	0.06	0.030	—	Bal
Udimet 710	18.0	15.0	3.0	1.5	—	2.5	5.0	—	—	—	0.07	0.020	—	Bal
Udimet 720	17.9	14.7	3.0	1.25	—	2.5	5.0	—	—	—	0.035	0.033	0.03	Bal
Udimet 720LI	16.0	15.0	3.0	1.25	—	2.5	5.0	—	—	—	0.025	0.018	0.05	Bal
Waspaloy	19.5	13.5	4.3	—	—	1.3	3.0	—	—	—	0.08	0.006	—	Bal

elements Al and Ti are generally lower, and Ta is not generally employed. Appreciable quantities of Fe are sometimes employed, and quantities of Nb are present. The Cr content is usually at least 15 wt%, higher than for the cast superalloys. Furthermore, additions such as Re and Ta are not generally made. In Chapters 3 and 4, the compositions of the cast and wrought forms of the superalloys are discussed in detail, and the differences rationalised.

1.3.3 Nickel as a high-temperature material: justification

Although the superalloys display excellent properties at elevated properties, it is not immediately obvious why nickel is a suitable solvent for a high-temperature alloy. Can nickel's role be justified?

Under conditions of high-temperature deformation, one might expect the creep shear strain rate, $\dot{\gamma}$, of a pure metal such as nickel to be proportional to the volume diffusivity, of activation energy Q_v and pre-exponential term $D_{0,v}$. This assumption is considered further in Chapter 2. Thus following ref. [16] one has

$$\dot{\gamma} \propto D_{0,v} \exp\left\{-\frac{Q_v}{RT}\right\} \tag{1.12}$$

In order to help in the comparison of different material classes, it is helpful to normalise T by the melting temperature, T_m, and $\dot{\gamma}$ by $D_{T_m}/\Omega^{2/3}$, where D_{T_m} is the diffusivity at the melting temperature and Ω is the atomic volume. Hence

$$\bar{\gamma} = \frac{\dot{\gamma}\Omega^{2/3}}{D_{T_m}} \propto \Omega^{2/3} \exp\left\{-\frac{Q_v}{RT_m}\left(\frac{T_m}{T}-1\right)\right\} \tag{1.13}$$

where $\bar{\gamma}$ is the dimensionless shear strain rate. One can see that the most important dimensionless groups are the homologous temperature, T/T_m, and the combination $Q_v/(RT_m)$. Also relevant is the normalising parameter, $D_{T_m}/\Omega^{2/3}$.

What does this analysis reveal? If one assumes for the moment that the combination $Q_v/(RT_m)$ is approximately the same for the different materials under consideration for this application, then for the best high-temperature performance (implying a small value of $\bar{\gamma}$) it is necessary to work at a low homologous temperature T/T_m; this emphasises the importance of materials which melt at high temperatures. Alternatively, if one accepts the need to work at a particular T/T_m, for example, $T = 0.8\,T_m$, then crystal classes which display the *maximum* $Q_v/(RT)$ and *minimum* $D_{T_m}/\Omega^{2/3}$ are the ones which will display the best properties.

We consider first the place of nickel in the family of metallic elements within the periodic table. The melting temperature of the elements shows a strong and remarkable correlation with atomic number (see Figure 1.19). If one focusses attention on the transition metals, then (i) there is a *maximum* in T_m somewhere towards the centre of each row of transition metals, for example, at V, Mo and W for the first, second and third rows, respectively, and (ii) the melting temperatures of the elements in each of the 3d, 4d and 5d rows increases generally as one descends a column from the 3d $\rightarrow$ 4d $\rightarrow$ 5d rows. However, due to the bonding characteristics induced by the increasing number of d electrons, the crystal structure also shows a strong correlation with position within the periodic table. For the transition

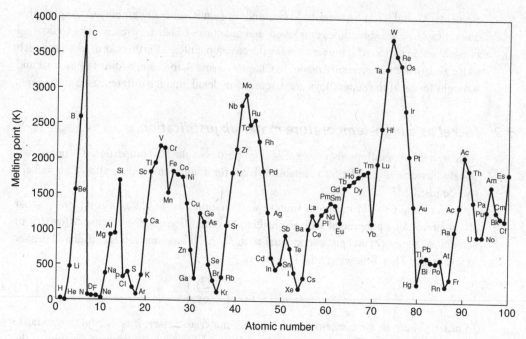

Fig. 1.19. Variation of the melting temperatures of the elements with atomic number.

BCC

IIIB	IVB	VB	VIB	VIIB	◄	—— VIIIB ——	►	IB	IIB
21	22	23	24	25	26	27	28	29	30
Sc	Ti	**V**	**Cr**	**Mn**	Fe	Co	**Ni**	**Cu**	Zn
44.956	47.90	50.942	51.996	54.9380	55.847	58.9332	58.71	63.54	65.37
39	40	41	42	43	44	45	46	47	48
Y	**Zr**	**Nb**	**Mo**	Tc	Ru	**Rh**	Pd	**Ag**	Cd
88.905	91.22	92.906	95.94	[99]	101.07	102.905	106.4	107.870	112.40
✱ 57	72	73	74	75	76	77	78	79	80
La	**Hf**	**Ta**	**W**	Re	Os	Ir	Pt	Au	Hg
138.91	178.49	180.948	183.85	186.2	190.2	192.2	195.09	196.967	200.59

Liquid

HCP   FCC

Fig. 1.20. Correlation of the crystal structures of the transition metals with position in the periodic table.

metals (see Figure 1.20), the face-centered cubic (FCC) metals lie towards the far east of the period (groups VIII and 1B), hexagonally close-packed (HCP) metals are at the centre (group VIIB) and the body-centered cubic (BCC) metals are at the far west (groups VB, VIB). Clearly, if one prefers for this application an FCC metal on the grounds that it will be ductile and tough, then the number of available metals is limited; moreover, the list includes

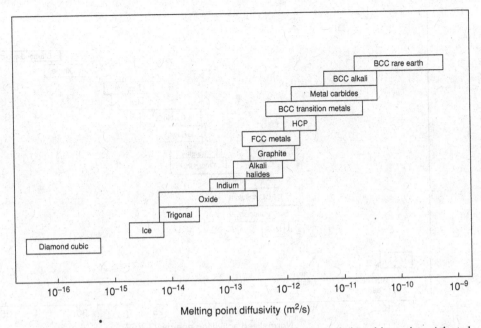

Fig. 1.21. Values of the diffusivities for the various crystal classes at their melting points. Adapted from ref. [17].

many of the platinum group metals (PGMs), which are characterised by their high density and significant cost.

In fact, it turns out that if one widens the search to include materials of different crystal classes, considering oxides and carbide ceramic systems for example, then there is quite a spread in the normalised activation energies, $Q_v/(RT_m)$, and the melting point diffusivity, D_{T_m}. Data for various crystal structures are given in Figures 1.21 and 1.22 for $Q_v/(RT_m)$ and D_{T_m}, respectively [17]. The FCC metals display very high and low values, respectively, and these are considerably better than for the BCC and the HCP metals, at $Q_v/(RT_m) = 18.4$ (compare 17.3 for HCP metals and 17.8 for the BCC metals) and $D_{T_m} = 5 \times 10^{-13}$ m^2/s (compare $\sim 2 \times 10^{-12}$ m^2/s for HCP and $10^{-12} - 10^{-11}$ m^2/s for the BCC transition metals). Other observations support these conclusions. For example, the BCC $\rightarrow$ FCC and BCC $\rightarrow$ HCP transformations in Fe and Ti, respectively, cause a 100-fold reduction in the measured diffusivity.

Given these considerations, one can list the reasons why nickel-based alloys have emerged as the ones chosen for high-temperature applications. First, nickel displays the FCC crystal structure and is thus both *tough* and *ductile*, due to a considerable cohesive energy arising from the bonding provided by the outer d electrons. Furthermore, nickel is stable in the FCC form from room temperature to its melting point, so that there are no phase transformations to cause expansions and contractions which might complicate its use for high-temperature components. Other metals in the transition metal series which display this crystal structure, i.e. the platinum group metals (PGMs), are dense and very expensive. Second, low rates of thermally activated creep require low rates of diffusion – as suggested by the correlation

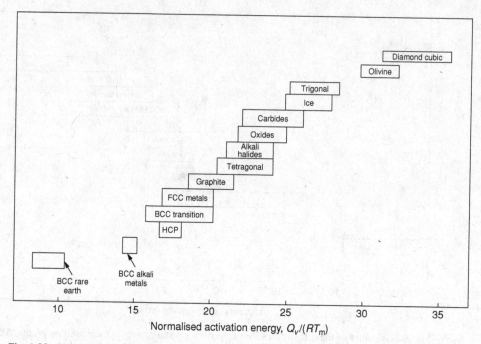

Fig. 1.22. Values of the normalised activation energy for diffusion for the various crystal classes. Adapted from ref. [17].

between the activation energies for self-diffusion and creep in the pure metals. Diffusion rates for FCC metals such as Ni are low; hence, considerable microstructural stability is imparted at elevated temperatures. Finally, consider other elements which possess different crystal structures. Of the HCP metals, only Co displays an acceptable density and cost; Re and Ru are PGMs and are therefore expensive; Os has an oxide which is poisonous; and Tc is radioactive. Co-based superalloys are, in fact, used for high-temperature applications; however, they tend to be more expensive than the nickel-based superalloys. The BCC metals such as Cr are prone to brittleness, and there is a ductile/brittle transition which means that the toughness decreases significantly with decreasing temperature.

1.4   Summary

With nickel as a solvent, but containing in excess of ten different alloying elements, the nickel-based superalloys have emerged as the materials of choice for high-temperature operation when resistance to creep, fatigue and environmental degradation is required. Their metallurgical development is linked inextricably to the history of the jet engine, for which the first superalloy grades were designed, although use has now spread to other high-temperature applications – notably to the turbines used for electricity generation. The technological incentive for further improvements being sought in alloy capability and processing developments arises from the enhanced fuel efficiency and reduced emissions expected of modern turbine systems; consistent with thermodynamic considerations, higher operating temperatures are then required. The risk of failure by fracture or fatigue necessitates that components fabricated from the superalloys are of the highest integrity, so that inspection and lifetime estimation procedures are critical to safe ongoing operation. A survey of the creep performance of available engineering materials confirms the superiority of the super-alloys for high-temperature applications. The use of nickel as the solvent for these materials can be justified on account of its face-centered cubic (FCC) crystal structure, its moderate cost and low rates of thermally activated processes.

Questions

1.1 What are the roles played by the fan, compressor, combustor and turbine arrangements in a typical gas turbine engine? How do these affect (i) the pressure and (ii) the average temperature of the gas stream? Explain why your findings justify the use of nickel-based superalloys in the combustor and turbine sections, but not in the compressor regions.

1.2 In an aeroengine, the low-pressure (LP) turbine blades are always very much longer than the high-pressure (HP) blading. Why?

1.3 Sketch the temperature–time cycle experienced by the first row of turbine blades for one take-off/landing cycle. Assume first that one is considering an aeroengine for a civil aircraft such as the Trent 800 or GE90. Discuss how this mission cycle would differ for (i) the turbine arrangement in a military jet and (ii) that in a land-based industrial gas turbine (IGT). Why do these considerations have implications for the alloy used, particularly the creep and fatigue properties?

1.4 A turbojet is operating with a turbine entry temperature (TET) of 1400 K, an overall pressure ratio (OPR) of 25 and an air mass intake, $\dot{M}$, of 20 kg/s. Assuming an ideal cycle such that the turbine work equals the work of compression, determine the *useful work* available to provide a propulsive force. Hence estimate the thermal efficiency, η_{th}. State your assumptions.

Take $c_p = 1000$ J/(kg K), intake temperature $= 288$ K, $\gamma = 1.4$ and atmospheric pressure $= 101\,325$ Pa. Assume adiabatic isentropic reversible processes such that

$$\frac{T}{T_{ref}} = \left(\frac{P}{P_{ref}}\right)^{(\gamma-1)/\gamma}$$

1.5 Using the concepts developed in Question 1.4, investigate how the thermal efficiency, η_{th}, depends upon the TET and the OPR. Which effect is stronger? Begin by making calculations with the OPR set at values of 15, 20 and 25 and the TET at 1200, 1300 and 1400 K. Then make calculations outside of these ranges. What is the relevance of your findings to turbine engine performance? Why, in practice, is there a stronger influence from the TET than predicted by the calculations?

1.6 From the definition of the thrust, F_N, as $\dot{M}(V_{jet} - V_0)$ and the propulsive efficiency, η_{prop}, as $F_N V_0 / \frac{1}{2}\dot{M}(V_{jet}^2 - V_0^2)$, where V_0 is the gas entry velocity, show that alternative expressions are

$$\eta_{\text{prop}} = \frac{2}{1 + V_{\text{jet}}/V_0} = \frac{1}{1 + F_N/(2V_0\dot{M})}$$

1.7 Develop 'performance diagrams' for the turbojet considered in Questions 1.4–1.6. First plot the thrust, F_N, against the specific fuel consumption, defined as $Z/(F_N C)$. The term C is the calorific value of the fuel (taken as 43 MJ/kg) and Z is the ideal work of combustion defined as $\dot{M}c_p\Delta T$, where ΔT is the difference between the turbine entry and compressor exit temperatures. Take the entry velocity, V_0, as 250 m/s. On your diagram, plot contours corresponding to specific values of the TET and OPR.

   As a second exercise, plot the overall efficiency, $\eta_{\text{overall}} = \eta_{\text{th}}\eta_{\text{prop}}$, versus OPR for various values of the TET. Comment on your findings.

1.8 One factor in setting the fan diameter of a modern turbofan such as the Trent 800 or GE90 is the clearance of the engine beneath the wing of the aircraft – this needs to be sufficient for take-off. Use these considerations to estimate (i) an approximate value of the fan diameter, (ii) the length of the fan blades, if the ratio of inner and outer diameter is constrained to be 0.37, (iii) the rate of mass intake into the engine, given your dimensions and a typical cruising speed and (iv) the mass intake (in kg/s) into the turbine section of the engine, if the bypass ratio (BPR) is 6:1.

   The shaft power developed by a turbine stage is given approximately by $\dot{M}c_p\Delta T$, where $\dot{M}$ is the mass flow (kg/s), c_p is the heat capacity of 1000 J/(kg K) and ΔT is the temperature drop across the blading. By estimating the temperature drop to be expected in the turbine, compare the power developed by the high-pressure (HP) and low-pressure (LP) stages. Take the turbine entry temperature (TET) to be 1700 K. If the HP and LP shafts spin at 1000 and 300 rad/s, respectively, estimate the power developed in the two shafts, and the associated torques. [*Hint*. The shaft power is torque multiplied by angular speed.]

1.9 The constant, C, in the Larson–Miller parameter, P, which is defined according to $P = T\{C + \log_{10} t_r\}$, is often taken to be 20. Give some justification for this.

1.10 The costs of superalloy components are often considered to be an important consideration. Estimate the relative costs of (a) a turbine blade cast in single-crystal, columnar and equiaxed forms and (b) a turbine disc fabricated from powder metallurgy and conventional cast-and-wrought processing.

1.11 The best choice of metal solvent for a high-temperature alloy is often considered to be that with the highest value of $Q_v/(RT_m)$, where Q_v is the activation energy for lattice diffusion, R is the gas constant and T_m is the melting temperature. On this basis, FCC metals (with a value of 18.4) perform better than BCC metals (17.8) which in turn are superior to HCP ones (17.3). An alternative measure is the diffusivity at the melting temperature.

   These suggestions, however, depend upon a normalisation against the melting temperature, T_m, whereas, for gas turbine applications, one could argue that (i) the absolute metal temperature is more relevant and (ii) the density and cost should be accounted for. Re-examine the properties of the transition metals with these considerations in mind. Suggest a merit criterion for turbine blade applications, and find the transition metal which maximises it. [*Hint.* Make use of the considerations of Brown and Ashby (see ref. [17]). For FCC metals, the pre-exponential factor is 5.4×10^{-5}; for BCC it is 1.6×10^{-4}, and for HCP it is 4.9×10^{-5}. All units are in m^2/s.]

References

[1] W. Betteridge and S. W. S. Shaw, Development of superalloys, *Materials Science and Technology*, **3** (1987), 682–694.

[2] C. T. Sims, N. S. Stoloff and W. C. Hagel, eds, *Superalloys II: High-Temperature Materials for Aerospace and Industrial Power* (New York: John Wiley and Sons, 1987).

[3] K. A. Green, T. M. Pollock, H. Harada *et al.*, eds., *Superalloys 2004, Proceedings of the Tenth International Symposium on the Superalloys* (Warrendale, PA: The Minerals, Metals and Materials Society (TMS), 2004).

[4] R. Schafrik and R. Sprague, The saga of gas turbine materials, *Advanced Materials and Processes*, **162** (2004), 3:33–36, 4:27–30, 5:29–33, 6:41–46.

[5] J. Stringer, The role of the coating and superalloy system in enabling advanced land-based combustion turbine development. In P. J. Maziasz, I. G. Wright, W. J. Brindley, J. Stringer and C. O'Brien, eds, *Gas Turbine Materials Technology* (Materials Park, OH: ASM International, 1999).

[6] J. C. Williams and E. A. Starke, Progress in structural materials for aerospace systems, *Acta Materialia*, **51** (2003), 5775–5799.

[7] G. Smith and L. Shoemaker, Advanced nickel alloys for coal-fired boiler tubing, *Advanced Materials and Processes*, **162** (2004), 23–26.

[8] Rolls-Royce plc, *The Jet Engine*, 4th edn (Derby, UK: The Technical Publications Department, Rolls-Royce plc, 1992).

[9] M. McLean, *Directionally Solidified Materials for High Temperature Service* (London: The Metals Society, 1983).

[10] N. A. Cumpsty, *Jet Propulsion: A Simple Guide to the Aerodynamic and Thermodynamic Design and Performance of Jet Engines* (Cambridge: Cambridge University Press, 1997).

[11] N. R. Blyth, *Examination of a Failed CFM56-3C-1 Turbofan Engine – Boeing 737-476, VH-TJN*, Technical Analysis Report 3/01 (Reference BE/200000023) (Canberra, Australia: The Australian Transport Safety Bureau, 2001).

[12] D. A. Cooper and D. F. King, *Report on the Accident to Boeing 737-237 Series 1 (G-BGJL) at Manchester International Airport on 22nd August 1985*, Aircraft Accident Report 8/88 (London: United Kingdom Department of Transport, Her Majesty's Stationery Office, 1988).

[13] D. Kondepudi and I. Prigogine, *Modern Thermodynamics – From Heat Engines to Dissipative Structures* (Chichester, UK: Wiley, 1998).

[14] F. R. N. Nabarro and H. L. de Villiers, *The Physics of Creep* (London: Taylor and Francis, 1995).

[15]   O. D. Sherby and P. M. Burke, Mechanical behaviour of crystalline solids at elevated temperatures, *Progress in Materials Science*, **13** (1967), 325–390.

[16]   H. J. Frost and M. F. Ashby, *Deformation-Mechanism Maps: The Plasticity and Creep of Metals and Ceramics* (Oxford: Pergamon Press, 1982).

[17]   A. M. Brown and M. F. Ashby, Correlations for diffusion coefficients, *Acta Metallurgica*, **28** (1980), 1085–1101.

2 The physical metallurgy of nickel and its alloys

Nickel is the fifth most abundant element on earth. The atomic number is 28, and this places it in the first row of the d block of transition metals, alongside iron and cobalt. The atomic weight is 58.71, the weighted average of the five stable isotopes 58, 60, 61, 62 and 64, which are found with probabilities 67.7%, 26.2%, 1.25%, 3.66% and 1.16%, respectively. The crystal structure is face-centred cubic (FCC; see Figure 2.1), from ambient conditions to the melting point, 1455 °C, which represents an absolute limit for the temperature capability of the nickel-based superalloys. The density under ambient conditions is 8907 kg/m^3. Thus, compared with other metals used for aerospace applications, for example, Ti (4508 kg/m^3) and Al (2698 kg/m^3), Ni is rather dense. This is a consequence of a small interatomic distance, arising from the strong cohesion provided by the outer d electrons – a characteristic of the transition metals.

In this chapter, some important aspects of the physical metallurgy of nickel and its alloys are considered. Section 2.1 is concerned with the compositions of the superalloys and the phases promoted by the presence of the alloying elements. Such composition–microstructure relationships have been established over many years, and considerable use of them is required when designing new grades of superalloy. Much of this quantitative information now resides in thermodynamic databases, which are capable of predicting the equilibrium fractions and compositions of the metallurgical phases as a function of mean chemistry and temperature; the background to these methods is presented. One of these – the ordered γ' phase – plays a pivotal role in superalloy metallurgy, and its characteristics are introduced here. In Section 2.2, attention is focussed on the different lattice defects which arise – it turns out that control of these is vital if significant high-temperature strength and creep resistance are to be achieved. The role of vacancies is emphasised since these play an important role in the mechanism of diffusion, together with the planar defects such as anti-phase boundaries and stacking faults which are involved in the dislocation dissociation reactions which determine the high-temperature properties. Section 2.3 is concerned with the role of the γ' phase in promoting high-temperature strength. First, theory for order strengthening by distributions of the γ' phase is presented. Next, the anomalous yielding effect – the origin of which has proved a controversial topic in superalloy metallurgy – is reviewed. Consideration is given to the role played by the γ' phase in promoting the high-temperature creep strength in these systems. Finally, in Section 2.4, attention is turned to the creep behaviour of nickel and its alloys.

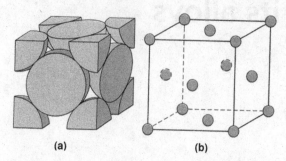

Fig. 2.1. The unit cell of the face-centred cubic (FCC) crystal structure, which is displayed by nickel.

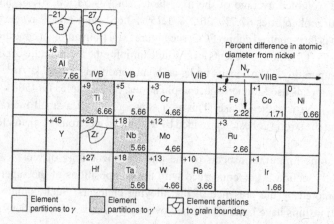

Fig. 2.2. Categories of elements important to the constitution of the nickel-based superalloys, and their relative positions in the periodic table. Adapted from ref. [1].

2.1   Composition–microstructure relationships in nickel alloys

The compositions of some important nickel-based superalloys have been given in Tables 1.3 and 1.4. One can see that the number of alloying elements is often greater than ten and consequently, if judged in this way, the superalloys are amongst the most complex of materials engineered by man. Although a wide variety of alloying elements are used, some broad rules are apparent. Most of the alloys contain significant amounts of chromium, cobalt, aluminium and titanium. Small amounts of boron, zirconium and carbon are often included. Other elements that are added, but not to all alloys, include rhenium, tungsten, tantalum and hafnium, from the 5d block of transition metals, and ruthenium, molybdenum, niobium and zirconium from the 4d block. Certain superalloys, such as IN718 and IN706, contain significant proportions of iron, and should be referred to as nickel–iron superalloys.

Thus, most of the alloying elements are taken from the d block of transition metals [1]. Unsurprisingly, the behaviour of each alloying element and its influence on the phase stability depends strongly upon its position within the periodic table (see Figure 2.2). A first class of elements includes nickel, cobalt, iron, chromium, ruthenium, molybdenum, rhenium and tungsten; these prefer to partition to the austenitic γ and thereby stabilise it. These elements have atomic radii not very different from that of nickel. A second group of elements, aluminium, titanium, niobium and tantalum, have greater atomic radii and these promote the formation of ordered phases such as the compound $Ni_3(Al, Ta, Ti)$, known as γ'. Boron, carbon and zirconium constitute a third class that tend to segregate to the grain boundaries of the γ phase, on account of their atomic sizes, which are very different from that of nickel. Carbide and boride phases can also be promoted. Chromium, molybdenum, tungsten, niobium, tantalum and titanium are particularly strong carbide formers; chromium and molybdenum promote the formation of borides.

The microstructure of a typical superalloy consists therefore of different phases, drawn from the following list [2].

(i) The gamma phase, denoted γ. This exhibits the FCC structure, and in nearly all cases it forms a continuous, matrix phase in which the other phases reside. It contains significant concentrations of elements such as cobalt, chromium, molybdenum, ruthenium and rhenium, where these are present, since these prefer to reside in this phase.

(ii) The gamma prime precipitate, denoted γ'. This forms as a precipitate phase, which is often coherent with the γ-matrix, and rich in elements such as aluminium, titanium and tantalum. In nickel–iron superalloys and those rich in niobium, a related ordered phase, γ'', is preferred instead of γ'.

(iii) Carbides and borides. Carbon, often present at concentrations up to 0.2 wt%, combines with reactive elements such as titanium, tantalum and hafnium to form MC carbides. During processing or service, these can decompose to other species, such as $M_{23}C_6$ and M_6C, which prefer to reside on the γ–grain boundaries, and which are rich in chromium, molybdenum and tungsten. Boron can combine with elements such as chromium or molybdenum to form borides which reside on the γ–grain boundaries.

Other phases can be found in certain superalloys, particularly in the service-aged condition, for example, the topologically close-packed (TCP) phases μ, σ, Laves, etc. However, the compositions of the superalloys are chosen to avoid, rather than to promote, the formation of these compounds.

2.1.1 The FCC phase

Since the FCC phase constitutes the matrix phase for the superalloys, it is desirable to quantify its stability with respect to other possible crystal structures. This is important since any phase transformations, either during thermal cycling or during extended periods of operation, will confer poor high-temperature properties. It is also relevant to the coupling of phase diagram information with thermochemical data, for the analysis of the microstructural stability of the superalloys. One can begin by estimating the driving force required for transformation of pure Ni to a crystal structure other than the FCC form. Consider the binary Ni–Cr binary phase diagram [3] (see Figure 2.3). The solute Cr displays the body-centered cubic (BCC) crystal structure, and extrapolation of the (liquid + BCC) two-phase field on the Cr-rich side of the binary gives a theoretical metastable liquid $\rightarrow$ BCC

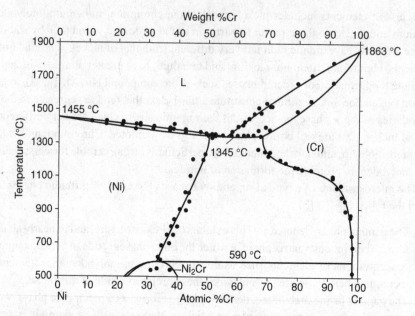

Fig. 2.3. The Ni–Cr binary phase diagram [3]. Note that extrapolation of the liquid + BCC two-phase region to pure Ni gives a metastable liquid to BCC transformation temperature of ~700 °C.

transformation temperature of ~700 °C, 750 °C below Ni's true melting temperature of 1455 °C. For most solidification reactions, the molar entropy of fusion has a magnitude of about 10 J/(mol K), as noted by Trouton's rule, which states that the absolute melting temperature is proportional to the molar enthalpy of fusion. The latent heat of melting is thus about $(1455 + 273) \times 10 \times 10^{-3} = 17.3$ kJ/mol. It follows that the lattice stability at 750 °C of the FCC phase with respect to BCC is approximately given by

$$\Delta G_{Ni}^{BCC \rightarrow FCC} = G_{Ni}^{FCC} - G_{Ni}^{BCC} = 750 \times (-10) = -7500 \text{ J/mol} \qquad (2.1a)$$

and that this is a rough estimate of the value at the melting temperature of pure nickel, 1455 °C, if Trouton's rule applies. The lattice stabilities of the FCC and BCC phases with respect to the liquid (L) form are thus given by

$$\Delta G_{Ni}^{L \rightarrow FCC} = G_{Ni}^{FCC} - G_{Ni}^{L} = -173\,00 + 10\,T \text{ J/mol} \qquad (2.1b)$$

and

$$\Delta G_{Ni}^{L \rightarrow BCC} = G_{Ni}^{BCC} - G_{Ni}^{L} = -9800 + 10\,T \text{ J/mol} \qquad (2.1c)$$

The lattice stability of the hexagonally close-packed (HCP) form of Ni can be estimated in a similar way, for example, by considering the Ni–Ru binary diagram. The lattice stabilities deduced in this simple way are in reasonable agreement with the values suggested in the literature [4] (see Figure 2.4). The values confirm the considerable solid-state stability of the FCC form.

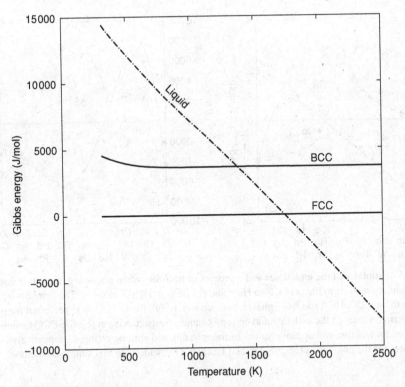

Fig. 2.4. Variation of the liquid and BCC lattice stabilities of Ni, with respect to the FCC form. Data taken from ref. [4].

It is possible to take these calculations further, to justify the stability of the FCC structure of Ni in the context of the crystal structures displayed by the transition metals; see Figure 1.20. To a zeroth-order approximation, the liquid, FCC, BCC and HCP phases can be modelled as ideal [5]. Defining $\Delta G_i^{L \to \phi} = G_i^{\phi} - G_i^{L}$ as the driving force for transformation for the pure constituent i, the two-phase (liquid + ϕ) region bounded by the liquidus and solidus is described by

$$\mu_i^{L} = G_i^{L} + RT \ln x_i^{L} = \mu_i^{\phi} = G_i^{\phi} + RT \ln x_i^{\phi} \qquad (2.2a)$$

or, after some manipulation,

$$\frac{x_i^{\phi}}{x_i^{L}} = \exp\left\{-\frac{\Delta G_i^{L \to \phi}}{RT}\right\} \qquad i = \text{solute, solvent} \qquad (2.2b)$$

consistent with the equality of chemical potential, μ_i^{ϕ}, required by equilibrium conditions. The term x_i^{ϕ} represents the mole fraction of component i in phase ϕ. Thus it is seen that, for the ideal case, each two-phase region on the binary phase diagram requires just four parameters for its description, two for each of the pure constituents: a transformation temperature T_m for the liquid to solid phase transformation (at which the free energy, $\Delta G_i^{L \to \phi}$, disappears) and an associated entropy of fusion, $\Delta S_i^{L \to \phi}$. Optimisation of the four

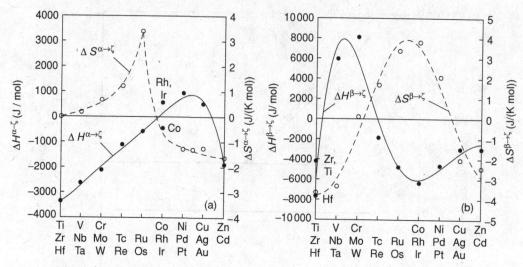

Fig. 2.5. Estimates of the enthalpies and entropies of transformation necessary to convert one mole of transition metal from the (a) FCC to HCP and (b) BCC to HCP forms [5]. The symbols α, ζ and β refer to the FCC, HCP and BCC phases, respectively. In (a), the FCC form is preferred for positive and negative values of the enthalpy and entropy changes, respectively; in (b), the BCC form is preferred for positive and negative values of the enthalpy and entropy changes, respectively. Estimates for the magnetic elements Mn and Fe cannot be made with this simplistic method.

parameters to give a best fit to the available phase diagram information can then be carried out, enabling estimates of $\Delta H_i^{\text{FCC}\rightarrow\text{HCP}}$, $\Delta S_i^{\text{FCC}\rightarrow\text{HCP}}$, $\Delta H_i^{\text{BCC}\rightarrow\text{HCP}}$ and $\Delta S_i^{\text{BCC}\rightarrow\text{HCP}}$ to be made for each transition metal i. It should be noted, however, that this is a lengthy exercise. The values derived in this way [5] are given in Figure 2.5. Predicted correctly is the stability of the BCC, HCP and FCC crystal structures as one proceeds from the far west to the far east of the d block of transition metals.

In practice, the solutions formed by Ni with the other elements are rarely ideal, so that it is seldom accurate to approximate the activity coefficients as unity – thus the above considerations will necessarily contain systematic errors. For example, the liquid phase in the binary Ni–Cr system displays a finite enthalpy of mixing of 2300 J/mol at $x_{\text{Cr}} = 0.5$ at 1873 K [3], when the standard states of both Ni and Cr correspond to the liquid phase. The activities of Cr show negative deviations from ideality up to about $x_{\text{Cr}} = 0.33$, and positive deviations thereafter [3]. This experimental information cannot be reproduced by the ideal solution model. Hence the greater complexity of the regular solution model is more appropriate, in which one includes a molar excess energy, $\Delta G_{\text{m}}^{\text{excess}}$, which is formally identified with the enthalpy of mixing, $\Delta H_{\text{m}}^{\text{mix}}$, and an interaction parameter, Ω, such that

$$\Delta G_{\text{m}}^{\text{excess}} = \Delta H_{\text{m}}^{\text{mix}} = \Omega x_i (1 - x_i) \qquad (2.3a)$$

The corresponding partial molar quantities can be shown to be

$$\Delta \overline{H}_i = \Delta \overline{G}_i^{\text{excess}} = \Omega (1 - x_i)^2 \qquad i = 2 \text{ (solute), and } 1 \text{ (solvent)} \qquad (2.3b)$$

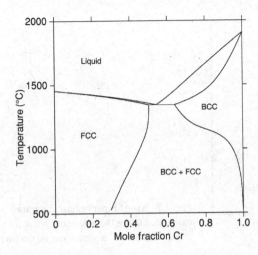

Fig. 2.6. Ni–Cr binary phase diagram calculated using the Thermo-Calc software and a database of thermodynamic parameters.

These must be added to the chemical potentials of Equation (2.2a), so that

$$\mu_i^L = G_i^L + RT \ln x_i^L + \Delta \overline{G}_i^{\text{excess, L}} = \mu_i^\phi = G_i^\phi + RT \ln x_i^\phi + \Delta \overline{G}_i^{\text{excess, }\phi} \qquad (2.3c)$$

After some manipulation, one obtains

$$\frac{\Delta G_1^{L \rightarrow \phi}}{RT} + \frac{\Omega^\phi}{RT}\left(x_2^\phi\right)^2 - \frac{\Omega^L}{RT}\left(x_2^L\right)^2 + \ln\left\{\frac{\left(1 - x_2^\phi\right)}{\left(1 - x_2^L\right)}\right\} = 0 \qquad (2.4a)$$

and

$$\frac{\Delta G_2^{L \rightarrow \phi}}{RT} + \frac{\Omega^\phi}{RT}\left(1 - x_2^\phi\right)^2 - \frac{\Omega^L}{RT}\left(1 - x_2^L\right)^2 + \ln\left\{\frac{x_2^\phi}{x_2^L}\right\} = 0 \qquad (2.4b)$$

Equations (2.4a) and (2.4b) represent a set of coupled equations which can be solved, iteratively, for the unknowns x_2^ϕ and x_2^L. The values of Ω^L and Ω^ϕ are chosen such that the available experimental information is reproduced accurately.

For the calculation of the phase diagrams of multicomponent nickel-based superalloys and the associated phase equilibria, there is now available thermodynamic software such as the Thermo-Calc package [6], which enables calculations to be performed quickly and with precision. Such software requires a thermodynamic database of parameters [7] which describe the lattice stabilities and the interaction parameters. These need to be chosen by careful assessment of experimental phase diagram information, using methods of the type described here but also others of greater sophistication. Figures 2.6 and 2.7 illustrate computed Ni–Cr and Ni–Al binary diagrams determined with these so-called CALPHAD (CALculation of PHAse Diagrams) methods. Note the very good agreement between the computed Ni–Cr diagram and the experimental one given in Figure 2.3.

The liquidus and solidus lines for the Ni–Re, Ni–W, Ni–Ru, Ni–Co, Ni–Al, Ni–Mo, Ni–Ta and Ni–Ti binary diagrams are plotted in Figure 2.8. One can see that the variation in the liquidus induced by each alloying element i does not correlate with its melting temperature. For example, W and Ta (which lie next to each other in the periodic table)

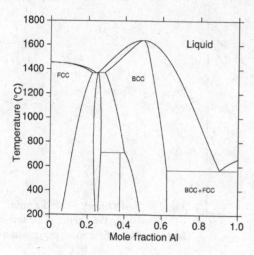

Fig. 2.7. Ni–Al binary phase diagram calculated using the Thermo-Calc software and a database of thermodynamic parameters.

increase and *decrease* the liquidus temperature by about $+5\,\mathrm{K/at\%}$ and $-5\,\mathrm{K/at\%}$ of solute, respectively. The element Re, however, is notable for its strong positive influence on the liquidus temperature. Consistent with the Hume–Rothery rule, the solubility of i in FCC nickel is found to depend strongly upon element i's atomic diameter; see Figure 2.9. The solubility is large only when atomic diameter is comparable to that of nickel, as for the platinum group metals Pt, Ir, Pd and Rh.

2.1.2 The gamma prime phase

Although Ni and Al possess the same crystal structure so that complete mutual solid solubility might be anticipated, the binary Ni–Al system exhibits a number of solid phases other than the FCC one, as the phase diagram confirms. These possess the following characteristics: (i) a significant degree of directional, covalent bonding such that precise stoichiometric relationships exist between the number of Ni and Al atoms in each unit cell, and (ii) crystal structures in which Ni–Al rather than Ni–Ni or Al–Al bonds are preferred. Thus a strong degree of chemical order is displayed, and consequently these phases are therefore referred to as *ordered* to distinguish them from the *disordered* solutions based on the FCC or BCC crystal structures. The chemical formulae are Ni_3Al, $NiAl$, Ni_2Al_3, $NiAl_3$ and Ni_2Al_9. For each compound, the enthalpy of ordering is significant. This is demonstrated by Figure 2.10, in which the enthalpies of formation at $25\,^\circ\mathrm{C}$ are plotted against the enthalpy of mixing of the FCC phase [8]. The enthalpy of formation is largest for the β–NiAl compound, which displays the CsCl crystal structure. However, of great significance is the gamma prime (γ') phase, Ni_3Al, particularly in view of the role it plays in conferring strength to the superalloys. It displays the primitive cubic, $L1_2$, crystal structure (see Figure 2.11), with Al atoms at the cube corners and Ni atoms at the centres of the faces. It is notable that each Ni atom has four Al and eight Ni as nearest neighbours, but that each Al atom is co-ordinated by twelve Ni atoms – thus Ni and Al have distinct site occupances. The data in Figure 2.10 indicate that the ordering energy with respect to the FCC phase is about $3\,\mathrm{kJ/mol}$.

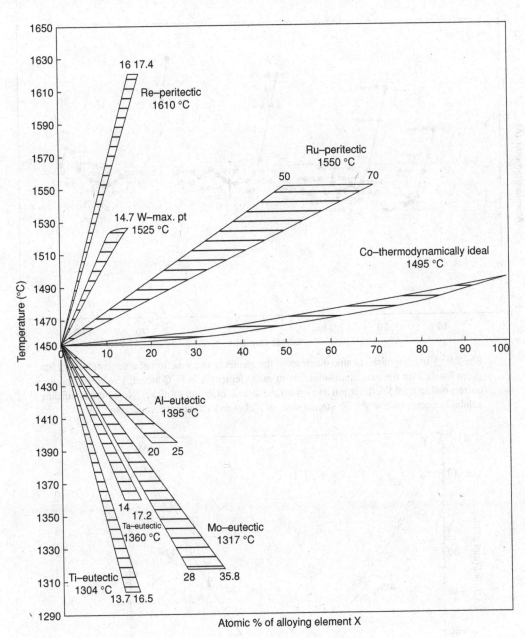

Fig. 2.8. The solidus and liquidus lines in the Ni–X binary systems.

It is interesting to examine the various pieces of evidence which confirm the ordered nature of the γ' phase. Consider first the ternary phase diagrams Ni–Al–X, where X = Co, Cr, Mo, W, etc. In each case, the γ' phase field is extended in a direction which depends upon the solubility of X in the γ' phase [9] (see Figure 2.12). Elements such as Co and Pt promote γ' phase fields which are parallel to the Ni–X axis on the ternary section, implying

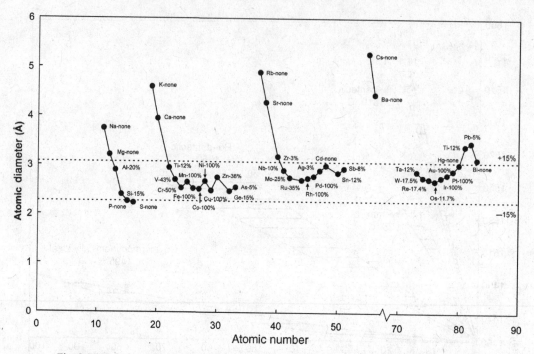

Fig. 2.9. Variation of the atomic diameter of the elements as a function of atomic number; also given are data for the maximum solubility of each element X in FCC nickel. The horizontal lines correspond to a ±15% deviation away from the atomic diameter of nickel. Note that significant solubility occurs only when the atomic sizes of nickel and solute are comparable.

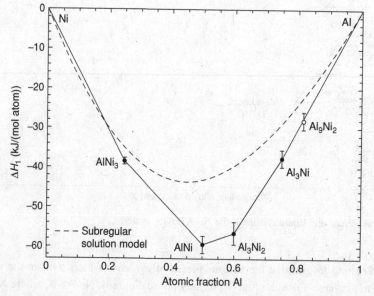

Fig. 2.10. Variation of the enthalpy of formation, ΔH_f, for the various intermetallic compounds in the binary Ni–Al system [8]. The broken line corresponds to the enthalpy of mixing of the disordered FCC phase, with respect to the pure Ni and Al constituents.

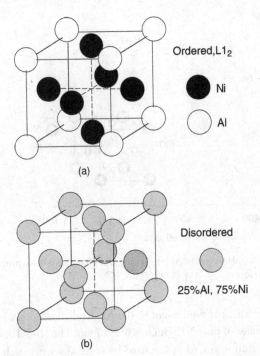

(a)

Ordered, L1$_2$

● Ni

○ Al

(b)

Disordered

25%Al, 75%Ni

Fig. 2.11. Arrangements of Ni and Al atoms in (a) the ordered Ni$_3$Al phase and (b) after disordering.

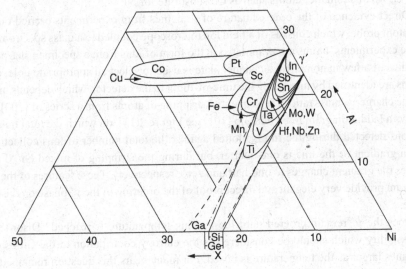

Fig. 2.12. Superimposed ternary phase diagrams Ni–Al–X [9], illustrating the great effect of X on the extent of the γ' phase field.

a constant Al fraction and providing confirmation that substitution for Ni on the first of the two sublattices is preferred. Elements such as Ti and Ta promote fields parallel to the Al–X axis, and thus they replace Al on the second sublattice. Only rarely, for example, for Cr, Fe and Mn, is mixed behaviour observed. The behaviour of the element X in this regard depends rather strongly on its size relative to Ni and Al. The lattice parameter, a, of γ'

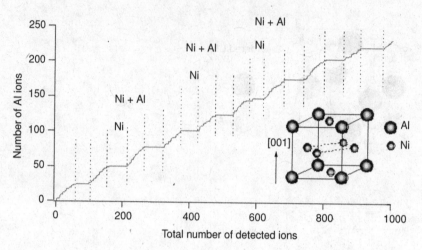

Fig. 2.13. Ladder diagram taken from a $\langle 001 \rangle$-orientated Ni_3Al single crystal using atom probe field ion microscopy. Courtesy of Kazihiro Hono.

at room temperature is 0.3570 nm, which is equivalent to the Al–Al distance; this is only $\sim$1.5% larger than the lattice parameter of pure Ni, which is 0.3517 nm. The Ni–Al distance is $a/\sqrt{2}$, or 0.2524 nm. Thus substitution for Al is favoured by large elements such as Ta and Ti, whereas smaller atoms such as Co substitute for Ni.

Direct evidence of the ordered nature of γ' comes from experiments carried out using the atom probe, which consists of a field ion microscope coupled to a mass spectrometer. In these experiments, atoms are evaporated in the form of ions from a specimen sharpened to a radius of a few nanometres; a channel plate is used to select an appropriate pole, and the atoms are identified by measuring the time-of-flight to the detector, which depends uniquely on the charge-to-mass ratio. The sequential stripping of atoms from a series of {200} planes allows a ladder diagram to be built up [10] (see Figure 2.13), on which the total number of Al ions detected during the run is plotted against the total number of ions collected. The mean gradient of the line is one-quarter, but during the stripping of mixed Ni/Al and Ni planes the gradient changes to one-half and zero, respectively. These features of the ladder diagram provide very elegant and direct proof of the ordering of the γ' phase on the atomic scale.

Does the γ' remain ordered until the melting temperature is reached? Disordering is a possibility which should be considered as the entropy contribution to the Gibbs energy becomes larger as the temperature is raised. For many years this question remained unanswered, but it is now known that the ordering temperature of Ni_3Al is roughly equivalent to its melting temperature of about 1375 °C, although it does depend strongly upon the degree of stoichiometry and the concentrations of any impurities. This has been demonstrated [11] by doping Ni_3Al with varying amounts of Fe, which weakens the ordering; this lowers the critical ordering temperature, which can then be measured using dilatometry (see Figure 2.14). The ordering temperature of pure stoichiometric Ni_3Al is then estimated by extrapolation of the data to zero iron content; the result of $\sim$1375 °C must be termed 'virtual' since it lies above the melting point. In fact, evidence of the disordering of Ni_3Al

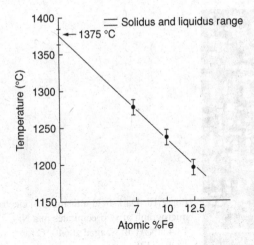

Fig. 2.14. Variation of the ordering temperature of Ni_3Al when doped with Fe [11]. Note that the ordering temperature of pure Ni_3Al can be deduced by extrapolation to zero Fe content.

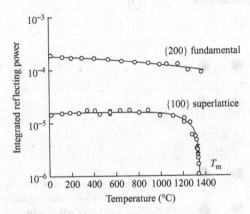

Fig. 2.15. Temperature dependence of the intensity of two diffraction lines from a Ni_3Al crystal, containing slightly less than 25 at%Al [12].

has been observed [12] using high-temperature neutron diffractometry; see Figure 2.15. In the vicinity of the melting temperature, the intensity of the {100} superlattice reflection drops sharply – implying the onset of disordering before the melting temperature is reached.

Binary Ni–Al alloys of composition consistent with a two-phase γ/γ' microstructure exhibit γ' precipitates which are often cuboidal in form [13] (see Figure 2.16). Analysis of these structures using transmission electron microscopy confirms that a distinct cube–cube orientation relationship exists between the γ' precipitates and γ matrix in which they reside (see Figure 2.17). Since the lattice parameters of the disordered γ and ordered γ' phases are very similar, the electron diffraction patterns exhibit maxima which are common, for example, from the {110}, {200}, {220} reflections, and others which are due only to γ', for example, {100}, {210} etc. The orientation relationship can be described by

$$\{100\}_\gamma //\{100\}_{\gamma'}$$
$$\langle 010 \rangle_\gamma //\langle 010 \rangle_{\gamma'} \tag{2.5}$$

which is referred to as the cube–cube orientation relationship. The γ/γ' interfaces have the $\langle 001 \rangle$ directions as their plane normals. Provided that the lattice misfit between the lattice

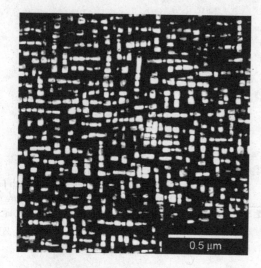

Fig. 2.16. Dark-field transmission electron micrograph of γ' precipitates in a Ni–13.4 at%Al alloy, aged at 640 °C for 1000 h [13].

(a)

γ

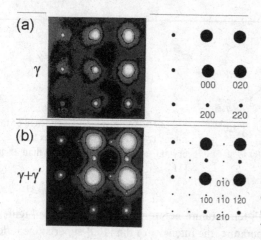

(b)

$\gamma + \gamma'$

Fig. 2.17. Selected area diffraction patterns taken in the transmission electron microscope of a single-crystal superalloy, with foil normal close to $\langle 001 \rangle$, (a) of the disordered γ and (b) of both the ordered γ and disordered γ', with superlattice reflections.

parameters of the γ and γ' phases is not too large, the γ/γ' interface remains coherent and the interfacial energy remains low. The γ' precipitates often align along the elastically soft $\langle 100 \rangle$ direction (see Figure 2.16).

The properties of the superalloys are found to depend critically on the coherency of the γ/γ' interface. This is favoured by small values of the lattice misfit, δ, defined according to

$$\delta = 2 \times \left[\frac{a_{\gamma'} - a_\gamma}{a_{\gamma'} + a_\gamma} \right] \qquad (2.6)$$

which is thus negative if $a_{\gamma'} < a_\gamma$ and positive if the converse is true. Consistent with Vegard's law, one has relationships such as $a_\gamma = a_\gamma^0 + \sum_i \Gamma_i^\gamma x_i^\gamma$ and $a_{\gamma'} = a_{\gamma'}^0 + \sum_i \Gamma_i^{\gamma'} x_i^{\gamma'}$ so that the lattice parameters are linear with the mole fractions of added solutes. The Vegard coefficients Γ_i^γ and $\Gamma_i^{\gamma'}$ show a strong dependence on the position of i in the periodic table (see Figure 2.18). Values are largest for elements from the far west and far east of

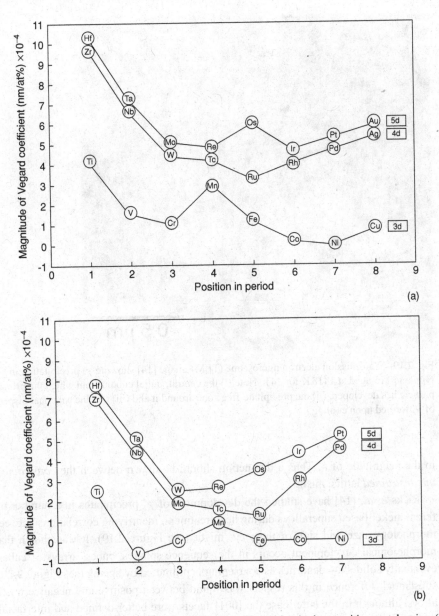

Fig. 2.18. Variation of the Vegard coefficients with position in the transition metal series: (a) for the disordered FCC phase, γ, and (b) for the ordered Ni_3Al phase, γ'. Data from ref. [3].

the transition metal block, when the discrepancy in atomic size is largest. Moreover, the lattice parameter of the γ phase shows the greater sensitivity to solute additions. The lattice misfit, δ, then depends on two factors: (i) the partitioning of solutes i between γ and γ' and (ii) the corresponding influence of the solutes i on the lattice parameters consistent with Vegard's law. Coarsening of the γ' particles leads to loss of coherency and an increase

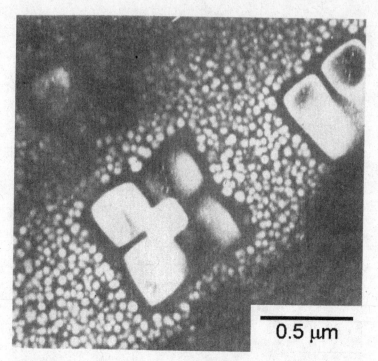

0.5 μm

Fig. 2.19. Transmission electron microscope (TEM) image [14] showing γ' precipitation in Nimonic 115 aged at 1418 K for 24 h. Note (i) the central, initial cuboid from which the primary γ' particle has developed, (ii) the precipitate-free zone around it and (iii) the fine spherical dispersion of γ' formed upon cooling.

in the magnitude of δ. Thus a distinction should be drawn between the *constrained* and *unconstrained* lattice misfit.

Ricks *et al.* [14] have studied the development of γ' precipitates in a number of different nickel-based superalloys during heat treatment, identifying correlations between the morphology, size and sign of the γ/γ' misfit (see Figure 2.19). It was shown that the morphological development occurs in the sequence spheres, cubes, arrays of cubes and eventually solid-state dendrites as coarsening is promoted by ageing (see Figure 2.20). No substantial difference in this respect was found between positive and negative misfitting alloys, although in the latter case the {001} facets were better defined and interfacial dislocation networks were found to develop at an earlier stage in the γ' growth sequence. Interestingly, both the size at which the γ' particles depart from the spherical morphology and the severity of the heat treatment required to form the cuboidal arrays were found to be sensitive functions of the lattice misfit; thus, for alloys of low misfit the size of the γ' particle needs to be larger before the effects of misfit strain are sufficient to influence the particle shape. For the Nimonic alloys 80A, 90, 105 and 115, for which convergent beam electron diffraction (CBED) was used to estimate the room-temperature misfits at +0.32%, +0.34%, −0.04% and −0.18%, respectively, the γ' precipitate sizes at which significant departure from the spherical shape was observed were found in turn to be 0.3, 0.3, 0.7 and

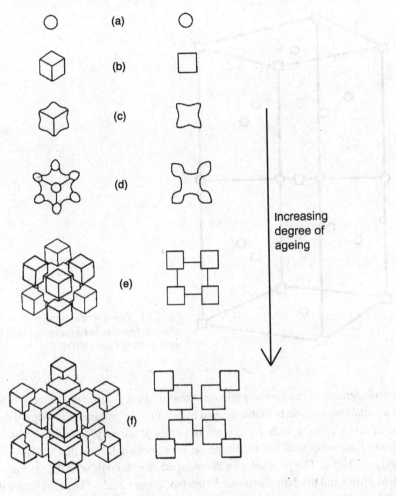

Fig. 2.20. Schematic diagram illustrating the development, from (a) to (f), of strain-induced, faceted γ' morphologies during ageing: left, projection along $\langle 111 \rangle$; right, projection along $\langle 001 \rangle$. Adapted from ref. [14].

0.5 µm. This suggests that when the magnitude of the misfit is small, the γ' particles must grow to a larger size before the cuboidal form is found.

2.1.3 Other phases in the superalloys

A The gamma double prime phase

In nickel–iron superalloys such as IN718 and IN706, which contain quantities of niobium, the primary strengthening precipitate is not γ' but instead a body-centred tetragonal (BCT) ordered compound; this displays the $D0_{22}$ crystal structure and a composition which can be represented approximately by Ni_3Nb. In IN718, the lattice parameters are approximately $c = 0.740$ nm and $a = 0.362$ nm [15]. The unit cell is shown in Figure 2.21. One can see

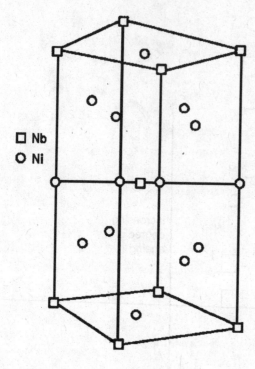

□ Nb
○ Ni

Fig. 2.21. The unit cell of the γ'' precipitate which is found in iron–nickel superalloys such as IN718 and IN706.

that the arrangement of the atoms is related to that of the L1$_2$ structure of the γ' phase; in fact the a parameter is close to that displayed by γ' but the c parameter is roughly doubled. Because of this similarity with γ', one refers to this phase as γ''. As with γ', a distinct orientation relationship with the matrix phase γ is displayed, such that $\langle 001 \rangle_{\gamma''}//\langle 001 \rangle_{\gamma}$ and $\langle 100 \rangle_{\gamma''}//\langle 100 \rangle_{\gamma}$. The γ'' displays a disc-shaped morphology, with the thickness often as small as 10 nm and the diameter about 50 nm (see Figure 2.22). The particles are usually found to be coherent with the matrix, with coherency strains equivalent to several per cent. The excellent high-temperature properties of IN718 – which is used very widely in polycrystalline cast-and-wrought form to operating temperatures approaching 650 °C on account of its relatively low cost – are due to the coherency strains so imparted and the limited number of available slip systems which operate in γ''. The kinetics of formation of γ'' are sluggish as a consequence of the high coherency strains.

Nickel–iron alloys which are strengthened by γ'' are susceptible to the formation of an orthorhombic δ phase in the overaged condition [16,17]. This is invariably incoherent with γ, and therefore does not confer strength even when present in significant quantities. It forms in the temperature range 650 to 980 °C, but the characteristics of its formation depend strongly upon temperature. Below 700 °C, nucleation of δ is observed at γ–grain boundaries and growth occurs at the expense of γ''. In the range 700 °C to 885 °C, formation of δ is accompanied by rapid coarsening of γ''; beyond the solvus temperature of 885 °C γ'' is no longer stable. At temperatures between 840 °C and 950 °C, the plates of δ form rapidly in times less than 24 h. The δ solvus temperature is ~1000 °C. It should be emphasised, however, that the kinetics of formation of δ are strongly accelerated if forging is carried

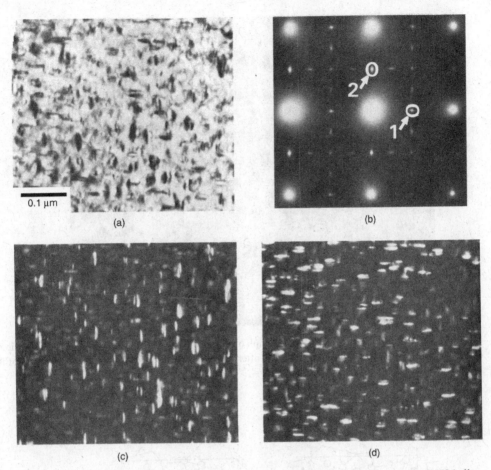

Fig. 2.22. Transmission electron micrographs [17] showing the γ' and γ'' phases in the IN706 alloy aged at 732 °C for 16 h: (a) bright-field image, (b) diffraction pattern, (c) dark-field image using spot 1 and (d) dark-field image using spot 2.

out below the δ solvus; the resultant δ-phase distribution can be used to control and refine the grain size to help optimise the tensile and fatigue properties. However, the formation of extensive amounts of δ during service leads to severe degradation of properties and is to be avoided.

B The TCP phases

Excessive quantities of Cr, Mo, W and Re promote the precipitation of intermetallic phases which are rich in these elements [18] (see Figure 2.23). The resulting phases have a number of distinct characteristics: (i) a high and uniform packing density of atoms, (ii) a degree of non-metallic, directional bonding and (iii) complex crystal structures, each built up of distinct tessellated layers consisting of arrays of hexagons, pentagons and triangles (see Figure 2.24) stacked into a limited number of so-called Kasper co-ordination polyhedra

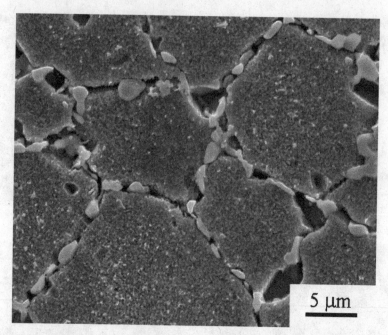

Fig. 2.23. Scanning electron micrograph of the superalloy RR1000 showing extensive precipitation of the σ phase at γ–grain boundaries, after a heat treatment of 5000 h at 750 °C. Courtesy of Rob Mitchell.

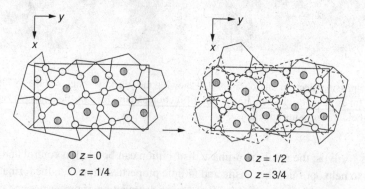

Fig. 2.24. The unit cell of the P phase showing the primary and secondary layers. For clarity, all atoms are represented by circles of the same size. Adapted from ref. [19].

[19]. This last fact gives rise to the name *topologically close-packed*, or TCP, phases. Generally speaking, the TCP phases have chemical formulae $A_x B_y$, where A and B are transition metals, such that A falls to one side of the group VIIB column defined by Mn, Tc and Re (see Figure 1.21) and B falls to the other [19]. The μ phase is based on the ideal stoichiometry $A_6 B_7$ and has a rhombohedral cell containing 13 atoms; examples include $W_6 Co_7$ and $Mo_6 Co_7$. The σ phase is based upon the stoichiometry $A_2 B$ and has a tetragonal cell containing 30 atoms; examples include $Cr_2 Ru$, $Cr_{61} Co_{39}$ and $Re_{67} Mo_{33}$. The P phase, for example, $Cr_{18} Mo_{42} Ni_{40}$ is primitive orthorhombic, containing 56 atoms per cell. Finally,

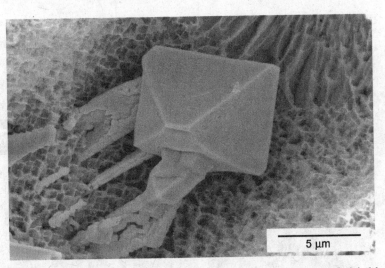

Fig. 2.25. Example of a blocky MC carbide formed in an experimental nickel-based single-crystal superalloy [20].

the R phase, for example, $Fe_{52}Mn_{16}Mo_{32}$, has a rhombohedral cell containing 53 atoms. It should be emphasised that the TCP phases can have very different compositions from those quoted here, and that the stoichiometry ranges are often very wide. However, the fact that the crystal structures are based upon A_xB_y, with A and B thus displaying different electronegativities, suggests that electronic factors are important in determining stability, in an analogous way to the 'electronic compounds' of Hume-Rothery in alloys which contain combinations of Cu, Ag and Au.

C Carbide and boride phases

Various carbide and boride species form in the superalloys, the type depending upon the alloy composition and the processing conditions employed. Some of the more important types include MC, M_6C, $M_{23}C_6$, M_7C_3 and M_3B_2, where M stands for a metal atom such as Cr, Mo, Ti, Ta or Hf. In many superalloys the MC carbide, which is usually rich in Ti, Ta and/or Hf since these are strong carbide formers, precipitates at high temperatures from the liquid phase. Consequently, the carbide is often found in interdendritic regions and no distinct orientation relationship with the matrix is displayed. A number of morphologies, for example, globular, blocky and script have been reported [20] (see Figure 2.25). Carbides such as $M_{23}C_6$ form at lower temperatures – around 750 °C – during protracted periods of service exposure, particularly in alloys which are rich in Cr. Their formation has been attributed to the breakdown of the MC carbides, via reactions of the type

$$MC + \gamma \rightarrow M_{23}C_6 + \gamma' \qquad (2.7)$$

The $M_{23}C_6$ carbide is usually found to precipitate on the γ–grain boundaries. The complex cubic structure exhibited by this phase is, in fact, closely related to that of σ – with the carbon atoms removed, the crystal structures resemble each other [18]. This fact explains

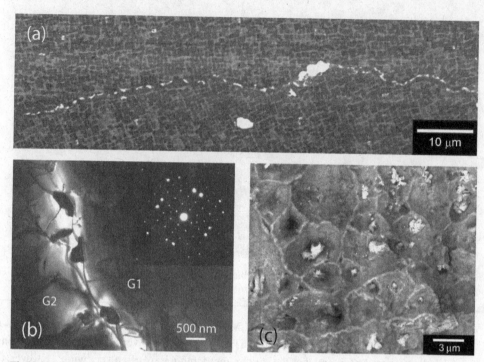

Fig. 2.26. Precipitation of the $M_{23}C_6$ carbide on the γ–grain boundary in an experimental second-generation single-crystal superalloy doped with carbon and boron and crept at 950 °C and 290 MPa [21]: (a) scanning electron micrograph; (b) transmission electron micrograph and diffraction pattern identifying the $M_{23}C_6$ carbide; (c) dimpled fracture surface of creep specimen, showing evidence of ductile failure mode.

why σ is found to nucleate sympathetically from $M_{23}C_6$ in polycrystalline turbine disc alloys such as Udimet 720Li and RR1000.

The role of carbides and borides has proved controversial, but it is now accepted that the high-temperature creep properties of the superalloys are improved in many circumstances if carbon and boron are present; the formation of these phases is then very likely. This is primarily due to the preferred location of carbon, boron, carbides and borides at the γ–grain boundaries, which has a potent effect on the rupture strength via the inhibition of grain-boundary sliding. Thus, carbon and boron are often referred to as grain-boundary strengtheners. This explains why carbon and boron levels are generally higher in polycrystalline or columnar-grained superalloys used in cast form, and the absence of these elements in the single-crystal superalloys, for which grain-boundary strengthening is unnecessary. The benefits of doping with carbon and boron when grain boundaries are present has been confirmed by Chen and Knowles [21], who tested an experimental second-generation single-crystal superalloy in creep at 950 °C and 290 MPa. When grain boundaries of misorientation equal to 10° were introduced, carbon and boron concentrations of 0.09 and 0.01 wt%, respectively, were sufficient to improve the rupture life from 10 h to 100 h. This effect was attributed to small, coherent and closely spaced $M_{23}C_6$ carbides precipitating on the grain boundaries (see Figure 2.26).

2.2   Defects in nickel and its alloys

A nickel alloy, despite being a crystalline solid, possesses a number of intrinsic defects present on the scale of the lattice. One can distinguish between (i) planar defects such as stacking faults, (ii) line defects such as dislocations and (iii) point defects, for example, vacancies, interstitials and anti-site defects. In turns out that the mechanical properties displayed by the superalloys are strongly dependent upon the type and concentration of defects which occur. In this section, consideration is given to the details of the defects which arise and to some of the implications of their presence.

2.2.1   Defects in the gamma (FCC) phase

The face-centred cubic (FCC) structure consists of close-packed planes, stacked with periodicity equal to three. Thus the lattice, if it is perfect, can be denoted ABCABCABCABC..., where each of A, B or C represents a close-packed layer; this contrasts with the hexagonally close-packed (HCP) structure, which is denoted ABABABABAB... Since the lattice vector is $a/\sqrt{2}$, where a is the length of the side of the unit cell, one can demonstrate that two consecutive close-packed layers are displaced by a vector of magnitude $a/\sqrt{6}$, measured within a close-packed plane. The spacing of the close-packed planes is $a/\sqrt{3}$.

A   Planar defects – the stacking fault

A stacking fault arises when the stacking sequence ABCABCABCABC... is interrupted [22]. The *removal* of one plane introduces an *intrinsic* stacking fault, for example, ABCACABCABC..., and causes two neighbouring planes to possess the HCP co-ordination locally. Conversely, the *addition* of one plane introduces an *extrinsic* stacking fault, for example, ABCACBCABCABC... Although two HCP planes are again introduced, these are separated by a single plane of FCC co-ordination, and are thus a distance $2a/\sqrt{3}$ apart – a spacing corresponding to two lattice planes.

B   Line defects – dislocations

In a macroscopic sense, the slip system in an FCC metal such as Ni is $a/2\langle 1\bar{1}0\rangle\{111\}$; hence the Burgers vector is $a/\sqrt{2}$. However, this statement does not properly respect the micromechanics of deformation; it is well established that the glide of an $a/2\langle 1\bar{1}0\rangle\{111\}$ dislocation occurs by the passage of two *partial* dislocations, which, although in close proximity to each other, are separated by a distance which depends upon the force of their elastic repulsion and the energy of the planar stacking fault so produced (see Figure 2.27). This implies that the $a/2\langle 1\bar{1}0\rangle\{111\}$ dislocations are dissociated; electron microscopy confirms that the reaction is of the form

$$\frac{a}{2}\langle 110\rangle\{\bar{1}11\} \rightarrow \frac{a}{6}\langle 211\rangle\{\bar{1}11\} + \frac{a}{6}\langle 12\bar{1}\rangle\{\bar{1}11\} \qquad (2.8)$$

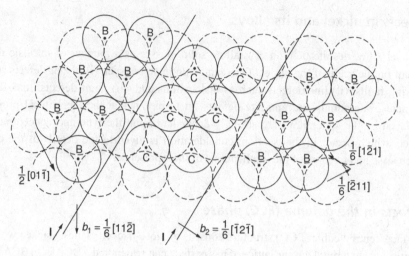

Fig. 2.27. Dislocation on a {111} plane of γ, dissociated into two Shockley partial dislocations [22].

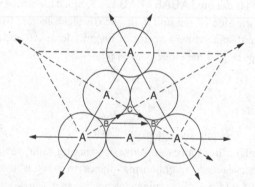

Fig. 2.28. Geometry of the close-packed plane, appropriate for the dissociation into Shockley partials; the large arrow corresponds to $\frac{a}{2}\langle 110 \rangle$, and the smaller arrows correspond to $\frac{a}{6}\langle 211 \rangle$ and $\frac{a}{6}\langle 12\bar{1} \rangle$.

where each dislocation on the right-hand side of the equation is known as a Shockley partial. The magnitude of the Burgers vector of each is $a/\sqrt{6}$, equivalent to the displacement of two neighbouring close-packed planes in the perfect FCC crystal (see Figure 2.28). Thus the passage of a single Shockley partial dislocation causes the introduction of an intrinsic stacking fault, which is removed by the action of the second – hence the width of the stacking fault depends upon the spacing of the partials. The reader can readily apply Frank's rule, which states that the elastic energy of a dislocation is proportional to the square of the magnitude of the Burgers vector, to confirm that the dissociation of Equation (2.8) is energetically favourable.

It is of interest to determine the relationship between the spacing of the partial dislocations and the stacking fault energy (SFE) – the energy per unit area of the planar defect between the two partials, denoted γ_{SF}. Consider the dissociation of an edge dislocation (Figure 2.29). The spacing of the partials is given the symbol d; each has a Burgers vector of magnitude $a/\sqrt{6}$, which can be resolved into an edge component, $a\cos\{\pi/6\}/\sqrt{6}$, and a screw component, $a\sin\{\pi/6\}/\sqrt{6}$, $\pi/6$ being the angle between the Burgers vectors of the dissociated and undissociated dislocations. It follows from the geometry of the situation, see Figure 2.29,

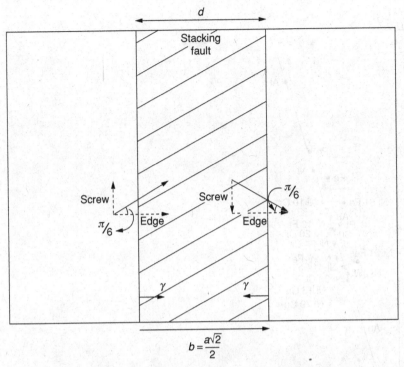

Fig. 2.29. Schematic illustration of the dissociation of an edge dislocation into two Shockley partial dislocations, which are in turn resolved into their edge and screw components.

that the screw components of the dissociated dislocations are of opposite sense so that they provide an attractive force complementing that due to the stacking fault energy. On the other hand, the edge components are of equivalent sign and thus repel each other; it is this effect which provides the driving force for dissociation.

The force balance consists therefore of three terms, the repulsive force per unit length of dislocation due to the edge components of the partials, balanced at equilibrium by attractive forces per unit length arising from the screw components and the stacking fault energy itself. It is given by

$$\gamma_{SF} + \frac{Gb_s^2}{2\pi d} = \frac{Gb_e^2}{2\pi d(1 - v)} \tag{2.9}$$

where b_e and b_s are the edge and screw components of the partials of magnitude $a\cos\{\pi/6\}/\sqrt{6}$ and $a\sin\{\pi/6\}/\sqrt{6}$, respectively, G is the shear modulus and v is Poisson's ratio. Note that isotropic elasticity is assumed to hold. The forces $Gb_s^2/2\pi d$ and $Gb_e^2/2\pi d(1 - v)$ arise from the shear stresses $Gb_s/2\pi d$ and $Gb_e/2\pi d(1 - v)$ acting at a distance d from the centres of perfect screw and edge dislocations, respectively, as given by elementary dislocation theory; the forces per unit length, F, are obtained by multiplying by the appropriate shear stress, τ, by the corresponding Burgers vector, b, consistent with $F = \tau b$. After some algebraic manipulation, one can show that the spacing of the partials

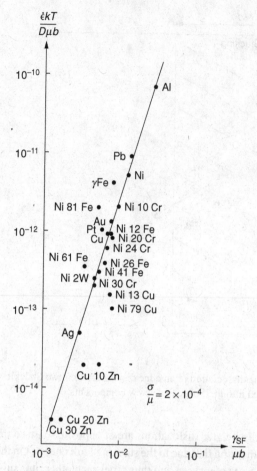

Fig. 2.30. Correlation between the normalised steady-state creep rate, $\dot{\epsilon}$, and the stacking fault energy, γ_{SF}, for a number of solid-solution-strengthened alloys [23].

is given by

$$d = \frac{Ga^2}{48\pi \gamma_{SF}} \left(\frac{2 + \nu}{1 - \nu} \right) \tag{2.10}$$

the important result being that the spacing, d, is inversely proportional to the stacking fault energy, γ_{SF}. All dislocations, whether edge, screw or mixed, are dissociated in this way. It can be demonstrated by the reader that if a screw rather than an edge dislocation is considered, the spacing, d, is reduced by a factor $(2 - 3\nu)/(2 + \nu)$, which is approximately one-half if $\nu = 0.33$.

For high-temperature applications, the magnitude of the stacking fault energy is an important consideration. This is because screw portions of perfect dislocations, dissociated into their Shockley partials, must constrict locally if they are to cross-slip. This is a common mechanism by which obstacles such as precipitates can be circumvented. On this basis, widely spaced partials arising from low values of γ_{SF} are advantageous to high-temperature properties; this has been confirmed by the work of Mohamed and Langdon [23], who measured the steady-state creep rate, $\dot{\epsilon}$, of a number of solid-solution-strengthened nickel alloys (see Figure 2.30). A correlation between the parameter $\dot{\epsilon}kT/(DGb)$ was found, where D is the diffusion coefficient, and the stacking fault energy, γ_{SF}, normalised by Gb. Transmission electron microscopy studies have confirmed a strong influence of alloying elements on the stacking fault energy. See, for example, the data [24] for Ni–Co alloys (Figure 2.31), which indicates that γ_{SF} decreases from about $250 \, \text{mJ/m}^2$ for pure Ni to less than half this value for an alloy of composition Ni–50wt%Co. It appears, however, that there have been no systematic attempts to correlate the influence of stacking fault energy on the type of solute, although Delehouzee and Deruyttere [25] have measured the stacking fault probability, α, a measure of the number of stacking faults existing in a metal, from the broadening and shifting of diffraction peaks using X-ray methods; this is expected to be inversely proportional to γ_{SF}. Their results are given in Figure 2.32. The slope of the lines correlate strongly with the valency difference between Ni and the solute being added. Given these results, on the basis of at% solute added one would expect elements such as Ti, which lie at the far left of the d block of transition metals, to have the greatest effect on γ_{SF} and the high-temperature properties which are influenced by it. Such alloying effects are important since Ni itself has one of the largest stacking fault energies of the FCC metals at $250 \, \text{mJ/m}^2$; by contrast, for Al, Cu, Au and Ag the values are approximately 200, 55, 50 and $22 \, \text{mJ/m}^2$, respectively. Thus the partial dislocations in unalloyed Ni are relatively closely spaced.

C Point defects – vacancies

Vacancies are present in the lattice of high-purity Ni, because the increase in free energy caused by the breaking of bonds is more than offset by the term due to the configurational entropy which then arises. From statistical mechanics one expects the equilibrium vacancy fraction, f_{Va}^{eq}, to be given by

$$f_{Va}^{eq} = \exp\left\{-\frac{\Delta G_{Va}^{f}}{RT}\right\} = \exp\left\{+\frac{\Delta S_{Va}^{f}}{R}\right\} \times \exp\left\{-\frac{\Delta H_{Va}^{f}}{RT}\right\} \qquad (2.11)$$

where ΔG_{Va}^{f}, ΔH_{Va}^{f} and ΔS_{Va}^{f} are, respectively, the Gibbs energy, enthalpy and entropy of formation of a vacancy. For most metals, the pre-exponential term given by $\exp\{+\Delta S_{Va}^{f}/R\}$ lies between 1 and 10, and is nearly invariant with temperature. Electrical resistivity measurements on quenched and quenched/tempered nickel [26] indicate that ΔH_{Va}^{f} is $\sim 160 \, \text{kJ/mol}$; taking the pre-exponential term as 5, this gives a fraction f of 10^{-28} at ambient temperature, rising to 8.5×10^{-5} at the melting temperature of $1455 \, °\text{C}$. This value is comparable to other FCC metals; for example, for Al, Au and Ag the enthalpies ΔH_{Va}^{f} are $65 \, \text{kJ/mol}$, $85 \, \text{kJ/mol}$ and $107 \, \text{kJ/mol}$, respectively, so that f_{Va}^{eq} is 9.4×10^{-4}, 7.2×10^{-5} and 1.7×10^{-5} at the melting temperature of the metal concerned.

The presence of vacancies can be inferred directly from measurements of the lattice spacing and thermal expansion coefficient, i.e. using a combination of diffractometry and dilatometry. This is because one atomic site is added each time a vacancy is created at a free surface, grain boundary or dislocation. The corresponding changes in the dimensions of the

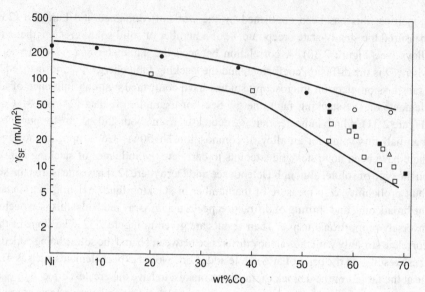

Fig. 2.31. Variation of the stacking fault energy, γ_{SF}, with composition for a number of binary Ni–Co alloys [24].

specimen will be detectable by dilatometry; however, if internal stresses can be ignored, the lattice spacing will not be altered substantially. To illustrate this, consider a one-dimensional analogy: a perfect row of N atoms of spacing a_0 which is heated from T_0 to T_1, causing (i) the formation of n vacancies such that $n \ll N$ and (ii) a change in length from L_0 to L_1. The expansion $\Delta L = L_1 - L_0$ has two contributions: the first is due to the increase in the periodicity from a_0 to a_1 due to the thermal expansion, denoted Δa, and the second is due to the introduction of the vacancies. Hence,

$$\Delta L = (N \times \Delta a) + (n \times a_1) \tag{2.12}$$

Dividing through by $Na_1 \sim L_1$ and rearranging, one has

$$f_{Va}^{eq} = \frac{n}{N} = \frac{\Delta L}{L_1} - \frac{\Delta a}{a_1} \tag{2.13}$$

For metals, this simple estimate is, in fact, in error by a factor of 3, since one has in reality a three-dimensional rather than a one-dimensional lattice; the fraction f_{Va}^{eq} is then given instead by $3(\Delta L/L_1 - \Delta a/a_1)$. Small errors are expected due to strain relaxation effects and the possibility of vacancy clustering to form divacancies; nevertheless, $3(\Delta L/L_1 - \Delta a/a_1)$ is usually found to be of the order 10^{-4} for most metals [27]; see Figure 2.33.

   The presence of vacancies in the nickel alloys is of great significance, since they mediate the diffusional flow required for a number of phenomena: creep occurring at a rate dependent upon diffusional rearrangements at dislocation cores, coarsening of the γ' precipitates on the scale of their periodicity, and oxidation, which occurs at a rate which is diffusion-controlled; clearly, the retardation of diffusional processes is important if the very best properties are to be attained. Consider the self-diffusion of nickel. The diffusion coefficient, D_{Ni}^{self}, depends on two factors:

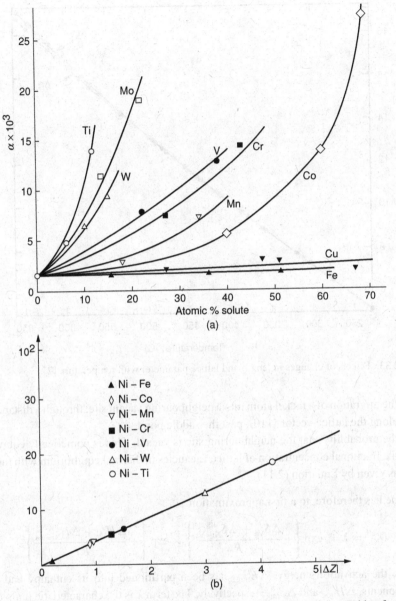

Fig. 2.32. For a number of binary nickel alloys [25]: (a) variation of the stacking fault probability with solute content, and (b) correlation of the slope of (a) at zero solute content, with the difference between the valency of nickel and solute denoted ΔZ.

(i) the rate of successful atom–vacancy exchanges, denoted Γ_{Va}, which is given by

$$\Gamma_{Va} = \nu_o \exp\left\{-\frac{\Delta E_{Ni-Va}^{mig}}{RT}\right\} \tag{2.14}$$

where ν_o is an attempt frequency and ΔE_{Ni-Va}^{mig} is an activation energy associated with

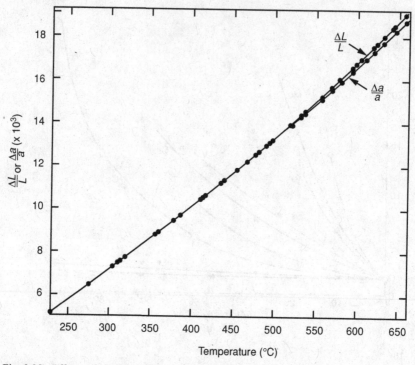

Fig. 2.33. Effects of changes in length and lattice parameter with temperature [27].

the migration of a nickel atom into a neighbouring vacant site, through a distance $a/\sqrt{2}$ along the lattice vector $\langle 110 \rangle$, past the saddle point;

(ii) the probability that the neighbouring site is vacant. This is numerically equivalent to the fractional concentration of lattice vacancies in thermal equilibrium with the lattice, as given by Equation (2.11).

One has therefore, to a first approximation,

$$D_{\mathrm{Ni}}^{\mathrm{self}} = \lambda^2 \nu_o \exp\left\{+\frac{\Delta S_{\mathrm{Va}}^f + \Delta S_{\mathrm{Ni-Va}}^{\mathrm{mig}}}{R}\right\} \times \exp\left\{-\frac{\Delta H_{\mathrm{Va}}^f + \Delta H_{\mathrm{Ni-Va}}^{\mathrm{mig}}}{RT}\right\} \qquad (2.15)$$

where the activation energy, ΔE_{mig}, has been partitioned into its enthalpy and entropy components, ΔH_{mig} and ΔS_{mig}, respectively. The term λ is the characteristic jump distance given by $a/\sqrt{2}$. One can see that the apparent activation energy for self-diffusion, Q_{self}, is composed of two terms, such that

$$Q_{\mathrm{self}} = \Delta H_{\mathrm{Va}}^f + \Delta H_{\mathrm{Ni-Va}}^{\mathrm{mig}} \qquad (2.16)$$

The experimental measurements indicate that $Q_{\mathrm{self}} \sim 280\,\mathrm{kJ/mol}$ [28]. Hence, $\Delta H_{\mathrm{Ni-Va}}^{\mathrm{mig}} \sim 120\,\mathrm{kJ/mol}$, given the value of ΔH_{Va}^f quoted above. This value is associated with the opening of the diffusion window along $\langle 110 \rangle$.

The presence of vacancies influences the rate of diffusion of substitutional solutes in nickel. The rate of interdiffusion of the 4d and 5d transition metals has been studied

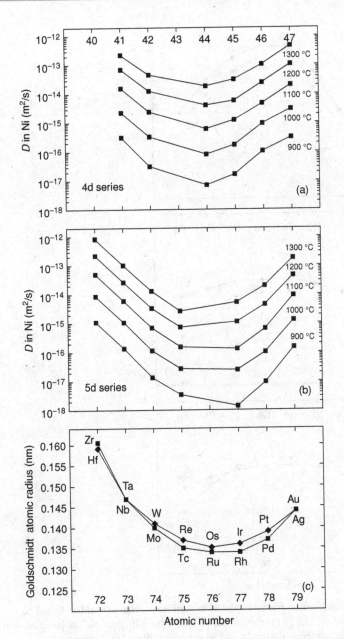

Fig. 2.34. Measurements of the interdiffusion coefficient in binary Ni/Ni–X diffusion couples: (a) for the 4d series of the d block of transition metals, (b) for the 5d series, and (c) variation of the Goldschmidt radius of the transition metals with atomic number [29,30].

experimentally [29,30], by analysing diffusion couples fabricated from Ni and Ni–X alloys, where X is a transition metal. Elements near the centre of the d block of the periodic table, for example, Re and Ru, were found to diffuse several orders of magnitude more slowly than those from the far west or far east of the d block, for example, Hf and Au; see Figure 2.34. These experimental results are interesting, since they disprove the widely held

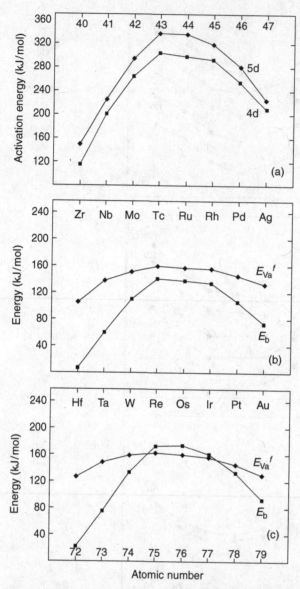

Fig. 2.35. Results of calculations from a model of the diffusion behaviour of the 4d and 5d transition metals in nickel [31,32]: (a) the activation energy for diffusion, and the partition of it into vacancy and barrier energy terms for the 4d and 5d solutes in (b) and (c), respectively.

view that diffusion is least rapid when the lattice misfit between solute and solvent atom radii is the greatest. These observations have been rationalised using quantum-mechanical modelling using first principles methods [31,32]; this has allowed an evaluation of the activation energies for solute–vacancy exchange, denoted E_b, and the vacancy formation energy for a solute–vacancy complex, denoted E^f_{X-Va}. The calculations indicate (see Figure 2.35) that E^f_{X-Va} does not vary strongly as one crosses the d block and by no more than 40 kJ/mol; in fact, the magnitudes of $E^f_{\text{Hf}-\text{Va}}$ and $E^f_{\text{Au}-\text{Va}}$ are close to 120 kJ/mol, and thus are close to the vacancy formation energy in pure Ni, implying that the solute–vacancy binding energy for elements such as Hf and Au is negligible. The Re–Va binding energy is about +40 kJ/mol. Thus, the dependence of $E^f_{\text{X}-\text{Va}}$ on atomic number is relatively weak. Instead, the major contribution to the variation of the interdiffusion coefficient with atomic number arises from the diffusion energy barrier, E_b. This is significant for elements such as Re and Ru. Alloying with these elements, which are amongst the densest in the d block owing to their electronic bonding and many unpaired electrons, causes directional and incompressible Ni–Re and Ni–Ru bonds which do not favour solute–vacancy exchanges. This effect dominates any influences of atomic radius, misfit strain and differences in the solute–vacancy binding energy. It is probable that this is the reason why Re has such a significant effect on the high-temperature properties of the superalloys.

2.2.2 Defects in the gamma prime phase

A Planar defects – the anti-phase boundary

An important planar defect found in the nickel alloys is the *anti-phase boundary*, the meaning of which can be understood in the following way. Consider two perfect crystals of γ' displaced by $a/2\langle\bar{1}01\rangle$; this vector links neighbouring Ni and Al atoms, but note that it is not a lattice vector. When bonded together, an interface known as the anti-phase boundary (APB) separates the otherwise perfect lattices; in the vicinity of the APB, the number of Ni–Al bonds is substantially reduced since 'forbidden' Ni–Ni and Al–Al bonds are formed there. The net effect is to cause an APB energy which is substantial – usually of order 100 mJ/m^2. The number of forbidden bonds is found to depend upon the crystallographic plane on which the APB resides [33] (see Figure 2.36); consequently, for this reason, the APB energy is anisotropic. This can be rationalised as follows. First, consider two {111} planes of the L1$_2$ structure stacked perfectly (Figure 2.37) – note that there are no instances in which two Al atoms are nearest neighbours. Upon displacing the second layer by $a/2\langle1\bar{1}0\rangle$, forbidden bonds are created: one is formed for each area, $a\sqrt{2}/2 \times a\sqrt{6}/2$ or $a^2\sqrt{3}/2$. The APB energy on the {111} plane is then $2\Omega/(a^2\sqrt{3})$, where Ω is the ordering energy, defined by $\Omega = \Omega_{\text{NiAl}} - \frac{1}{2}(\Omega_{\text{NiNi}} + \Omega_{\text{AlAl}})$, where Ω_{NiAl}, Ω_{AlAl} and Ω_{NiNi} are the Ni–Al, Al–Al and Ni–Ni bond energies, respectively. Consider next the {110} plane (Figure 2.38). One forbidden bond is formed per area $a \times a\sqrt{2}/2$ or $a^2\sqrt{2}/2$ so that the APB energy is $2\Omega/(a^2\sqrt{2})$, which is about 23% higher than for an APB on the {111} plane. Finally, consider the APB energy on the {001} plane. In this case, Figure 2.36 confirms that a translation of $a/2\langle1\bar{1}0\rangle$ does not cause forbidden bonds; one way to realise this is to consider the {200} planes, every second one of which consists solely of Ni atoms. It follows then that Al atoms are never brought into nearest-neighbour positions, so that the APB energy will be close to zero. In practice, the APB energy on the {001} plane is non-zero; it should be remembered that here only first-order nearest-neighbour bonding has been accounted for in these arguments.

The planar fault known as the APB is of profound importance in superalloy metallurgy. This is because the γ' phase deforms very differently from the disordered γ matrix in which

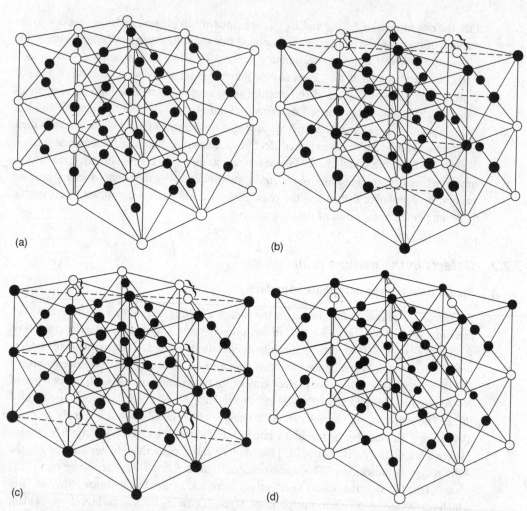

(a)   (b)

(c)   (d)

Fig. 2.36. Structure of the perfect L1$_2$ lattice, and anti-phase boundary (APB) structures caused by the APB fault lying on different planes. (a) The ordered L1$_2$ structure; (b) (111)[110] anti-phase boundary in the L1$_2$ structure; (c) (110)[110] anti-phase boundary in the L1$_2$ structure; (d) (001)[110] anti-phase boundary in the L1$_2$ structure. Adapted from ref. [33].

it resides. In γ, slip deformation is due to dislocation glide on the $a/2\langle 1\bar{1}0\rangle\{111\}$ system; the $a/2\langle 1\bar{1}0\rangle$ lattice vector is the shortest one available, and it lies in the close-packed plane. However, in the L1$_2$ structure, the close-packed planes are again $\{111\}$, but the shortest lattice vectors, $a\langle 100\rangle$, do not reside in them. Consequently, at first sight it is not clear whether the preferred slip system should be $a\langle 100\rangle\{001\}$, $a/2\langle 1\bar{1}0\rangle\{001\}$ or $a\langle 1\bar{1}0\rangle\{111\}$; in fact, at low temperatures it is the last one which is observed. This causes the following effect. A single $a/2\langle 1\bar{1}0\rangle\{111\}$ dislocation in γ, although a perfect dislocation in that phase, cannot enter γ' without an APB fault being formed and therefore without a substantial energy penalty – this is because the closure vector required to restore the γ' lattice to its perfect state is *twice* the Burgers vector of a single dislocation in γ. It follows that $a/2\langle 1\bar{1}0\rangle\{111\}$ dislocations

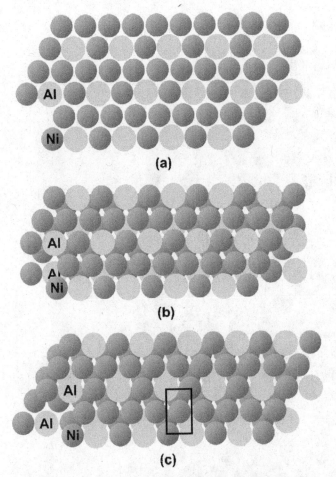

Fig. 2.37. (a) Representation of the {111} plane of the L1$_2$ crystal structure of Ni$_3$Al. (b) Two such planes stacked according to the crystal structure. (c) Situation after displacing the second plane by $a/2\langle1\bar{1}0\rangle$ on {111}. One forbidden bond is formed in the area marked.

must travel in pairs through γ'. Each dislocation is referred to as a *superpartial*, and a pair of two such dislocations is called a *superdislocation*. Since each superpartial is an imperfect dislocation in γ', two superpartials are linked by a faulted strip, or *anti-phase boundary*. Since the associated APB energy is relatively high, and particularly when the γ' fraction is significant, the superpartials usually reside in the same γ' precipitate and are then said to be 'strongly coupled'. When the γ' fraction is smaller, so-called 'weak coupling' becomes possible. The consequences of 'strong' and 'weak' coupling of superpartial dislocations are discussed in more detail in Section 2.3.

   Three other sorts of planar defect have been identified in Ni$_3$Al; see, for example, the pioneering papers of Kear and co-workers [34,35]. It is helpful to consider the stacking of successive {111} planes (see Figure 2.39) [36]; three is sufficient since this corresponds to the periodicity along $\langle111\rangle$, although the vector displacement of one layer above another is now twice that for the disordered γ phase. The {111} planes are the appropriate ones to

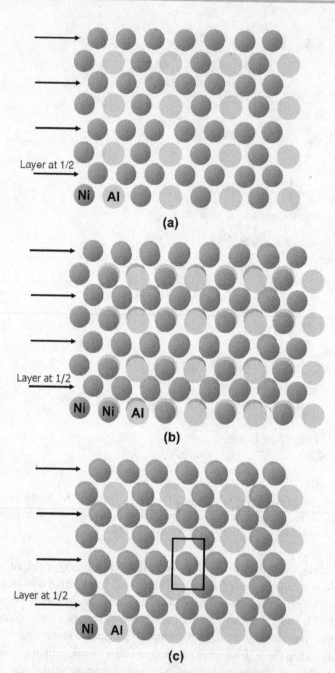

Fig. 2.38. (a) Representation of the {110} plane of the $L1_2$ crystal structure of Ni_3Al. (b) Two such planes stacked according to the crystal structure. (c) Situation after displacing the second plane by $a/2\langle 1\bar{1}0 \rangle$ on {110}. One forbidden bond is formed in the area marked.

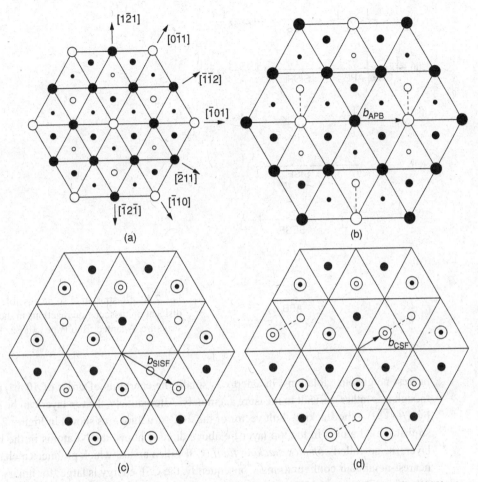

Fig. 2.39. Illustration of various faults possible in the $L1_2$ structure formed by shearing {111} planes past one another: (a) three successive {111} planes, unfaulted; (b) after shearing the top layer by $a/2\langle\bar{1}01\rangle$ to form an anti-phase boundary (APB); (c) after shearing by $a/3\langle\bar{2}11\rangle$ to form a superlattice intrinsic stacking fault (SISF); (d) after shearing by $a/6\langle\bar{1}\bar{1}2\rangle$ to form a complex stacking fault (CSF). Note that the large, medium and small circles represent atoms in the top, middle and bottom planes, respectively [36].

consider since planar faults in Ni_3Al arise only as a consequence of mechanical deformation; thermal treatment is insufficient – there have been no recorded observations of their occurrence due to recrystallisation, for example. If the top layer is sheared by $a/2\langle\bar{1}01\rangle$, as shown in Figure 2.39(b), one has an APB on {111} as discussed above, and there are changes in the nearest-neighbour bonds; for example, Al–Al bonds are formed. But other displacement vectors are possible. If one shears the top layer by $a/3\langle\bar{2}11\rangle$, see Figure 2.39(c), the Al atoms in the top layer now lie directly above those in the bottom one. The stacking sequence changes from ABCABCABCABC... to ABCACABCABC..., which is equivalent to a single {111} being removed; hence, by analogy with the intrinsic stacking fault in γ, this is referred to as a *superlattice intrinsic stacking fault* (SISF). Since it produces no

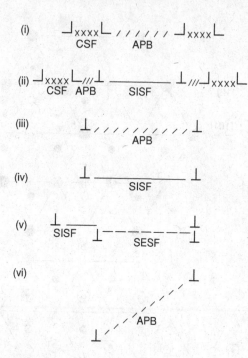

Fig. 2.40. Illustration of some possible dislocation dissociation reactions in the L1$_2$ structure. Adapted from Pope and Ezz [36].

nearest-neighbour violations, its energy is small; the energy is of order of 10 mJ/m^2, although it is rather difficult to measure accurately. Alternatively, the top layer can be shifted by $a/6\langle \overline{1}12 \rangle$ – the stacking fault vector of the FCC structure – as shown in Figure 2.39(d), so that the Al atoms in the top layer lie above directly above the Ni atoms in the bottom layer; the so-called *complex stacking fault* (CSF) then arises, which produces a change in nearest-neighbour configurations. Consequently, the CSF energy is large; for binary Ni$_3$Al it has been estimated [37] to be ~250 mJ/m^2, varying by perhaps 50 mJ/m^2 across the range of stoichiometry exhibited by the compound. Finally, the *superlattice extrinsic stacking fault* (SESF) – not shown in Figure 2.39 – is produced by displacing the top layer by $a/3\langle \overline{2}11 \rangle$. In this case, the shearing can be shown to be equivalent to the introduction of an extra {111} layer. As with the SISF fault, no nearest-neighbour violations are produced, so the fault energy is low.

B   Line defects – dislocations

Dislocations in the L1$_2$ structure of the γ' phase dissociate into partial dislocations, just as for a $a/2\langle 1\overline{1}0 \rangle\{111\}$ dislocation in γ. However, the Burgers vectors of dislocations in γ' are expected to be longer than in γ as a consequence of the ordering reaction; possible closure vectors are of length $a\sqrt{2}$ or $a\sqrt{6}$, as can be seen by analysis of Figure 2.39. Hence, there is greater scope for the dissociations to be complex. The following possibilities are suggested in the review by Pope and Ezz [36], as depicted in Figure 2.40:

(i) dissociation by CSF, APB and CSF on {111} planes, according to

$$a[\bar{1}01] \rightarrow \frac{a}{6}[\bar{1}\bar{1}2] + \frac{a}{6}[\bar{2}11] + \frac{a}{6}[\bar{2}11] + \frac{a}{6}[\bar{1}\bar{1}2] \tag{2.17}$$

(ii) dissociation by APB, CSF and SISF on {111} planes, according to

$$a[\bar{1}01] \rightarrow \frac{a}{6}[\bar{1}\bar{1}2] + \frac{a}{6}[\bar{2}11] + \frac{a}{6}[\bar{1}2\bar{1}] + \frac{a}{6}[1\bar{2}1] + \frac{a}{6}[\bar{1}\bar{1}2] + \frac{a}{6}[\bar{2}11] \tag{2.18}$$

(iii) dissociation by APB on {111} planes, according to

$$a[\bar{1}01] \rightarrow \frac{a}{2}[\bar{1}01] + \frac{a}{2}[\bar{1}01] \tag{2.19}$$

(iv) dissociation by SISF on {111} planes, according to

$$a[\bar{1}01] \rightarrow \frac{a}{3}[\bar{2}11] + \frac{a}{3}[\bar{1}\bar{1}2] \tag{2.20}$$

(v) dissociation by SISF and SESF on {111} planes according to

$$a[\bar{2}11] \rightarrow \frac{a}{3}[\bar{1}2\bar{1}] + \frac{a}{3}[\bar{1}\bar{1}2] + \frac{a}{3}[\bar{2}11] + \frac{a}{3}[\bar{2}11] \tag{2.21}$$

(vi) dissociation by APB on {010} planes, according to

$$a[\bar{1}01] \rightarrow \frac{a}{2}[\bar{1}01] + \frac{a}{2}[\bar{1}01] \tag{2.22}$$

Extensive use has been made of transmission electron microscopy to study the dislocation dissociation reactions which occur in γ'. Scheme (v) has been observed in single-crystal alloys at temperatures of around 750 °C if the applied stress level is high enough to cause primary creep; this is considered in greater detail in Section 3.3 of Chapter 3. Of the other reactions given, no evidence has been found to support the dissociations given by (ii) and (iv) [36,37]. Dissociation by (i) and (iii) can occur when γ' or $\gamma + \gamma'$ alloys are deformed at the strain rates characteristic of a tension test. However, in practice, the dissociations are usually even more complicated than those given. This is because any given superdislocation is found to dissociate on both {111} and {010}, with distinct segments of it lying on each of the two planes. For example, consider scheme (i). It is possible for the APB, or segments of it, to cross-slip to the cube plane {010}, whilst the two CSFs remain on {111}. This configuration is expected to be sessile, or 'locked', since the cube plane is not a glide plane for γ' at low temperatures. This configuration is termed a Kear–Wilsdorf lock, after those who discovered it in L1$_2$ crystal structures [38]. The behaviour of the Kear–Wilsdorf lock is central to our understanding of the anomalous yield effect displayed by L1$_2$ compounds and Ni$_3$Al in particular. This is considered in greater detail in Section 2.3.

C Point defects

The binary Ni–Al phase diagram indicates that the Ni$_3$Al compound is stable across a compositional range from approximately 23 to 27 at% Al. Thus, only a relatively small deviation from exact stoichiometry is possible, certainly very much smaller than that for the neighbouring β–NiAl compound, which displays the ordered B2 structure. Such deviations

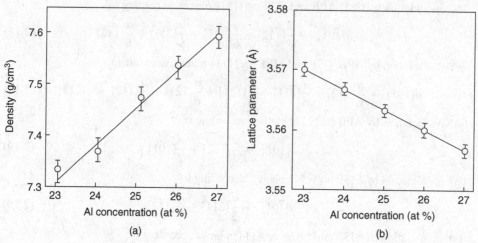

Fig. 2.41. Measurements [39] of (a) the density and (b) the lattice parameter with composition for the compound Ni_3Al.

from exact stoichiometry must be accommodated by point defects such as constitutional vacancies or anti-site defects. There are four distinct types of point defect which are conceivable, either (i) anti-site defects, i.e. Al atoms located on Ni sites, or alternatively Ni atoms on Al sites, or (ii) constitutional vacancies, i.e. vacancies on either the Ni or Al lattices. Studies have indicated that the Ni_3Al compound might more accurately be written as $Ni_{3+x}Al_{1-x}$ since non-stoichiometry is always accommodated by anti-site defects: Ni atoms on Al sites and Al atoms on Ni sites for Ni-rich and Ni-lean compositions, respectively. This is confirmed by measurements [39] of the density and lattice parameter, both of which vary monotonically as the composition passes through the point of exact stoichiometry; see Figure 2.41. In constructing a thermodynamic model for the Ni_3Al compound, Numakura et al. [40] have estimated that the density of constitutional vacancies at 950 K is very small, at $\sim 10^{-8}$ and $\sim 10^{-10}$ for the Ni and Al lattices, respectively, at the point of exact stoichiometry, varying by no more than an order of magnitude across the solubility range. Once again, this behaviour is in contrast to that of β–NiAl, for which Ni constitutional vacancies exist on the Al-rich side and Ni anti-site defects exist on the Ni-rich side [41].

The formation of anti-site defects in Ni_3Al has been inferred from the mechanical loss (internal friction) spectrum in $Ni_3(Al, Ta)$ single crystals, by measuring the free decay of vibrations in a torsional pendulum [42]. The stresses set up in this way cause time-dependent, but recoverable, *anelastic* strains, due to the movement of the defects from one site to another. The internal friction shows a sharp maximum at frequencies between 0.03 Hz and 3 Hz, the exact value depending upon the temperature; see Figure 2.42. The activation energy is then about 3.0 eV, very much greater than that for the formation of a constitutional vacancy, but a reasonable value for the sum of the activation energies for vacancy formation and migration of an anti-site defect. Since the anelastic strains necessary for this effect must be due to a defect which breaks the cubic symmetry, the lattice strain introduced by a Ni atom on an Al site cannot be responsible since it is isotropic; on the other hand, an

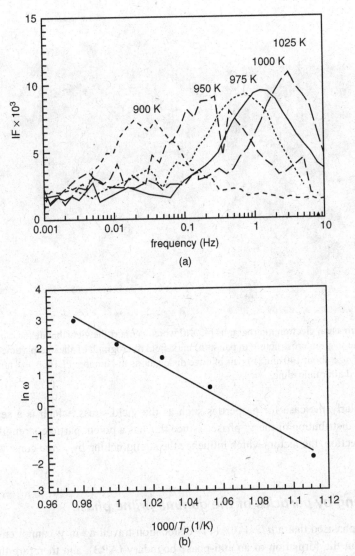

Fig. 2.42. Internal friction (IF) measurements [42] on polycrystalline $Ni_3(Al, Ta)$. (a) The variation of the IF signal as a function of frequency and temperature. (b) Arrhenius plot corresponding to (a), from which an activation energy H_{act} of 290 kJ/mol is deduced.

Al atom on a Ni site causes the tetragonal distortion that is required for the mechanical damping to occur.

2.3 Strengthening effects in nickel alloys

The mechanical properties of nickel alloys depend strongly upon the state of microstructure, which, in turn, is controlled by the chemical composition and the processing conditions.

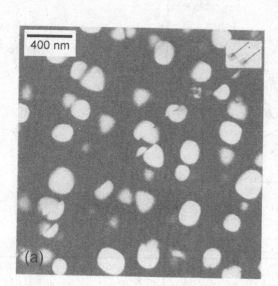

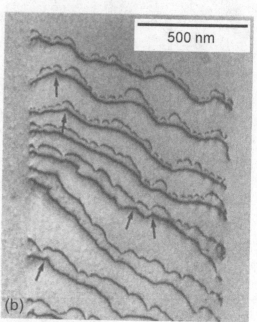

Fig. 2.43. Transmission electron micrographs [45,46] which support the view that dislocations travel through the γ/γ' microstructure in pairs: (a) dark-field micrograph of sheared particles in Nimonic 105, of size about 140 nm; (b) pair of edge dislocations in Nimonic PE16, $r = 8$ nm and $f = 0.09$, imaged after unloading.

This is particularly the case for properties such as the yield stress, which is a sensitive function of the distribution of the γ' phase – since this has a potent particle-strengthening effect. In this section, the factors which influence the strengthening by γ' are considered in detail.

2.3.1 Strengthening by particles of the gamma prime phase

It has been emphasised that a $a/2\langle1\bar{1}0\rangle\{111\}$ dislocation travelling in γ cannot enter the γ' phase without the formation of an anti-phase boundary (APB), and therefore that the dislocations must travel through the γ/γ' structure in pairs, with a second $a/2\langle1\bar{1}0\rangle\{111\}$ dislocation removing the anti-phase boundary introduced by the first. This is supported by evidence (see, for example, refs. [43]–[46]) from transmission electron microscopy (TEM); see Figure 2.43. The associated anti-phase boundary energy, γ_{APB}, represents a barrier which must be overcome if particle cutting is to occur. Although detailed calculations are required for an exact estimate, the particle-cutting stress is expected to be of the order γ_{APB}/b, where b is the Burgers vector. TEM experiments [47] indicate that $\gamma_{APB} \sim 0.1$ J/m^2; consequently, with $b = 0.25$ nm, the cutting stress is approximately $0.1/0.25 \times 10^{-9}/10^6$, or 400 MPa. This value is considerable, and it indicates that substantial *order strengthening* is expected. Indeed, in the nickel-based alloys, the contribution from order strengthening outweighs other contributions due to differences of modulus, stacking fault energy, interfacial energy and Orowan strengthening.

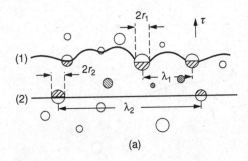

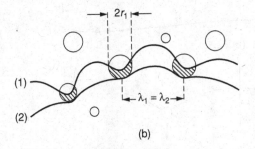

Fig. 2.44. Particles of ordered γ' being sheared by pairs of dislocations: (a) weak pair-coupling and (b) strong pair-coupling. Adapted from ref. [49].

A The case of weakly coupled dislocations

In order to develop expressions for the strengthening expected from γ' particles of volume fraction f, the case of 'weakly coupled' dislocations [48] is considered first; see Figure 2.44(a). By 'weak' one means that the spacing of the two paired dislocations is large in comparison with the particle diameter; consequently, the second trailing dislocation is some way behind the first, leaving faulted particles between the two. This situation is applicable to the case of the under-aged condition. To simplify matters, the γ' particles are assumed to be spherical and the volume fraction of γ' is taken to be small. The particles intersecting the first and second dislocations are assumed to be spaced at intervals λ_1 and λ_2, respectively, and each is known as a 'Friedel spacing'. The radius of the particles along the first and second lines are r_1 and r_2.

The onset of particle shearing is controlled by the following forces which act on the pair of dislocations: (i) the forces $\tau b \lambda_i$ $(i = 1, 2)$ driving the particle-shearing process, due to the applied shear stress; (ii) the elastic force, R, repulsive in sign since the dislocations are of the same sense, acting to keep the pair separated; and (iii) the pinning forces, F_i $(i = 1, 2)$, in magnitude equal to $2\gamma_{APB}r_i$, which are a consequence of the APB energy. The force balances may be written as follows [48,49]:

$$\tau b \lambda_1 + R + F_1 = 0 \tag{2.23a}$$

for the first dislocation and

$$\tau b \lambda_2 - R + F_2 = 0 \tag{2.23b}$$

for the second, where $F_1 = -2\gamma_{APB}r_1$ and $F_2 = +2\gamma_{APB}r_2$. The terms F_1 and F_2 are of opposite signs since the first dislocation introduces the APB, whereas the second dislocation removes it. The net shear stress required for cutting, denoted τ_c, is found by elimination of

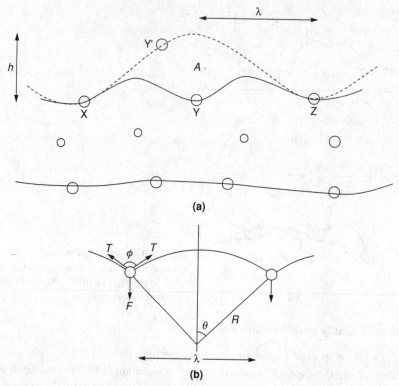

Fig. 2.45. For weakly coupled dislocations, (a) illustration of the unzipping of the leading dislocation from the pinning point Y to Y' – an area A on the slip plane is swept out; (b) geometry for force balance calculation between the dislocation line tension and pinning force due to order hardening, in which the particles are assumed to be point sources.

R from Equations (2.23a) and (2.23b); this yields [49]

$$\tau_c = \frac{\gamma_{APB}}{b}\left(\frac{2r_1 - 2r_2}{\lambda_1 + \lambda_2}\right) \simeq \frac{\gamma_{APB}}{2b}\left[\frac{2r_1}{\lambda_1} - \frac{2r_2}{\lambda_2}\right] \qquad (2.24)$$

the second expression following from the approximation that λ_1 and λ_2 are expected to be very similar. Equation (2.24) indicates that τ_c depends critically upon the difference between the ratios $2r_1/\lambda_1$ and $2r_2/\lambda_2$, and therefore the way in which the dislocations interact with the γ' particles. The factor of $\frac{1}{2}$ in the first term of Equation (2.24) can be thought of as arising from the presence of the second dislocation.

Figure 2.43 indicates that the onset of particle shearing is controlled by the behaviour of the leading dislocation, which will, in practice, be held up by closely spaced, larger particles. It must bow in order to overcome the pinning forces due to the particles, i.e. to develop a component of the line tension in a direction opposite to that in which the pinning forces act. On the other hand, the trailing dislocation can be assumed to be straight; thus, to a good approximation, $2r_2/\lambda_2 = f$. The challenge therefore is to estimate the ratio $2r_1/\lambda_1$. This can be done by considering the conditions necessary for the leading dislocation to break away or 'unzip' from a particle; see Figure 2.45(a). The leading dislocation (initially

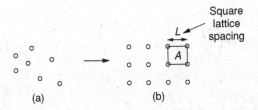

Fig. 2.46. Representation of a random distribution of γ' particles on the slip plane, (a) by a 'square lattice array', (b) such that the number density, N_s, is identical in both cases.

pinned at X, Y and Z) breaks away from Y and becomes pinned at Y', bowing out to another configuration compatible with the stress field and sweeping out an area A on the slip plane. The new configuration has a radius of curvature denoted R.

In practice, the curvature of the leading dislocation is caused by the resistance imparted by the particles – therefore the force balance in the vicinity of a particle needs to be considered carefully (see Figure 2.45(b)). Each particle is considered to be a point obstacle providing a pinning force $F = 2\gamma_{APB}r$, the largest that can be exerted, consistent with the 'unzipping' condition being analysed. The condition for static equilibrium is given by resolving forces in the vertical direction [48]; it is

$$2T \cos\{\phi/2\} = F \tag{2.25}$$

where $\phi = \pi - 2\theta$ and T is the line tension, which is approximately $\frac{1}{2}Gb^2$. The angles ϕ and θ are defined in Figure 2.45(b). Also by geometry

$$\lambda_1 = 2R \sin\theta \tag{2.26}$$

In order to proceed to an estimate of $2r_1/\lambda_1$, a relationship is required between λ_1 and the area, A, swept out by the unzipping process. To derive this, one models the random array of particles on the slip plane as a 'square lattice' of spacing L, such that $A = L^2$; see Figure 2.46. Then the number of particles per unit area, N_s, equals L^{-2}, where $L = A^{-1/2}$; the term N_s is equal to $N_V \times 2r$, where N_V is the number of particles per unit volume, consistent with $N_V = f/\frac{4}{3}\pi r_1^3$. It follows that

$$L = \left(\frac{2\pi}{3f}\right)^{1/2} r_1 \tag{2.27}$$

It is assumed that A can be approximated by $h\lambda_1$, where h is defined by the semicircle passing through X, Y' and Z (see Figure 2.45(a)); this has radius R. From the property of a circle, one has $R^2 = \lambda_1^2 + (R - h)^2 = \lambda_1^2 + R^2 - 2Rh + h^2$; if $h \ll R$, then $\lambda_1^2 = 2Rh$. Consequently,

$$\lambda_1^3 = 2Rh \times \frac{L^2}{h} = 2L^2 R = L^2\lambda_1/\cos\{\phi/2\} \tag{2.28a}$$

and thus

$$\left(\frac{\lambda_1}{L}\right)^2 = \frac{1}{\cos\{\phi/2\}} = \frac{2T}{F} \tag{2.28b}$$

so that

$$\lambda_1 = \left(\frac{2T}{F}\right)^{1/2} \times L \tag{2.28c}$$

or, making use of Equation (2.27),

$$\lambda_1 = \left(\frac{2T}{F}\right)^{1/2} \times \left(\frac{2\pi}{3f}\right)^{1/2} \times r_1 \tag{2.28d}$$

Equations (2.28c) and (2.28d) are the critical equations for the Friedel spacing, λ_1; as expected, it is proportional to the square lattice spacing, L, which is, in turn, inversely proportional to the square root of the unzipped area, A.

An estimate of the critical resolved shear stress, τ_c, is derived by inserting Equation (2.28d) into Equation (2.24) and placing F equal to $2\gamma_{APB}r_1$. The result is as follows:

$$\tau_c = \frac{\gamma_{APB}}{2b}\left[\left(\frac{6\gamma_{APB}fr}{\pi T}\right)^{1/2} - f\right] \tag{2.29}$$

where r_1 has been replaced by r since the result will be valid for the general case of a uniform distribution of particles.

One can see that the factor $\frac{1}{2}\left[(6\gamma_{APB}fr/(\pi T))^{1/2} - f\right]$ modifies the first estimate of γ_{APB}/b for the cutting stress. It is instructive to introduce some numbers for the quantities in Equation (2.29). With $\gamma_{APB} \sim 0.1\,\mathrm{J/m^2}$, $f = 0.3$, $r = 25\,\mathrm{nm}$, $G = 80\,\mathrm{GPa}$ and $b = 0.25\,\mathrm{nm}$, the second factor $\left[(6\gamma_{APB}fr/(\pi T))^{1/2} - f\right]$ in Equation (2.29) is ~ 0.46, i.e. not substantially different from unity. A final estimate of τ_c is therefore 90 MPa, a factor of ~ 4 different from the 400 MPa determined from the ratio γ_{APB}/b.

In practice, for large particles, $(6\gamma_{APB}fr/(\pi T))^{1/2} \gg f$ so that, to a good approximation, $\tau_c \propto (fr)^{1/2}$. Thus, significant fractions of larger particles are expected to promote hardening. Data [48] for the age hardening of monocrystalline Ni–Al alloys tested in compression have confirmed the $r^{1/2}$ dependence expected as $f \rightarrow 0$; see Figure 2.47. However, as $r^{1/2}$ becomes large, the linear dependence breaks down; there are a number of reasons for this behaviour. First, the point obstacle approximation is no longer valid, since the volume fraction, f, of γ' is then large. Second, the dislocation pairs become more strongly coupled; an analysis of the kind given in the following section is required.

B   The case of strongly coupled dislocations

When the γ' particles are large, as will be the case for the over-aged condition, the spacing of the dislocation pairs becomes comparable to the particle diameter. Thus, any given particle may contain a pair of dislocations which are now 'strongly coupled'; see Figure 2.44(b). In this case, the behaviour becomes critically dependent upon the elastic repulsive force, R, since this must be overcome if the trailing dislocation is to enter the γ' particle. The analysis proceeds as follows [49]. The Friedel spacings, λ_1 and λ_2, are now equal, and can be approximated by the square lattice spacing, L. In these circumstances, Equations (2.23a) and (2.23b) can be manipulated to yield [49]

$$2\tau bL + (F_1 + F_2) = 0 \tag{2.30a}$$

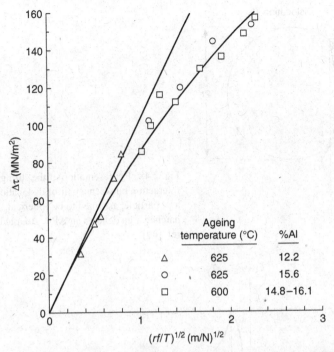

Fig. 2.47. Illustration of the agreement between the degree of ordering hardening, $\Delta\tau$, measured during room-temperature compression testing of Ni–Al single crystals and $(rf/T)^{1/2}$, where r is mean particle radius, f is volume fraction and T is dislocation line tension [48].

and

$$2R + (F_1 - F_2) = 0 \tag{2.30b}$$

Thus, $\tau = -\frac{1}{2}(F_1 + F_2)/(bL)$, where both F_1 and F_2 depend critically upon the line length, $\ell\{x\}$, of the corresponding dislocation inside the particle; see Figure 2.48. The symbol x denotes the distance to which a particle is penetrated, measured from the point of entry. Elementary geometry indicates that the function $\ell\{x\}$ is given by $2(2rx - x^2)^{1/2}$, if the particles are spherical. The forces F_1 and F_2 are of opposite sign; therefore force–distance plots (see Figure 2.49) confirm that the maximum value of $(F_1 + F_2)$ occurs when the trailing dislocation just touches the particle, at which point $x = x_m$ defines the position of the leading dislocation. Then $F_2 = 0$, and Equations (2.30a) and (2.30b) reduce to

$$2\tau bL = -F_1\{x_m\} = \gamma_{APB}\ell\{x_m\} = 2R \tag{2.31}$$

Thus one can see that the magnitude of the repulsive force R determines the critical value of τ required for particle shearing.

An estimate of R can be made by appealing to dislocation theory [50]. The repulsive force acting between two dislocations of the same sign, separated by a distance x_m, is approximately $Gb^2/(2\pi x_m)$ per unit length of the dislocation line. To account for small uncertainties, a dimensionless constant, w, is introduced which is expected to be of the

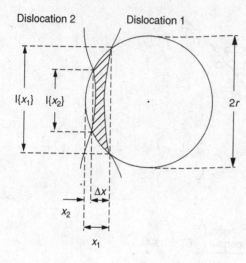

Fig. 2.48. Representation of the strong interaction between a pair of dislocations and a γ' particle, assumed to be spherical. The hatched area denotes an APB. Adapted from ref. [49].

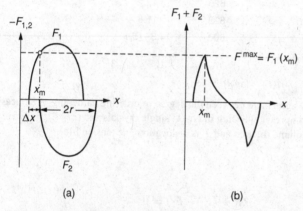

(a)   (b)

Fig. 2.49. Force–distance profiles of dislocations cutting an ordered γ' particle, assumed spherical: (a) variation of the forces F_1 and F_2 acting on the first and second dislocations with distance x, measured from the point of penetration; (b) net force on the pair, their spacing assumed to be constant. Adapted from ref. [49].

order of unity; R is then approximately given by

$$R = w \left(\frac{Gb^2}{2\pi} \right) \frac{\ell\{x_m\}}{x_m} \tag{2.32}$$

A first estimate of τ for the case of strongly coupled dislocations is derived by combining Equations (2.31) and (2.32), yielding

$$\tau = \frac{1}{2} \left(\frac{Gb}{L} \right) \frac{w}{\pi} \frac{\ell\{x_m\}}{x_m} \tag{2.33}$$

Clearly, an estimate of the spacing x_m is required. It follows also from Equations (2.31) and (2.32), giving

$$x_m = \frac{2w}{\gamma_{APB}} \left(\frac{Gb^2}{2\pi} \right) \tag{2.34}$$

so that [51]

$$\tau_c = \frac{1}{2} \left(\frac{Gb}{L} \right) \frac{2w}{\pi} \left(\frac{2\pi \gamma_{APB} r}{w G b^2} - 1 \right)^{1/2} \tag{2.35}$$

Finally, one can substitute into Equation (2.35) the expression for L given by Equation (2.27), to identify the dependence upon r and f; this is

$$\tau_c = \sqrt{\frac{3}{2}} \left(\frac{Gb}{r} \right) f^{1/2} \frac{w}{\pi^{3/2}} \left(\frac{2\pi \gamma_{APB} r}{w G b^2} - 1 \right)^{1/2} \tag{2.36}$$

A number of interesting points arise from this analysis. First, the factor of $\frac{1}{2}$ arising in Equation (2.35) can again be traced to the pair-coupling of the dislocations. Taking $\gamma_{APB} \sim 0.1$ J/m^2, $f = 0.3$, $r = 25$ nm, $G = 80$ GPa and $b = 0.25$ nm, as before, one finds that $\tau_c = 141$ MPa. This is about 50% lower than the Orowan bowing stress given by Gb/L, which features in the expression for τ_c in Equation (2.35) – suggesting that the Orowan mechanism is unlikely to be the major cause of strengthening in nickel alloys, at least at room temperature. Finally, as $r \to \infty$, an $r^{-1/2}$ dependence for τ_c is expected, so that the resistance provided by strongly coupled dislocations disappears at large particle sizes. This is in contrast to the $r^{1/2}$ dependence for weakly coupled dislocations, which is found provided that f is small; clearly in that case the strengthening increases monotonically as r increases. In fact, it can be shown that the transition from 'weak' to 'strong' coupling occurs when $r \sim 2T/\gamma_{APB}$, yielding a peak-strength of $\sim \frac{1}{2} \gamma_{APB} f^{1/2}/b$.

That the optimum hardening in the nickel-based superalloys occurs for a γ' particle size which lies at the transition from weak to strong coupling has been confirmed by a number of studies. Reppich et $al.$ [43,44,52] have measured the critical resolved shear stress at room temperature for the nickel-based superalloys PE16 and Nimonic 105, aged such that the volume fraction, f, and mean particle diameter, r, varied substantially. These systems show a degree of solid-solution strengthening, estimates for which were subtracted from the total strengthening displayed. The results are given in Figure 2.50. For PE16, an optimum r was found to be in the range 26–30 nm. For Nimonic 105, the range was found to be 55–85 nm. Jackson and Reed [51] have used these concepts to design a heat treatment schedule for the Udimet 720Li superalloy which places the γ' particles at a size which confers hardening at the transition from weak to strong coupling – and thus the maximum which can be achieved.

2.3.2 Temperature dependence of strengthening in the superalloys

The nickel alloys exhibit a remarkable characteristic: the yield stress, by which one means the stress necessary to cause the onset of plastic deformation by dislocation flow, does not decrease strongly with increasing temperature, as is the case for most other alloy systems. In fact, for many superalloys the yield stress $increases$ with increasing temperature, typically until temperatures of about 800 °C are reached. Figure 2.51 shows some typical data for a number of single-crystal alloys, tested along the $\langle 001 \rangle$ direction. The peak stress of beyond 1000 MPa is about 50 times greater than the flow stress of pure Ni – which confirms that considerable metallurgical strengthening effects are at play. For temperatures beyond 800 °C,

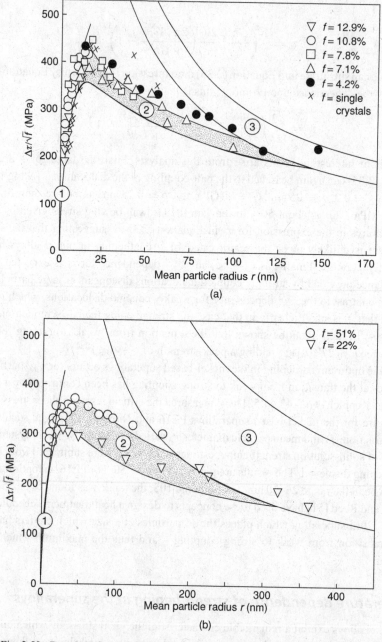

Fig. 2.50. Data [52] for the increase in critical resolved shear stress (divided by the square root of the volume fraction of the particles) as a function of the mean particle radius for two nickel-based superalloys: (a) Nimonic PE16 ($\gamma_{APB} = 0.125$ J/m^2) and (b) Nimonic 105 ($\gamma_{APB} = 0.110$ J/m^2). In regime 1, hardening is by weakly coupled pairs; in regime 2 it is by strongly coupled pairs; and in regime 3 it is by the Orowan mechanism. Note that the peak strength is associated with the transition from weak to strong coupling.

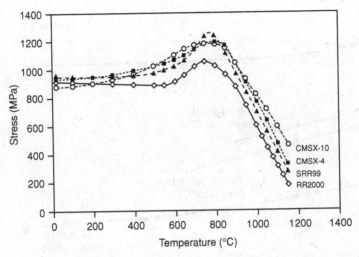

Fig. 2.51. Variation of the yield stress of a number of single-crystal superalloys with temperature.

the yield stress decreases quickly; by about 1200 °C little resistance to plastic deformation is displayed.

In a first attempt to rationalise this behaviour, Davies and Stoloff [53] studied the yield properties of Ni–Al binary alloys between −200 °C and 800 °C; see Figure 2.52. The Ni–8%Al alloy, which is alloyed insufficiently for the precipitation of the γ' phase, thus benefits only from a solid-solution strengthening of between 50 and 100 MPa, depending upon the temperature. The Ni–14%Al alloys displayed precipitation strengthening, but to an extent dependent upon the fraction, size and distribution of the γ' particles, and hence on the heat treatment applied. Quenching from 1000 °C introduced only very fine γ' precipitates which were barely resolvable in the transmission electron microscope (TEM). Ageing at 850 °C introduced particles spaced by 200–250 nm, significantly coarser than the 40–60 nm range when heat treatment was carried out at 700 °C. The TEM microscopy indicated that the micromechanism of deformation was Orowan looping when ageing was carried out at 850 °C, but particle cutting was observed for material subjected to the 700 °C treatment. This is strong evidence that the rather constant yield stress displayed by the Ni–14%Al alloy heat treated at 700 °C is due to microscopic deformation of γ'.

Further work reported by Piearcey et al. [54] sheds light on the role of γ' in the deformation of the superalloys. The Mar-M200 alloy, which contains about 60% of γ', was tested in single-crystal form (see Figure 2.53); this revealed a curve very similar to those of the more modern single crystals given in Figure 2.51. Also tested was single-crystal cube-orientated Ni$_3$Al, alloyed such that its composition matched that found in Mar-M200. The results (see Figure 2.53) confirm that at and beyond the peak stress, the behaviour of the alloy is determined by the strength of the γ' phase. Furthermore, from ambient to the temperature associated with the peak stress, the γ' imparts an increasing fraction of the strength as the temperature increases. Similar experiments were carried out by Beardmore et al. [55]. The temperature dependence of the yield stress was determined for a number of alloys with varying amounts of γ'. Rather ingeniously, a series of alloys lying on the tie-line

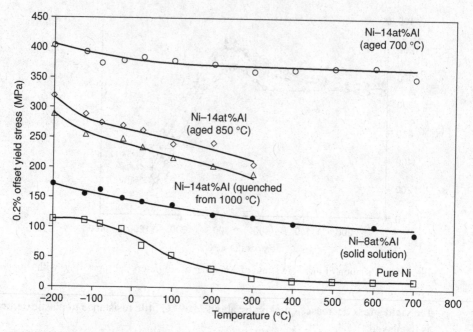

Fig. 2.52. Data for the yield stress of alloys in the binary Ni–Al system as a function of composition and temperature. Adapted from ref. [53].

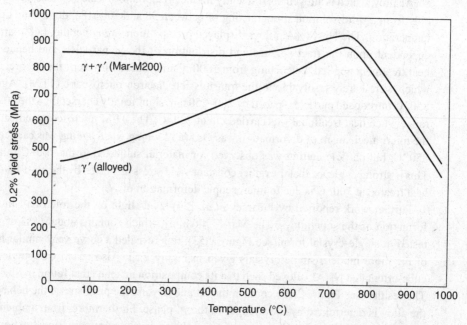

Fig. 2.53. Variation with temperature of the yield stress of the Mar-M200 alloy in single-crystal form, and a monolithic Ni$_3$Al alloy of composition equivalent to the γ' particles in it. In both cases, testing was along $\langle 100 \rangle$. Adapted from ref. [54].

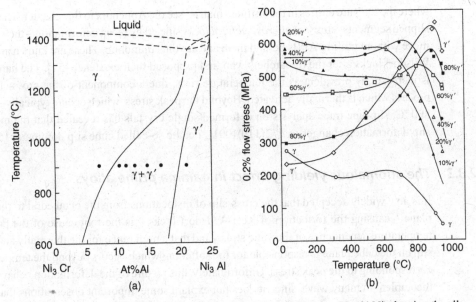

Fig. 2.54. (a) Section through the Ni–Cr–Al ternary phase diagram at 75 at% Ni, showing the alloys used in experiments by Beardmore *et al.* [55], and (b) variation of the yield stress with temperature for the alloys so produced.

joining Ni_3Al and the Ni–25at%Cr alloy were considered; see Figure 2.54(a). Once again, the alloy consisting of 100% γ' displayed a strong positive dependence on the temperature, until about 800 °C; see Figure 2.54(b). Above the temperature corresponding to the peak stress, the yield stress of a two-phase $\gamma + \gamma'$ alloy obeys a rule of mixtures, i.e. it corresponds to the weighted average of the values for γ and γ'. This contrasts strongly with the behaviour at low temperatures, where two-phase alloys are very much stronger than the rule of mixtures would predict.

The anomalous yielding effect is discussed in detail in the following section, but, broadly speaking, it arises for the following reason. Upon deformation, the applied stress, the anisotropy of the anti-phase boundary energy and further contribution from the elastic anisotropy combine to promote the *cross-slip* of segments of the γ' superpartial dislocations from the {111} slip plane to the cross-slip plane {001}. The force per unit length of dislocation which promotes cross-slip is equal to [56]

$$\gamma_{111}\left[\frac{1 + f_1\sqrt{2}}{\sqrt{3}}\right] - \gamma_{100} \tag{2.37}$$

where γ_{111} and γ_{100} are the APB energies on the octahedral and cube planes, respectively, and f_1 is given by

$$f_1 = \sqrt{2}\left[\frac{A-1}{A+2}\right] \tag{2.38}$$

where A is Zener's anisotropy factor, defined according to

$$A = \frac{2C_{44}}{C_{11} - C_{12}} \tag{2.39}$$

where the C_{ij}'s are entries in the stiffness matrix; see the Appendix to this chapter. The cross-slipped segments represent *microstructural locks* since they resist deformation – they cannot move without trailing APBs behind them and are thus immobile. These are known as Kear–Wilsdorf locks after the researchers who first proposed their existence [57]. The hardening is increasingly prevalent as the temperature rises, due a component of the cross-slipping process which is thermally activated. Beyond the peak stress, which occurs typically around 800 °C, slip-line trace analysis on deformed single crystals has revealed that the mode of slip deformation changes to $a/2\langle1\bar{1}0\rangle\{001\}$, i.e. the so-called cube slip dominates [58].

2.3.3   The anomalous yielding effect in gamma prime alloys

It is now widely accepted that the cross-slip of dislocations from the octahedral to the cube plane – causing the formation of Kear–Wilsdorf locks – is the root cause of the positive temperature dependence of the yield stress, and that a preference for the thermally activated slip on the cube plane is responsible for the softening which occurs beyond the temperature corresponding to the peak stress. Unfortunately, this picture of the deformation behaviour is incomplete in many ways, since it does not explain some important observations that have emerged from recent research carried out on this topic [59,60]. In particular, for most L1$_2$-based nickel alloys the yield stress increases with temperature from about 77 K [61] – too low for a phenomenon which relies purely upon thermal activation. Furthermore, the behaviour is found to be reversible; thus, prior deformation at an elevated temperature exerts only a very modest effect on the low temperature behaviour, and only a small component of the work hardening increment at the higher temperature is inherited. These observations are consistent with a strain-activated phenomenon; thus cross-slip is obviously heavily influenced by the process of deformation. The work hardening rate is also found to be abnormally high – at a strain of 0.2%, it is one or two orders of magnitude greater than the value typically observed for single slip in FCC metals – and it increases in magnitude over some, if not all, of the temperature range of the anomaly. Furthermore, the strain-rate sensitivity is not at all pronounced, as would be expected for a diffusion-controlled phenomenon [62]. Other interesting effects include a significant tension/compression asymmetry, the magnitude of which varies with crystallographic orientation, so that Schmid's law is violated [63,64]. For single crystals aligned along $\langle001\rangle$, the critical resolved shear stress (CRSS) on $[\bar{1}01]\{111\}$ is greater in tension than in compression; conversely, for alignment at orientations lying along the $[011]/[\bar{1}11]$ boundary of the stereographic triangle, the CRSS is greater in compression rather than tension; see Figure 2.55. Such effects have been rationalised [65] by considering combinations of Schmid factors on the cube cross-slip (for the direction $\langle110\rangle$) and on the primary octahedral plane (along $\langle110\rangle$ and $\langle112\rangle$).

Microstructural observations using transmission electron microscopy have helped to clarify the micromechanical effects occurring during deformation, although one should note that considerable controversy still exists over their interpretation [60]. Post-mortem examinations reveal that long, straight screw superdislocations are prevalent in samples deformed in the anomalous regime, indicating that screws are very much less mobile than non-screws. The dislocation microstructure is remarkably homogeneous – there is little apparent tendency to form the sub-grains or cell structures found in pure metals [66] – so

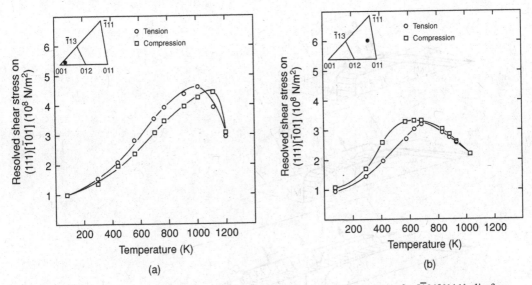

Fig. 2.55. Variation with temperature of the critical resolved shear stress for $[\bar{1}01]\{111\}$ slip for $Ni_3(Al, Nb)$ single crystals, measured in tension and compression, for two orientations within the stereographic triangle [63]; note the considerable tension/compression asymmetry which violates Schmid's law.

that it seems likely that the macroscopic properties are controlled to a good approximation by the behaviour of individual dislocations. The superdislocations are indeed found to be dissociated completely onto the cube plane, i.e. into the Kear–Wilsdorf (KW) configuration, but incomplete KW locks are also present in which a portion of the APB resides on the octahedral plane. Incomplete KW locks, which are mechanically unstable and thus prone to relax back to the complete configuration, are preferred, particularly after deformation at low temperatures. *In-situ* TEM observations using straining stages have unambiguously confirmed the dynamical locking of screw dislocations by cross-slip. However, instead of being curled between pinning points, as might have been expected, mobile superpartials are found to lock themselves over significant lengths almost instantaneously and to remain rather rectilinear during the locking and unlocking processes; when dissociated on the octahedral plane their velocity is extremely high, but their motion is jerky. Veyssiere and Saada [60] have constructed diagrams which are helpful in visualising the situation; see Figure 2.56. As the leading superpartial cross-slips onto the cube plane, it does so by the lateral spreading of a pair of jogs in opposite directions; this must occur with the lateral displacement of a pair of macrokinks on the octahedral plane. It has been pointed out that the macrokinks can be of two types – simple macrokinks and switch-over macrokinks – the distinction depending upon whether a single superpartial remains consistently the first in a ribbon built up by multiple cross-slip events; see Figure 2.57. The situation then resembles the images [66] seen in the TEM; see Figure 2.58. Careful analyses of the distributions of the heights of macrokinks have been made as a function of the temperature of deformation [67]; heights no greater than ten times the APB width, λ_0, are observed, the frequency increasing exponentially as the height reduces to zero, apart from a strong preference for heights of

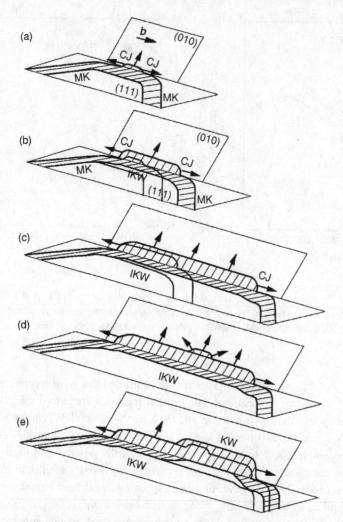

Fig. 2.56. Steps in the transformation of a screw superdislocation by cross-slip onto the cube plane [60]. CJ stands for complex jog and MK stands for macrokink; these lie on the cube and octahedral planes, respectively. KW and IKW are complete and incomplete Kear–Wilsdorf segments, respectively, a distinction which depends upon whether the superdislocation is totally or partially dissociated on the cube plane.

one to two λ_0. At this height, they are christened 'elementary kinks'. Other research has emphasised the dissociation of the leading and trailing superpartials and thus the formation of complex stacking faults (CSFs); estimates of the CSF fault energy have been possible through the use of image simulation techniques. A strong influence of Hf on the CSF energy has been reported [47], which explains the effect of Hf alloying on the values of the peak temperature, stress and work hardening rate. The elements Ti and Ta behave similarly.

An adequate quantitative model for the macroscopic yielding behaviour of Ni_3Al which accounts for the macroscopic and microscopic observations – especially the influence of alloying – is not available as yet. In constructing it, one will need to decide whether the

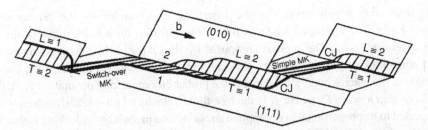

Fig. 2.57. Schematic representation of a ribbon of cross-slipped superdislocations, displaying macrokinks of both the simple and switch-over types [60] – the distinction depends upon whether the leading superpartial switches to the trailing one on either side of the cross-slipped segment.

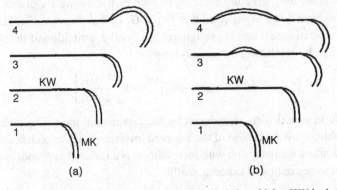

Fig. 2.58. Transmission electron micrograph [66] of a typical superdislocation in the deformed Ni$_3$Al compound. KW and MK refer to Kear–Wilsdorf locks and macrokinks, respectively.

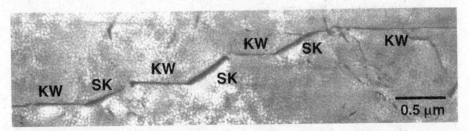

Fig. 2.59. Schematic illustration of the methods by which a KW lock may be overcome [60]: (a) bypassing by an expanding macrokink (MK), and (b) unlocking by the formation of a glissile double-kink configuration on the locked KW segment.

improvement in the flow stress with increasing temperature is due to an increase in the strength of the obstacles, or else to a reduction in their spacing; moreover, the extent to which the interaction of superdislocations is responsible for the work hardening behaviour (if at all) needs to be established. One can say that a successful theory is likely to be built on one or two foundations (see Figure 2.59): either (i) a treatment of locking/unlocking by an 'unzipping' process on the scale of the cores of the superpartial dislocations, which seems most likely to explain the considerable composition dependence displayed by the Ni$_3$Al compounds doped with Hf, Ta, Ti, etc. and ones away from the stoichiometric composition, and

(ii) treatments which emphasise the bypassing of obstacles by the propagation of macrokinks. The motivated reader is referred to ref. [60] for a full critique of the models developed so far. The approach proposed by Hirsch [68], which focusses attention on the macrokinks, is one of the more physically realistic models developed so far. Here, the height of the macrokinks, h, is related to the probability of locking per unit time, P_l, by the relationship $h = v_f/P_l$, where v_f is the free-flight velocity of glissile dislocations – this is expected to be proportional to the applied stress, σ. The probability of locking is then given by $P_l \propto l \exp\{-G_l/(kT)\}$, where G_l is an activation energy of locking and l is the spacing of macrokinks along the dislocation line; since this is assumed to vary as the inverse of σ, it follows that $h \propto \sigma^2 \exp\{+G_l/(kT)\}$. The *average* dislocation velocity, v, which is governed by the movement of macrokinks and thus the difference between the rates of pinning and unpinning events, is then given by

$$v = v_0 \left[\exp\left\{ -\frac{G_{ul}}{kT} \right\} - \exp\left\{ -\frac{G_l}{kT} \right\} \right] = v_0 \exp\left\{ -\frac{G_{ul} - G_l}{kT} \right\} \qquad (2.40)$$

where v_0 is a constant, and G_{ul} is the activation energy for unlocking; this is assumed to depend upon the effective stress, σ, necessary to cut pinning points, taken to be the ends of the macrokinks; thus

$$G_{ul} = {}^0G_{ul} - \sigma V^* \qquad (2.41)$$

where ${}^0G_{ul}$ is the activation energy in the absence of the applied stress and V^* is an activation volume – this is of order hb^2, since the breaking of pinning points sweeps an area which is proportional to h. Note that it is expected that $G_{ul} > G_l$ so that the rate of locking will be greater than the rate of unlocking. If one ignores the small contribution of the Orowan stress, Gb/h, one has, by combining the various terms,

$$\sigma \propto \left[{}^0G_{ul} - kT \ln\{v_0/v\} - G_l \right]^{1/3} \exp\left\{ -\frac{G_l}{3kT} \right\} \qquad (2.42)$$

so that the increase in σ with T is reproduced by the exponential term. The factor of 3 present in it is notable; it arises because of the assumed inverse square dependence of h on σ. Unfortunately, whilst a decrease in h with increasing σ is a reasonable proposition, this dependence has yet to be confirmed experimentally.

2.4   The creep behaviour of nickel alloys

So far, it has been assumed that a unique value of the plastic strain is produced when a nickel alloy is loaded under any given conditions of temperature and applied stress. In practice, this is not precisely the case – particularly at high temperatures, plastic strain is accumulated over time by the process of creep. This effect is pertinent to the turbine blading in the aeroengine, which is machined to tight tolerances with respect to the engine housing; consequently, excessive creep deformation during service constitutes 'failure' of the component since it is no longer able to satisfy the function for which it was designed. It follows that creep-resistant alloys are required for this application.

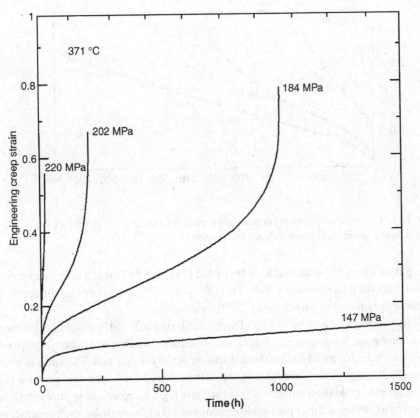

Fig. 2.60. Creep curves (constant load conditions) for nominally pure nickel tested at 371 °C [69].

2.4.1 The creep behaviour of nickel

Creep strengthening in polycrystalline nickel alloys arises both from solid-solution strength-ening due to the presence of solute atoms and from precipitation hardening due to phases such as γ'. Consider Figure 2.60, in which creep data [69] are given for nominally pure nickel tested under constant load conditions at 371 °C and stresses between 147 and 220 MPa. It is clear that the evolution of creep strain is remarkably sensitive to the stress applied. More-over, for each creep curve a rapid initial transient (known as a period of *primary* creep) gives way to a regime in which, to a good approximation, the creep strain rate is constant – this steady-state is referred to as *secondary* creep. The onset of failure is associated with a divergence from the steady-state behaviour; during this *tertiary* regime the creep strain rate increases with increasing strain before fracture finally occurs. Figure 2.61 shows additional creep data [70] for polycrystalline nickel dispersion-strengthened with a small fraction of ThO$_2$ particles, under constant stress conditions of 28 MPa over a range of temperatures. One can see that the creep strain is sensitive to the temperature employed – a rise of 25 °C is sufficient to double the creep strain. The fitting of the steady-state creep data in ref. [70] to Equation (1.7) yields an estimate of the activation energy for creep, $Q = 270$ kJ/mol, which is close to the value for the self-diffusion of nickel. When one collates the various

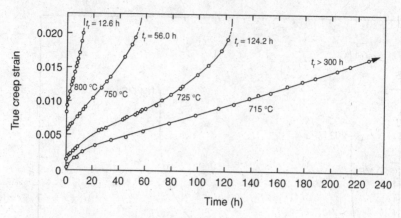

Fig. 2.61. Creep curves for nickel doped with a small fraction of ThO$_2$ particles [70]; $\sigma = 28$ MPa. The time to rupture, t_r, is given for each creep curve.

creep data which are available for pure nickel [71], one finds that a stress exponent, n, of about 5 fits the data reasonably well. This fifth-power law dependence on the applied stress is commonly found in pure metals [72,73]; see Figure 2.62.

The steady-state creep behaviour of nickel can be rationalised by assuming that the modes of softening and hardening, i.e. dislocation annihilation and multiplication, are operating at rates which are balanced so that the dislocation density is constant. The various 'recovery-controlled' creep models available in the literature are based upon this premise [73–75]. Then, provided that the temperature is sufficiently high, creep occurs in pure metals such as nickel by a combined climb-plus-glide mechanism – if a gliding dislocation becomes pinned, then a small amount of climb can release it, allowing further glide to occur. Dislocation glide is then responsible for the majority of the deformation, although the average velocity is determined by the climb step [74]; for most metals creeping above $0.6T_m$, climb occurs by lattice-diffusion (rather than, for example, by pipe diffusion along a dislocation core), and therefore this process is rate-controlling. In the steady-state regime, well-established dislocation cell sub-structures with associated sub-grain boundaries are observed using transmission electron microscopy – within the grain boundaries; see Figure 2.63. In general, the formation of sub-grains arises principally as a result of polygonisation due to the climb of edge dislocations since cross-slip is easy [72] – studies suggest that the sub-grain size and misorientation reach a constant value during primary creep consistent with the strain rate arising. The sub-grain size is found to be inversely proportional to the applied stress; moreover, the same sub-grain size is developed in steady-state creep, regardless of its previously work hardened or pre-crept condition. If the steady-state regime is interpreted in this way, then primary creep arises because a constant sub-structure is yet to be established; the rate of dislocation multiplication then outweighs the rate of the annihilation processes. Conversely, during tertiary creep other degradation mechanisms are occurring – for example, loss of net cross-section due to creep cavitation at grain boundaries.

Nevertheless, the σ^5-dependence of the steady-state strain rate needs to be rationalised. This can be done in the following way, the analysis being inspired by the work reported in refs. [76] and [77]. Assume that the characteristic dimension of the dislocation sub-structure

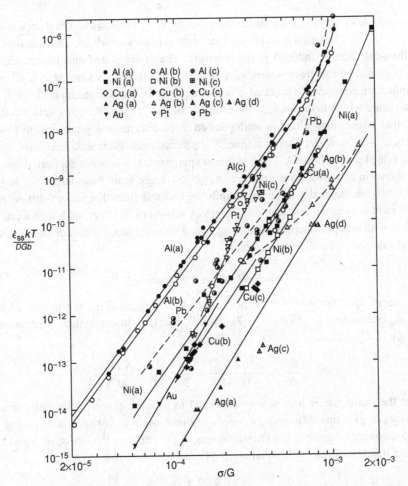

Fig. 2.62. Effect of stress on the normalised steady-state creep rates of the face-centred cubic (FCC) metals [72]. Note that the Al(b), Ni(a), Ag(a) and Ag(d) tests were not performed at constant stress.

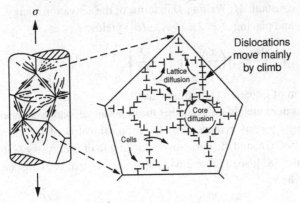

Fig. 2.63. Schematic illustration of the formation of cells by climb during power-law creep. After ref. [71].

in steady-state creep is λ; if the dislocation density is ρ, it follows to a first approximation that $\lambda = 1/\sqrt{\rho}$. The terms λ and ρ are then state variables which are characteristic of the metallurgical damage induced in the structure. The accumulated dislocation density, ρ, is proportional to the creep strain, ϵ, such that $\dot{\rho} = M\dot{\epsilon}$ [78], where M is a dislocation multiplication constant. To proceed, a law is required for the coarsening of the dislocation cell structure which is responsible for creep softening; consistent with a lattice-diffusion-controlled coarsening, which is analogous to three-dimensional grain growth [79], one expects $\lambda^3 - \lambda_0^3 = KDt$, where t is time, D is a diffusion coefficient, K is a kinetic constant and λ_0 is the initial value of λ. There is some experimental evidence in support of this; it has been shown in ref. [80], for example, that, for low-angle grain boundaries, the activation energy for grain-boundary migration is indistinguishable from that for self-diffusion.

A steady-state creep rate implies a constant sub-grain structure such that ρ and λ are invariant with time; the increments $d\rho_{\text{hardening}}$ and $d\rho_{\text{softening}}$ due to the increments of strain, $d\epsilon$, and sub-grain size, $d\lambda$, are such that

$$d\rho_{\text{total}} = d\rho_{\text{softening}} + d\rho_{\text{hardening}} = 0 \tag{2.43}$$

and hence of equal magnitude but opposite sign. Now, $d\rho_{\text{hardening}}$ is equivalent to $M\,d\epsilon$ and $d\rho_{\text{softening}}$ is equal to $-2\rho\sqrt{\rho}\,d\lambda = -[2\rho\sqrt{\rho}KD/(3\lambda^2)]\,dt$. Inserting these expressions into Equation (2.43), one has

$$\frac{d\epsilon}{dt} = \frac{2\rho\sqrt{\rho}KD}{3\lambda^2 M} = \frac{2}{3} \times \frac{1}{\lambda^5} \times \frac{KD}{M} \tag{2.44}$$

so that the strain rate is inversely proportional to the fifth power of the sub-grain size. The stress to give this strain rate, $\dot{\epsilon}$, needs to overcome the obstacles at a spacing of λ; it is approximately equal to the Orowan stress, Gb/λ. Placing the stress, σ, equal to this estimate, and combining with Equation (2.44), one can show that

$$\frac{d\epsilon}{dt} = \frac{2}{3}\left(\frac{\sigma}{G}\right)^5 \frac{KD}{Mb^5} \tag{2.45}$$

so that one arrives at the fifth power for the stress dependence. Note that the creep strain rate is then proportional to the diffusion coefficient, D, and inversely proportional to the dislocation multiplication constant, M. Writing D in terms of the activation energy, Q, and pre-exponential term, D_0, and placing $A = 2KD_0/(3Mb^5)$ yields

$$\frac{d\epsilon}{dt} = A\left(\frac{\sigma}{G}\right)^5 \exp\left\{-\frac{Q}{RT}\right\} \tag{2.46}$$

which has the familiar form of Equation (1.7) from Chapter 1.

In fact, the evidence for pure metals suggests that the constant, A, shows a dependence on the stacking fault energy, γ_{SF}, as evidenced by the proportionality of the normalised secondary creep rate, $\dot{\epsilon}_{ss}kT/DGb$, and the normalised stacking fault energy, γ_{SF}/Gb, when plotted on logarithmic scales [81]; see Figure 2.64. This indicates that the constant, A, of Equation (2.46) can be written

$$A = A'\left(\frac{\gamma_{SF}}{Gb}\right)^3 \tag{2.47}$$

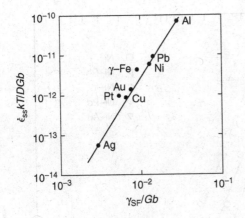

Fig. 2.64. Variation of the function $\dot{\epsilon}_{ss}kT/DGb$ with the normalised stacking fault energy, γ_{SF}/Gb, for various pure metals [81].

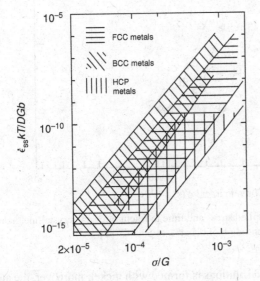

Fig. 2.65. Variation of the function $\dot{\epsilon}_{ss}kT/DGb$ with modulus-compensated stress for FCC, BCC and HCP metals, at a test temperature of 500 °C [81].

so that

$$\frac{d\epsilon}{dt} = A'\left(\frac{\gamma_{SF}}{Gb}\right)^3 \left(\frac{\sigma}{G}\right)^5 \exp\left\{-\frac{Q}{RT}\right\} \tag{2.48}$$

Although there is some scatter in the data for the pure metals, the steady-state creep rates of those which display the FCC crystal structures are, on average, about one or two orders of magnitude lower than for the BCC case; see Figure 2.65 [72,81]. The HCP metals are even lower still, presumably on account of the limited number of slip systems available. These findings lend further support to the choice of nickel as the solvent for the superalloys.

2.4.2 Creep strengthening in nickel alloys by solid-solution strengthening

Dennison et al. [82] have investigated the influence of solute additions on the creep resistance of polycrystalline nickel, of grain size ~80 μm. Alloying was carried out with gold, palladium, rhodium or iridium at concentrations of 0.1 at% – in each case, either an

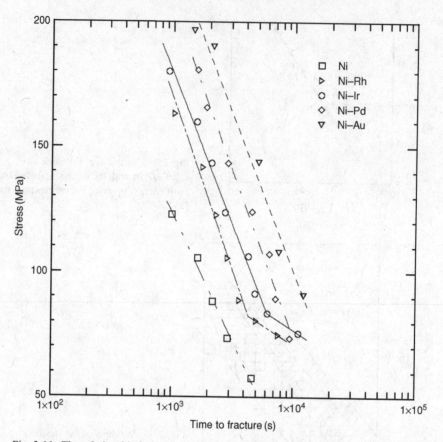

Fig. 2.66. The relationship between applied stress and time to fracture for polycrystalline nickel doped with various elements at a concentration of 0.1 at%, at 500 °C [82].

extensive or complete range of solid solutions is formed with nickel; moreover, the atomic sizes of the solutes are greater than that of the solvent. Testing was carried out at 500 °C and 600 °C under constant load conditions. The solutes were observed to have a profound effect on the creep resistance (see Figure 2.66), with Au and Rh being the most and least potent strengtheners, respectively. The amount of primary creep strain was found to increase in the order rhodium (4%), iridium (5%), palladium (6%) and gold (6%) and compared with nickel at 2%; the values did not depend strongly upon the stress level applied. The data were collated with those from previous work which involved doping with Co, Si and Sn; see Figure 2.67. The results confirm unambiguously that a strong correlation exists between the creep-strengthening increment and the percentage difference in atomic size between the solute and solvent (nickel) atoms. As expected, the degree of creep strengthening increases with the concentration of solute added; see Figure 2.68 [83].

Sherby and Burke [84] introduced a classification for solid-solution-strengthened materials which can be used to interpret the effects produced by alloying. It was realised that a great number of binary alloys behave similarly to pure metals at all concentrations across the phase diagram, but that this is not always so. Alloys whose behaviour is different

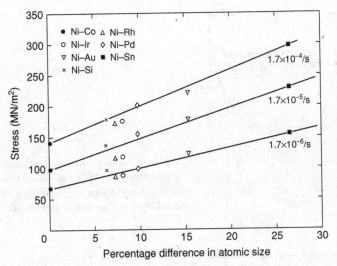

Fig. 2.67. The relationships between the percentage difference in atomic size of solute and solvent atoms and the creep resistance (defined as the stress for a given steady-state creep strain rate) of various dilute nickel alloys at 500 °C [82].

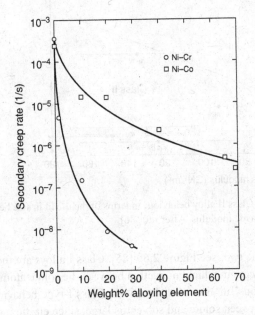

Fig. 2.68. Variation of the secondary creep rate of nickel with cobalt and chromium content during creep at 500 °C. Data taken from ref. [83].

from pure metals were termed Class I alloys and those behaving similarly designated Class II. Thus, Class II alloys exhibit a power-law dependence of secondary creep rate with n equalling five; these alloys are found to show significant primary creep and are influenced by changes in the stacking fault energy. Conversely, Class I alloys were found to exhibit an exponent n of about three; these do not show significant primary creep and are not influenced by the stacking fault energy. Thus, in the Ni–Au binary system, for example, which shows complete solid solubility, the stress exponent changes from five for the pure-end constituents

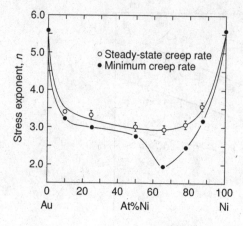

Fig. 2.69. Value of the stress exponent versus alloy content for Ni–Au alloys measured during compression creep; constant stress conditions at a test temperature of 860 °C [85].

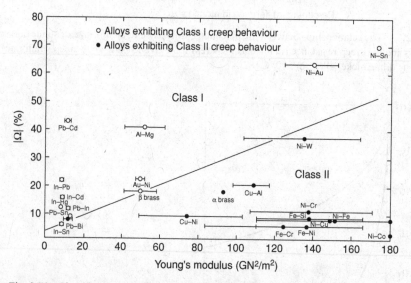

Fig. 2.70. Classification of Class I and Class II alloy behaviour in terms of the misfit in size between solute and solvent atoms, Ω, and the elastic modulus. After ref. [86].

to about three for alloys between the two; see Figure 2.69 [85]. Class I alloys are thought to deform by a glide-controlled process in which interaction between the solute atoms and dislocations leads to a 'linear viscous' motion of dislocations. Class I-type behaviour is expected when the atomic misfit between solute and solvent is large, since elastic interactions with dislocations are then favoured. The σ^3 dependence of the linear viscous glide behaviour can be rationalised by noting that the net dislocation velocity, v, is expected to be proportional to the applied stress, σ; i.e. $v \propto \sigma$. As in the case of pure metals, $\lambda = 1/\sqrt{\rho}$ and $\sigma = Gb/\lambda$, so that the dislocation density, ρ, is expected to be proportional to σ^2. The third-power dependence then follows from Orowan's equation, $\dot{\epsilon} = \rho b v$. This is the essence of Weertman's model of linear viscous creep; see ref. [75].

With these ideas in mind, Cannon and Sherby [86] have noted that Class I behaviour is expected when the size misfit, Ω, is high and the Young's modulus, E, is low; see Figure 2.70.

This observation can be rationalised as follows. For Class II behaviour – in which the climb rate is expected to be rate-controlling – the creep strain rate, $\dot{\epsilon}$, can be represented by $A''(\sigma/E)^5$. For Class I behaviour – controlled by linear viscous glide – one can take $\dot{\epsilon}$ equal to $K(\sigma/E)^3$ by analogy. Here, A'' and K are constants which depend upon temperature and other factors such as the diffusion coefficient. Treating the two phenomena of glide and climb as sequential processes, one can demonstrate that the net strain rate, $\dot{\epsilon}_{net}$, is given by

$$\dot{\epsilon}_{net} = \frac{KA''\sigma^5}{E^3(KE^2 + A''\sigma^2)} \tag{2.49}$$

so that one recovers Class I or Class II behaviour when $KE^2 \ll A''\sigma^2$ or $KE^2 \gg A''\sigma^2$, respectively. The dependence upon E follows immediately from this argument; since a large Ω acts to decrease K, all is consistent. Mohamed and Langdon [87] have developed a more sophisticated version of this model which takes the misfit more explicitly into account and includes also the influence of the stacking fault energy, γ_{SF}, in the expression for Class I behaviour.

It is interesting to consider the location on Figure 2.70 of the various alloying elements pertinent to commercial superalloys. The Young's modulus, E, is, in fact, not substantially altered by alloying; hence the data points are clustered around a value of about 130 GPa. The elements taken from the 3d row of transition metals (for example, Fe, Co, Cr) have atomic radii which are not substantially different from that of nickel; consequently the misfit, Ω, is small and therefore Class II behaviour is observed in Ni–Fe, Ni–Co and Ni–Cr binary alloys. The largest misfit is for Ni–Au, which displays Class I behaviour. Ni–W lies in the Class II regime but close to the Class I/Class II boundary. On the basis of the atomic size of their solute, Ni–Mo, Ni–Re and Ni–Ru alloys are expected to lie below Ni–W in the Class II regime, but this remains to be demonstrated.

2.4.3 *Creep strengthening in nickel alloys by precipitation hardening*

The presence of the γ' phase, which is promoted by alloying with Al, Ti and Ta, has a profound influence on the creep resistance of the nickel alloys. Gibbons and Hopkins [88] have studied the creep behaviour of a series of polycrystalline alloys based upon Nimonic 101 but possessing varying Al and Ti levels and thus volume fractions of the γ' phase. The results (see Figure 2.71, in which the minimum creep rate and life to rupture are plotted against the applied stress) indicate that, provided the stress level is not too high (in which case the γ' particles are sheared), the performance is improved substantially. This strong precipitation-hardening effect was exploited by the International Nickel Company in a series of wrought superalloys used for the turbine blading of early jet engines produced in the 1940s and 1950s; these were strengthened by progressively increasing the fraction of the γ' phase. For example, the Nimonic 80A, 90, 105 and 115 alloys, the compositions of which are given in Table 2.1, have γ' fractions of approximately 0.17, 0.18, 0.32 and 0.62, respectively; when creep resistance is measured with respect to the temperature for 1000 h creep life at 137 MPa, the last of these alloys is better than the first by about 120 °C. The data are plotted on Figure 1.17, which illustrates the evolution of these early alloys.

Gibbons and Hopkins [88] found also that the γ' strengthening effect was most potent when the grain size was large; see Figure 2.72. This effect occurs because the prominent

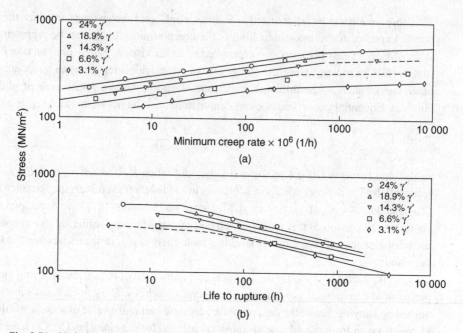

Fig. 2.71. Variation of the (a) minimum creep rate and (b) life to rupture for a number of polycrystalline superalloys with varying fractions of the γ' phase [88].

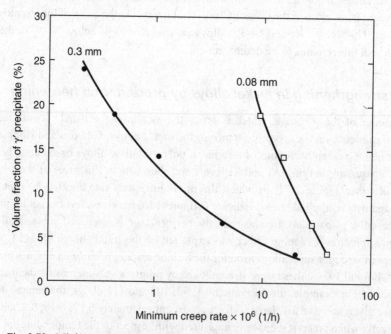

Fig. 2.72. Minimum creep rate of coarse- and fine-grained superalloys as a function of volume fraction of precipitate at 800 °C and 150 MPa [88].

Table 2.1. *The nominal compositions of the Nimonic series of superalloys, in weight%*

Alloy	C	Cr	Ti	Al	Co	Mo	Ni
Nimonic 80A	0.07	19.5	2.4	1.4	—	—	Bal
Nimonic 90	0.08	19.5	2.4	1.4	17	—	Bal
Nimonic 105	0.13	15.0	4.7	1.3	20	5.0	Bal
Nimonic 115	0.15	14.5	3.8	5.0	13	3.3	Bal

form of creep damage in polycrystalline nickel alloys is creep cavitation at the γ–grain boundaries – this becomes less influential as the grain size increases. This effect is exploited in the single-crystal superalloys; a full discussion of these is presented in Chapter 3, but at this point it should be acknowledged that the single-crystal superalloys behave under creep conditions in a very different manner from the polycrystalline alloys from which they were derived. The differences have been summarised succinctly by McLean [89]. First, the shape of the creep curve of the single-crystal superalloys is different; over wide ranges of temperature and applied stress there is no substantial secondary or 'steady-state' regime, but instead, over most of the life, the creep rate increases progressively. However, there is no obvious sign of the cavitation or cracking which causes tertiary creep in simple metals. Second, the values of the stress exponent, n, and creep activation energy, Q, derived by fitting the power law of Equation (1.7) to the creep data are typically greater than 8 and about three times the activation energy for self-diffusion of nickel – these findings imply a breakdown of power-law creep so that other theories are required. These are presented in Chapter 3. Third, single-crystal superalloys show a progressive increase in the dislocation density over all of the creep life; there is no evidence that a steady-state is reached, and indeed the dislocation densities are very low compared with simple metals. Finally, pre-straining of single-crystal superalloys weakens them in creep. This is in contrast to the simple metals, which are strengthened if the material is pre-conditioned in this way. This characteristic is, however, not limited to the single-crystal superalloys – the polycrystalline superalloys, such as Nimonic 80A, which possess a significant fraction of γ' behave in this way too [90]; see Figure 2.73. Despite the fact that grain-boundary cavitation is the predominant form of creep damage near the end of life, this confirms that it is the mobile dislocation density which controls the evolution of creep strain at short times, even in polycrystalline superalloys.

2.5 Summary

The physical metallurgy of nickel and its alloys has been reviewed. The high-temperature properties depend crucially upon the combination of alloying elements chosen, since these promote precipitation hardening (most often by the ordered phase γ', but also by γ'' in iron-rich alloys) and solid-solution strengthening. Additions of Al, Ti and Ta – the important γ' forming elements – are particularly critical if adequate strength levels are to be achieved.

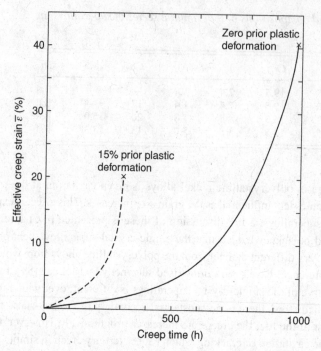

Fig. 2.73. Torsional creep data determined from hollow specimens of the Nimonic 80A polycrystalline superalloy, tested at 750 °C and an effective stress of 234 MPa [90]. The dashed line corresponds to a specimen which had been pre-strained plastically to 15% prior to creep testing. Note that creep softening is induced by the prior deformation.

The γ matrix phase possesses a cube–cube orientation relationship with the γ' precipitate phase which forms from it; these display the face-centred cubic (FCC) and L1$_2$ crystal structures, respectively. Provided that the elemental partitioning between γ and γ' is optimal such that the γ and γ' lattice parameters are balanced, the lattice misfit is small, the γ/γ' interfaces are coherent and the elastic strain energy is minimised. Other phases of importance are the carbides MC, M$_{23}$C$_6$ and M$_6$C, which are rich in Cr, Hf, Mo, Ti, Ta and W, and the topologically close-packed (TCP) phases μ, P, σ and R, which contain substantial concentrations of Cr, Mo and Re.

Defects are present – despite the crystalline nature of the phases – which control the mechanical properties exhibited. These are of various types: point defects such as vacancies; line defects such as dislocations (usually in dissociated form); and planar defects such as stacking faults. The stacking fault energy of nickel is one of the largest of all FCC metals at 250 mJ/m^2, so that partial dislocations in unalloyed γ are closely spaced. Dissociation is promoted by alloying additions, which decrease the stacking fault energy such that creep processes are hindered. The activation energy for solute–vacancy exchange depends strongly upon the solute species, with Re possessing a value (in excess of 160 kJ/mol) larger than for all others in the d block of transition metals; thus, Re is expected to retard diffusion-controlled processes in nickel alloys. Owing to the ordered structure, defects in γ' are complex. The anti-phase boundary (APB) is one type of planar defect which separates otherwise perfect crystals displaced by $a/2\langle\bar{1}01\rangle$, which is no longer a lattice vector; the associated APB energy is strongly anisotropic, being highest on the {110} plane and lowest on {001}. Intrinsic, extrinsic and complex stacking faults have also been identified. The energies of these planar defects control the mechanism of dislocation dissociation in the γ' phase.

Since a $a/2\langle1\bar{1}0\rangle\{111\}$ dislocation travelling in γ cannot enter the γ' phase without an APB forming, cutting of the γ/γ' structure requires pairs of them; this is the origin of a substantial precipitation-hardening effect known as order strengthening, which is relevant to static, tensile properties. A pair of dislocations (each one of which is known as a superpartial) may be either weakly or strongly coupled, depending upon whether their spacing is large or small in comparison with the γ' particle diameter. Cutting by the weak-coupling mechanism is applicable when the γ' particles are in the under-aged condition, but the strong-coupling one is appropriate for over-aged structures. Analysis indicates that the degree of hardening is approximately proportional to the square root of the particle radius, r, for the weak-coupling mechanism and inversely proportional to r for strong coupling; hardening is maximised for a γ' particle size at the transition between the two. For γ/γ' alloys, the yield stress does not decrease strongly as the temperature increases, provided that one stays below about 700 °C; this arises because of the anomalous yielding effect displayed by the γ' phase. The cause of this is the cross-slip of segments of the γ' superpartials from {111} to {001}, promoted by the applied stress, the anisotropy of the APB energy and a further contribution arising from elastic anisotropy.

The creep behaviour displayed by nickel and its alloys depends strongly upon the metallurgical structure, i.e. the grain size, γ', particle size and the concentrations of alloying additions. For polycrystalline nickel, a fifth-power-law dependence of the applied stress for the steady-state 'secondary' regime is found in common with other pure metals; this can be rationalised by balancing dislocation annihilation and multiplication processes (consistent with the 'recovery-controlled' model of creep deformation) provided that the sub-structure is assumed to coarsen with the cube root of time. Alloying additions have a strong strengthening effect on the creep of polycrystalline nickel, with strengthening being most potent when the atomic size misfit of solute and solvent is large; a distinction is made between 'Class I' behaviour, leading to a linear viscous motion of dislocations and a power-law stress dependence of three, and 'Class II' behaviour, with a dependence of five, which is analogous to the pure metals. Analysis indicates that most important alloying additions such as Fe, Co, Cr, W, Mo, Re and Ru promote Class II behaviour. The γ' phase fraction is found to have a profound influence on creep life, with the potency being greatest when the grain size is large. This effect occurs because the prominent form of creep damage in polycrystalline nickel alloys is creep cavitation at γ–grain boundaries, which becomes less influential as the grain size increases. This effect is exploited in the single-crystal superalloys, which are the topic of the Chapter 3.

Appendix. The anisotropic elasticity displayed by nickel

Nickel is *elastically anisotropic* when in single-crystal form. Hence the stiffness – a measure of the stress necessary to maintain a given strain – is dependent upon the crystallographic

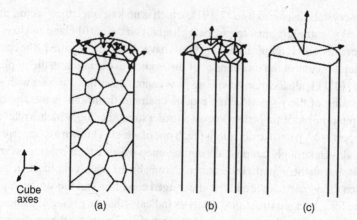

Cube axes

(a) (b) (c)

Fig. 2.74. Schematic illustration of the equiaxed, columnar and single-crystal structures used for turbine blade aerofoils.

orientation relative to the loading configuration. Consequently, specimens machined at random from a large single crystal will display different elastic properties. As turbine blading is often used in *single-crystal* or *columnar-grained* ($\langle 100 \rangle$-textured) form, see Figure 2.74, it is important to consider this effect. Moreover, the elastic anisotropy influences the cross-slip of dislocations, particularly in γ'.

A treatment of the elastic anisotropy of nickel requires an acknowledgement that the quantity relating the stress, σ, and strain, ϵ, both second-rank tensors, will then be a fourth-rank tensor [91]. Consider a cubic single crystal with orthogonal axes 1, 2, 3 lying along the [100], [010] and [001] directions, respectively. The elastic properties are then completely defined by *three* elastic stiffness constants C_{11}, C_{12} and C_{44}, such that

$$\sigma_{11} = C_{11}\epsilon_{11} + C_{12}\epsilon_{22} + C_{12}\epsilon_{33}$$
$$\sigma_{22} = C_{12}\epsilon_{11} + C_{11}\epsilon_{22} + C_{12}\epsilon_{33}$$
$$\sigma_{33} = C_{12}\epsilon_{11} + C_{12}\epsilon_{22} + C_{11}\epsilon_{33}$$
$$\sigma_{23} = C_{44}(\epsilon_{23} + \epsilon_{32}) = 2C_{44}\epsilon_{23}$$
$$\sigma_{31} = C_{44}(\epsilon_{31} + \epsilon_{13}) = 2C_{44}\epsilon_{31}$$
$$\sigma_{12} = C_{44}(\epsilon_{12} + \epsilon_{21}) = 2C_{44}\epsilon_{12} \tag{A1}$$

where use has been made of the reduced notation of Nye [91]. Note that the cubic symmetry reduces the number of constants in C_{ij} to three, from the 21 required for a triclinic crystal. Then $C_{11} = C_{22} = C_{33}$, $C_{44} = C_{55} = C_{66}$ and $C_{12} = C_{13} = C_{23}$; furthermore, the terms C_{14}, C_{15}, C_{16}, C_{24}, C_{25}, C_{26}, C_{34}, C_{35}, C_{36}, C_{45}, C_{46}, C_{56} are equal to zero.

An alternative representation is via the compliance constants S_{11}, S_{12} and S_{44}, such that

$$\epsilon_{11} = S_{11}\sigma_{11} + S_{12}\sigma_{22} + S_{12}\sigma_{33}$$
$$\epsilon_{22} = S_{12}\sigma_{11} + S_{11}\sigma_{22} + S_{12}\sigma_{33}$$
$$\epsilon_{33} = S_{12}\sigma_{11} + S_{12}\sigma_{22} + S_{11}\sigma_{33}$$

$$\epsilon_{23} = \frac{1}{2} S_{44} \sigma_{23}$$

$$\epsilon_{31} = \frac{1}{2} S_{44} \sigma_{31}$$

$$\epsilon_{12} = \frac{1}{2} S_{44} \sigma_{12} \tag{A2}$$

The stiffness and compliance constants are inter-related. For example, the compliance constants can be evaluated via

$$S_{11} = \frac{C_{11} + C_{12}}{(C_{11} - C_{12})(C_{11} + 2C_{12})} \qquad S_{12} = \frac{C_{12}}{(C_{11} - C_{12})(C_{11} + 2C_{12})} \qquad S_{44} = 1/C_{44} \tag{A3}$$

It is instructive to consider the magnitude of the constants, which have been determined by careful measurements on nickel single crystals. For pure nickel at room temperature, typical values are $S_{11} = 0.799 \times 10^{-5}/\text{MPa}$, $S_{12} = -0.312 \times 10^{-5}/\text{MPa}$ and $S_{44} = 0.844 \times 10^{-5}/\text{MPa}$. The stiffness along the $\langle 100 \rangle$ crystallographic direction is then derived by setting $\sigma_{22} = \sigma_{33} = 0$, so that

$$E_{\langle 100 \rangle} = \frac{\sigma_{11}}{\epsilon_{11}} = \frac{1}{S_{11}} \tag{A4}$$

For the pure nickel data, this gives $E_{\langle 100 \rangle} = 125\,\text{GPa}$. This is much less than the value for *polycrystalline* nickel, which, in the absence of texture, is expected to be isotropic, since the many grains average out the elastic properties over the different orientations. In fact, polycrystalline nickel has an elastic modulus of about 207 GPa, about 66% higher than that along the $\langle 100 \rangle$ direction. For nickel, it turns out that the $\langle 100 \rangle$ direction is the least stiff, and this is true for most other cubic metals.

How does one go about evaluating the elastic modulus at different crystallographic orientations, for example, along a direction with direction cosines l_1, l_2, l_3 defined relative to the 1, 2, 3 axes? To do this it is necessary to recognise that S is a fourth-rank tensor. Making use of the reduced notation, the new value of S_{11}, denoted S'_{11}, is given by

$$S'_{11} = \begin{bmatrix} +l_1^4 S_{11} & +2l_1^2 l_2^2 S_{12} & +2l_1^2 l_3^2 S_{13} & +2l_1^2 l_2 l_3 S_{14} & +2l_1^3 l_3 S_{15} & +2l_1^3 l_2 S_{16} \\ & +l_2^4 S_{22} & +2l_2^2 l_3^2 S_{23} & +2l_2^3 l_3 S_{24} & +2l_1 l_2^2 l_3 S_{25} & +2l_1 l_2^3 S_{26} \\ & & +l_3^4 S_{33} & +2l_2 l_3^3 S_{34} & +2l_1 l_3^3 S_{35} & +2l_1 l_2 l_3^2 S_{36} \\ & & & +l_2^2 l_3^2 S_{44} & +2l_1 l_2 l_3^2 S_{45} & +2l_1 l_2^2 l_3 S_{46} \\ & & & & +l_1^2 l_3^2 S_{55} & +2l_1^2 l_2 l_3 S_{56} \\ & & & & & +l_1^2 l_2^2 S_{66} \end{bmatrix} \tag{A5a}$$

for the general case of a monoclinic crystal. Note that the various 2's are present due to the symmetry of S. For a cubic crystal such as Ni, $S_{11} = S_{22} = S_{33}$, $S_{44} = S_{55} = S_{66}$ and $S_{12} = S_{13} = S_{23}$ with the other terms zero. Hence,

$$S'_{11} = S_{11}(l_1^4 + l_2^4 + l_3^4) + 2S_{12}(l_1^2 l_2^2 + l_1^2 l_3^2 + l_2^2 l_3^2) + S_{44}(l_1^2 l_2^2 + l_1^2 l_3^2 + l_2^2 l_3^2) \tag{A5b}$$

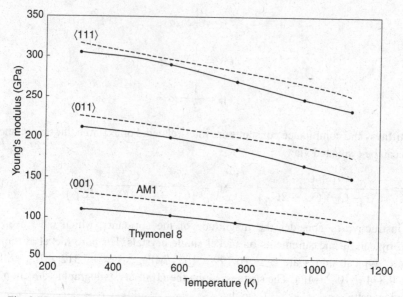

Fig. 2.75. Variation of the elastic modulus of single-crystal Thymonel and AM1 with crystallographic orientation and temperature [92].

Making use of the identities $l_1^2 + l_2^2 + l_3^2 = 1$ and $l_1^4 + l_2^4 + l_3^4 = 1 - 2(l_1^2 l_2^2 + l_1^2 l_3^2 + l_2^2 l_3^2)$, the transformation to the new axis is accomplished using

$$\frac{1}{E} = S'_{11} = S_{11} - 2\left(S_{11} - S_{12} - \frac{1}{2}S_{44}\right)\left[l_1^2 l_2^2 + l_2^2 l_3^2 + l_3^2 l_1^2\right] \tag{A6}$$

Thus, the degree of anisotropy depends upon the term $\left[l_1^2 l_2^2 + l_2^2 l_3^2 + l_3^2 l_1^2\right]$, which has a minimum value of zero along the cube axes, a maximum value of one-third along $\langle 111\rangle$ and a value of one-quarter along $\langle 110\rangle$. Using the values of S_{ij} for pure nickel, one can show that the $\langle 100\rangle$ direction is the least stiff, $\langle 111\rangle$ the stiffest (about 2.4 times stiffer than $\langle 100\rangle$), with $\langle 110\rangle$ between the two limits. The values are $E_{\langle 100\rangle} = 125$ GPa, $E_{\langle 110\rangle} = 220$ GPa and $E_{\langle 111\rangle} = 294$ GPa. Single-crystal superalloys display a similar degree of anisotropy; see Figure 2.75 [92].

What can be said about the resistance to shearing, for example, in the 23 plane, which has as its normal the solidification direction $\langle 100\rangle$? From the elasticity equations, one can see that this is largely controlled by the value of S_{44}; however, the resistance to shearing does vary with angle in the 23 plane. One can transform S_{44} to S'_{44} using similar methods used to determine S'_{11} above; one can show that the rotation through an arbitrary angle θ is accomplished via

$$S'_{44} = S_{44} + (2S_{11} - 2S_{12} - S_{44}) \times 4\sin^2\theta\cos^2\theta = S_{44} + (2S_{11} - 2S_{12} - S_{44}) \times \sin^2 2\theta \tag{A7}$$

Thus, when $\theta = 0$, G is the reciprocal of S_{44} and about 119 GPa. When $\theta = \pi/4$, G is the reciprocal of S'_{44} and is about 26 GPa. For isotropic, polycrystalline nickel, the value of G has been determined to be about 82 GPa, not far from the numerical average of 119 GPa and 26 GPa. More strictly, the value averaged over all loading configurations in the 23 plane is given by

$$\frac{4}{\pi} \int_0^{\pi/4} \frac{d\theta}{S_{44} + (2S_{11} - 2S_{12} - S_{44}) \times \sin^2 2\theta} = [2S_{44}(S_{11} - S_{12})]^{-1/2} \qquad (A8)$$

For the nickel data, this gives about 73 GPa. This is the appropriate value for columnar-grained nickel grown by directional solidification, with the long axis of each grain equivalent to the $\langle 100 \rangle$ growth direction.

For polycrystalline alloys, such as wrought superalloys or equiaxed castings, the elastic properties are expected to be isotropic and thus invariant with crystallographic orientation. Examination of Equation (A5) reveals that the condition for isotropy is

$$S_{11} - S_{12} - \frac{1}{2}S_{44} = 0 \qquad (A9)$$

In this case, the Young's modulus, E, Poisson's ratio, v, and shear modulus, G, are given by the identities

$$S_{11} = \frac{1}{E} \qquad S_{12} = -v/E \qquad \text{and} \qquad 2(S_{11} - S_{12}) = 1/G \qquad (A10)$$

Since $S_{44} = 2(S_{11} - S_{12})$, only two of E, v and G are independent; the third can be calculated from a knowledge of the other two, for example, via identities such as $G = E/(2(1 + v))$. Measurements on polycrystalline nickel at room temperature have yielded $E = 207$ GPa, $v = 0.26$, and thus $G = 82$ GPa. For polycrystalline nickel one has therefore $S_{11} = 4.83 \times 10^{-12}$ m^2/N, $S_{12} = -1.26 \times 10^{-12}$ m^2/N and $S_{44} = 12.17 \times 10^{-12}$ m^2/N.

Given the above considerations, it is interesting to determine whether the *elastic anisotropy* of nickel is greater or less than that displayed by other cubic metals. It is helpful to define an 'anisotropy factor', A, according to

$$A = \frac{2C_{44}}{C_{11} - C_{12}} \qquad (A11)$$

which will be unity in the case of isotropy. The value for Ni is 2.44, which compares with Cu at 3.19, Ag at 2.97, Au at 2.90, Ge at 1.65, Si at 1.57 and Al at 1.22, for other cubic metals. Thus, whilst Ni displays a fair degree of anisotropy, it is not the most anisotropic of FCC metals.

To summarise, Ni in single-crystal form is elastically anisotropic. When designing single-crystal blading from the nickel-based superalloys, it is important to recognise this fact. Compare, for example, the elastic properties of polycrystalline and directionally solidified nickel, which displays a strong $\langle 100 \rangle$ growth texture. Polycrystalline nickel displays values of E and G of 207 GPa and 82 GPa, respectively. Directionally solidified nickel displays a stiffness in the growth direction, denoted $E_{\langle 100 \rangle}$ of 125 GPa, i.e. only 60% of the isotropic value. The shear modulus for shearing in the plane whose normal is the growth direction is 73 GPa, about 89% of the isotropic case.

Questions

2.1 Nickel has an atomic weight of 58.71, a number which arises from the relative proportions of isotopes of weights 58, 60, 61, 62 and 64. Why is there little contribution from the isotopes of weight 59 and 63?

2.2 Nickel exhibits the FCC crystal structure, with a well-defined melting temperature of 1455 °C. But suppose that it was possible to supercool nickel liquid rapidly below the melting temperature, suppressing the formation of γ. At what approximate temperature might one expect the metastable HCP solid to form? Answer this question by investigating a binary phase diagram Ni–X, where X is an element from the d block which displays the HCP crystal structure. Given that the latent heat of melting of nickel is about 10 J/(mol K), estimate the difference in the Gibbs energy of the FCC and HCP forms at the temperature of solidification of the metastable HCP form.

Kaufman has carried out pioneering work on the lattice stabilities of the pure elements. His data for nickel (relative to the ferromagnetic FCC form) are: (i) liquid: $17\,614 - 10.299\,T$; (ii) HCP: $1046 - 1.255\,T$; (iii) BCC: $5564 - 1.06\,T$, where T is in Kelvin and the expressions give data in J/mol. Plot out these expressions as a function of temperature. Are Kaufman's data in reasonable agreement with your findings?

2.3 Consider a hypothetical binary phase diagram which displays complete solid solubility at low temperatures and a (solid + liquid) two-phase region below the liquid phase field. If both phases are ideal, demonstrate that just four parameters are sufficient for the calculation of the phase diagram. Using a spreadsheet, investigate how the shape and width of the two-phase region depends upon different combinations of the two parameters $\Delta S_1^{L\rightarrow\phi}$ and $\Delta S_2^{L\rightarrow\phi}$. Give a physical explanation for your findings.

In the Ni–Cr binary system, the liquid phase shows considerable deviation from ideality. For example, the Gibbs free energy of mixing at $x_{Cr} = 0.5$ is -13.1 kJ/mol at 1873 K. Using this information, modify the spreadsheet developed to describe the thermodynamic equilibria displayed by the Ni–Cr phase diagram. Demonstrate that the phase diagram is reproduced reasonably accurately.

2.4 Saunders has developed a thermodynamic model for the γ/γ' equilibria in the binary Ni–Al system. The molar Gibbs energy, $G_m^\varphi\{T\}$, of each phase, φ, as a function of temperature (T) is represented by

$$G_m^\varphi\{T\} - \sum_i x_i^\varphi\,{}^0H_i\{298.15\,\text{K}\} = G_m^{\text{ref},\varphi} + G_m^{\text{ent},\varphi} + G_m^{\text{E},\varphi}$$

where x_i^φ is the mole fraction of component i in phase φ and ${}^0H_i\{298.15\,\text{K}\}$ is the molar enthalpy of i in its standard state at 298.15 K. The terms $G_m^{\text{ref},\varphi}$, $G_m^{\text{ent},\varphi}$ and $G_m^{\text{E},\varphi}$ are the reference, configurational entropy and excess contributions to the molar Gibbs energy. The disordered γ phase is modelled with a single sub-lattice, with

$$G_m^\gamma = x_{Al}^\gamma\left[G_{Al}^\gamma\{T\} - {}^0H_{Al}\{298.15\,\text{K}\}\right] + x_{Ni}^\gamma\left[G_{Ni}^\gamma\{T\} - {}^0H_{Ni}\{298.15\,\text{K}\}\right]$$

$$G_m^{\text{ent},\gamma} = RT\left[x_{Al}^\gamma \ln\{x_{Al}^\gamma\} + x_{Ni}^\gamma \ln\{x_{Ni}^\gamma\}\right]$$

and

$$G_m^{E,\gamma} = x_{Al}^\gamma x_{Ni}^\gamma L_{Al,Ni}^\gamma$$

where the term $L_{Al,Ni}^\gamma$ is an interaction parameter. The Ni_3Al phase can be modelled as a point compound, where

$$G_m^{\gamma'} = G_{Ni:Al}^{\gamma'}$$

such that there are no entropy and excess contributions.

Given the thermodynamic data given below, plot out the variation of G_m^γ and $G_m^{\gamma'}$ with temperature. Using graphical means or otherwise, determine the position of the $\gamma/(\gamma + \gamma')$ phase diagram as a function of temperature. Furthermore, estimate the ordering energy as a function of temperature.

The following thermodynamic data are in J/mol:

$$G_{Al}^\gamma = -11\,278.378 + 188.684\,153\,T - 31.748\,192\,T\ln T - 1.231 \times 10^{28}\,T^{-9};$$
$$G_{Ni}^\gamma = -5179.159 + 117.854\,T - 22.096\,T\ln T - 0.004\,840\,7\,T^2$$
$$L_{Al,Ni}^\gamma = -168\,750 + 16\,T + 30\,600(x_{Al}^\gamma - x_{Ni}^\gamma) + 41\,700(x_{Al}^\gamma - x_{Ni}^\gamma)^2$$
$$G_{Ni:Al}^{\gamma'} = 0.75\,G_{Ni}^\gamma - 40\,000 + 3\,T + 0.25\,G_{Al}^\gamma$$

2.5 Using published ternary phase diagrams for the Ni–Al–Cr system, investigate the effect on the solid-state equilibria of increasing the Cr content from zero to 25 at%, in a series of alloys containing (i) a constant Al content of 15 at% and (b) a constant Ni content of 75 at%. Use Thermo-Calc to help validate your conclusions.

2.6 Examination of Ni–Al–X ternary phase diagrams, where X is a ternary addition such as Ti, Nb, Pt, Re, etc., indicates that the γ' phase field is often parallel to either (i) the Ni–X axis or alternatively (ii) the Al–X axis. What causes this to happen? Very rarely, this is not the case: for Cr, for example. Why should such elements not conform this to this general rule?

2.7 In Figure 2.10 (taken from ref. [8]) the enthalpies of formation of the intermetallic compounds in the Ni–Al system are compared with the enthalpy of mixing of the FCC phase. Since the data correspond to the low temperature of 25 °C, it is claimed that the entropy contribution is small, so that ΔH_f behaves like ΔG_f.

Using the Thermo-Calc software, investigate whether this is a good assumption. Estimate a value of (i) the ordering energy of β–NiAl and (ii) the enthalpy of formation at 25 °C, measured with respect to the disordered BCC phase.

It is found that, of all the intermetallic phases in the Ni–Al system, the ordering energy is largest for β–NiAl. Remembering that this displays the CsCl crystal structure, give a physical explanation for this.

2.8 The lattice parameters of the γ and γ' phases in the superalloys depend upon the alloy chemistry. However, the lattice parameter of γ is, in general, slightly more sensitive to solute additions than γ'; see Figure 2.18. Give a rationale for this observation.

2.9 The atom probe can be used to build up a ladder diagram in which the total number of atoms detected is plotted against the total number of atoms detected. For a specimen of pure Ni_3Al, explain the features of the curve which are observed, and estimate the slopes of the different regimes.

2.10 The lattice parameter of the Ni_3Al phase at room temperature is 0.357 00 nm. Given that it displays the $L1_2$ structure, estimate the atomic radii of Ni and Al in this compound. Is there any evidence that the bonding is directional in nature? How does the value for Ni compare with that of pure Ni, which has a lattice parameter of 0.351 67 nm?

2.11 Investigate how the radius of the transition group metals varies as one crosses any one period of the periodic table. Repeat this exercise so that you have information for the first three series. What patterns emerge? And do the γ'-forming elements display significantly different values from the elements which form solid solutions with γ–nickel?

2.12 In the intermetallic compounds γ'–Ni_3Al and β–NiAl, various sorts of point defects are possible. Distinguish between the four most common of these.

2.13 Whilst both Frenkel and Schottky defects can arise in compounds such as $\dot{\beta}$–NiAl, only one of these can occur in a pure elemental crystal such as Ni. Identify which of these defects may form in pure Ni, and explain why this is not possible for the other defect.

2.14 Bradley and Taylor [41] made measurements of the lattice spacing and density of the NiAl intermetallic compound at compositions at and around the point of exact stoichiometry (Figure 2.76). Use their data to prove conclusively that (i) on the Ni-rich side, the excess Ni atoms are accommodated as anti-site defects, and (ii) on the Al-rich side, the excess Al atoms are not accommodated as anti-site defects and that there must be a significant proportion of constitutional vacancies present. What is the fraction of constitutional vacancies present for an alloy on the phase boundary on the Al-rich side?

2.15 Theoretical expressions for the self-diffusion of pure metal indicate that the rate is proportional to Z, the number of neighbours nearest to any given atom. For nickel, which possesses the cubic F structure, determine Z and the corresponding lattice vector. How would your answers differ if nickel exhibited (i) the cubic P and (ii) the cubic I lattices?

2.16 Explain why the tracer diffusion coefficient of Al in the γ' phase (which has the $L1_2$ structure) is considerably smaller than that of Ni.

2.17 Comparison of the rates of interdiffusion of the transition group metals (the solutes) with nickel (the solvent) indicates that (i) the interdiffusion rate increases with increasing misfit strain between solvent and solute and (ii) the activation energy for interdiffusion decreases with increasing misfit strain. Why might these observations be contrary to expectation? How might this apparent anomaly be rationalised?

2.18 The interdiffusion coefficient of nickel with aluminium in the β–NiAl intermetallic compound is strongly dependent upon concentration. The minimum value is near the point of 50:50 stoichiometry, but not quite so. Why is this?

2.19 Self-diffusion in nickel is mediated by the exchange of atoms and vacancies on the lattice – so-called 'vacancy-assisted diffusion'. The micromechanism involves the

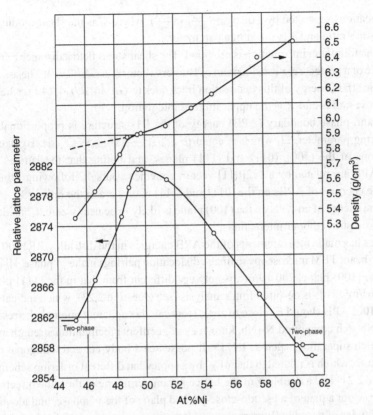

Fig. 2.76. Variation of the density and relative lattice parameter of NiAl around the point of stoichiometry [41].

movement of an atom by a distance $a/\sqrt{2}$ in the direction $\langle 110 \rangle$. Show that the size of the diffusion window is $(\sqrt{3} - 1)D_{Ni}$ at the saddle point, where D_{Ni} is the hard-sphere diameter of a Ni atom.

2.20 A rod of Ni is heated to its melting point, at which the concentration of vacancies is 3.3×10^{-5} per atomic site. The rod is then quenched so that all vacancies generated at the melting point are trapped on the lattice. Upon annealing at a higher temperature, the rod is observed to shrink. Why? If the total contraction is 5.1×10^{-6}, what change in lattice parameter should be observed during the annealing process?

2.21 Show that it is favourable for a perfect dislocation with Burgers vector $\frac{a}{2}[110]$ slipping on $(1\bar{1}1)$ in a cubic close-packed (CCP) metal such as nickel to dissociate into two partial dislocations. Give the Burgers vectors of the partial dislocations. What factors determine their equilibrium separation?

Cobalt is often added to the nickel-based superalloys. Explain why the addition of cobalt is beneficial. Why might the effects of other elements be more potent?

2.22 Show that, on the basis of dislocation mechanics and isotropic elasticity, a perfect screw dislocation gliding in nickel will dissociate into two Shockley partial dislocations separated by a distance $\frac{Ga^2}{48\pi\gamma}\left(\frac{2-3\nu}{1-\nu}\right)$, where G is the shear modulus, ν is Poisson's ratio and a is the lattice parameter.

The following information is provided. The shear stress field a distance r from the centre of a perfect screw dislocation of Burgers vector b is $Gb/(2\pi r)$. The associated elastic strain energy of this dislocation integrates to $Gb^2 \ln\{R/r_0\}/(4\pi)$, where r_0 is the core radius and R is the upper limit of integration.

2.23 The anti-phase boundary (APB) energy of the L1$_2$ structure is proportional to the ordering parameter, Ω, which is equal to $\Omega_{NiAl} - \frac{1}{2}(\Omega_{NiNi} + \Omega_{AlAl})$. By sketching sections of the (100), (011) and (111) planes, and noting that the lattices across the APB are related by a $1/2[01\bar{1}]$ vector, convince yourself (following Flinn) that (i) the energies of APBs on the (011) and (111) planes are in the ratio $\sqrt{3} : \sqrt{2}$, and (ii) the energy of an APB on the (100) plane is likely to be nearly zero. Consider only the nearest-neighbour interactions.

2.24 Stobbs has made measurements of the APB energies in the disc alloy RR1000, using weak-beam TEM microscopy to image dislocation pairings in the γ' phase. His figure for the {100} plane, 200 mJ/m^2, is not very different from that in the {111} plane, at 300 mJ/m^2. This is despite Flinn's analysis (see Question 2.23), which indicates that the {100} APB should be close to zero. How can this apparent anomaly be resolved?

2.25 The Nb-rich compound Ni$_3$Nb, known as γ'', confers precipitation strengthening to iron-rich superalloys such as IN718. It possesses a body-centred tetragonal crystal structure, which is related to that of γ' by a somewhat different ordering scheme (see Figure 2.21) and a doubling of the lattice parameter in one of the cube directions.

Draw out a plan of a pseudo-close-packed plane of the γ'' phase, and identify the magnitudes of possible Burgers vectors. Hence convince yourself that in only one of three directions will order be restored as a result of the movement of two $a/2\langle110\rangle$ matrix dislocations cutting through γ'', and that, in the other two directions, four such dislocations are required. Thus explain why γ''-strengthened alloys possesses extremely high strength levels at elevated temperatures.

2.26 Oblak, Paulonis and Duvall [93] studied the flow characteristics of the iron–nickel superalloy IN718, in single-crystal form, with the tensile axis along $\langle001\rangle$. When tested at 300 °C, the yield stress was 300 MPa; however, if the specimens were first aged at 790 °C for 5 h, the yield stress was altered by $+25\%$ and -6% for tensile and compressive loads, respectively. Give an explanation for these findings.

2.27 Each of the alloying elements present in a superalloy is expected to make a contribution to the level of solid-solution strengthening, such that the yield stress is increased by an increment $\Delta\sigma$. To sum the contributions from the concentrations, c_i, of each element i, it has been suggested that one uses

$$\Delta\sigma = \left(\sum_i k_i^2 c_i\right)^{1/2}$$

Tensile testing at room temperature indicates that suitable values for the k_i's are Al $= 225$, Ti $= 775$, Zr $= 2359$, Hf $= 1401$, Cr $= 337$, Mo $= 1015$, W $= 977$, Mn $= 448$, Fe $= 153$ and Co $= 39$, where the units are in MPa $\times$ at.fraction$^{-1/2}$.

Use this relationship and the data provided to quantify, for the single-crystal superalloy SRR99, the relative contributions of the different elements to the solid-strengthening levels to be expected in this alloy.

2.28 In order to evaluate the temperature dependence of the shear yield stress, τ_y, of a solid-solution-strengthened alloy, one can appeal to the trough model, which predicts

$$T\left(\frac{\tau_y}{G}\right)^{1/2} = K_1 - K_2\tau_y$$

where T is temperature and G is the shear modulus. The terms K_1 and K_2 are constants, which are in the range 5.7 to 27 K (K_1) and 0.22 to 0.26 K/MPa (K_2).

Using the expression, the data provided and a suitable estimate for G, investigate how the shear yield stress, τ_y, is expected to vary as a function of temperature. What is the effect of incorporating a temperature dependence into G?

2.29 The expression derived in Section 2.3 for the strengthening arising from strongly coupled dislocation pairs is dependent upon the function $\ell\{x\}$. For spherical particles of radius r, $\ell\{x\}$ is given by $2(2rx - x^2)^{1/2}$; the results in a $r^{-1/2}$ dependence for the strengthening, when r is large.

Suppose instead one considers cuboidal particles. What then is the appropriate shape of the cross-section, if the slip plane is $\{111\}$? Derive an expression for $\ell\{x\}$ for this case. Is the $r^{-1/2}$ dependence preserved? If not, is the dependence on r stronger or weaker than for spherical particles?

2.30 In nominally pure nickel in the annealed state, the movement of dislocations during plastic deformation is not limited by the Peierls–Nabarro stress, which is of very small magnitude (in common with all FCC materials). Evidence from this comes from TEM studies, which indicate that the dislocations are wavy with little evidence of pinning being displayed. This being the case, what phenomenon controls dislocation movement? How would the situation be different if nickel were to be heavily worked?

2.31 The shear strain rate, $\dot{\gamma}$, arising due to low-temperature plasticity has been modelled in a phenomenological sense by Kocks, Argon and Ashby according to

$$\dot{\gamma} = \dot{\gamma}_0\left(\frac{\sigma_s}{\mu}\right)^2 \exp\left\{-\frac{\Delta F}{kT}[1 - (\sigma_s/\tau_p)^p]^q\right\}$$

where σ_s is the deviatoric stress, μ is the shear modulus, τ_0 is the shear stress at 0 K and ΔF is an activation energy required either for (i) overcoming discrete obstacles or else (ii) formation of an isolated pair of kinks. The other terms, $\dot{\gamma}_0$, p and q, are constants; theory suggests that $0 \leq p \leq 1$ and $1 \leq q \leq 2$, with $\dot{\gamma}_0$ about 10^6/s.

Make a fit to the flow stress–temperature data for nickel (see Figure 2.52), including data down to cryogenic temperatures. Do your fitted parameters reveal much about the factors controlling slip? What is revealed about the strain-rate sensitivity? Convince yourself that the strain-rate sensitivity is low compared with, for example, that seen during power-law creep.

2.32 The elastic modulus of nickel is highly anisotropic; the Young's moduli along $\langle 111 \rangle$, $\langle 110 \rangle$ and $\langle 100 \rangle$ at ambient temperature are in the ratio 2.80:1.71:1. By determining the interplanar spacings in the cubic close-packed (CCP) crystal structure, rationalise the observation that the $\langle 111 \rangle$ and $\langle 100 \rangle$ directions are the most and least stiff, respectively.

2.33 The variation of the Young's modulus around the $\langle 100 \rangle$ pole is of profound importance in the design of turbine blading, since resonance can cause high-cycle fatigue failure. Using the expressions given in the Appendix to this chapter, investigate how the Young's modulus varies within $20°$ of $\langle 100 \rangle$. For CMSX-4 blading at $700\,°C$, take $E_{100} = 104\,$GPa, $E_{110} = 190\,$GPa and $E_{111} = 262\,$GPa.

2.34 In the context of the elasticity of monocrystalline nickel, prove the following identities:

$$S_{11} = \frac{C_{11} + C_{12}}{(C_{11} - C_{12})(C_{11} + 2C_{12})} \quad S_{12} = \frac{C_{12}}{(C_{11} - C_{12})(C_{11} + 2C_{12})} \quad S_{44} = 1/C_{44}$$

where the S_{ij} and C_{ij} are components of the compliance and stiffness matrices, respectively.

2.35 In Section 2.4, the σ^5 stress dependence of steady-state creep in pure nickel was justified on the basis that time scales with the third power of the characteristic length of the dislocation sub-structure λ. But suppose instead that a fifth power law (corresponding to coarsening of the the sub-grains by pipe diffusion) applies. Demonstrate that the exponent in the steady-state creep law is then equal to 7.

2.36 In the climb-plus-glide creep model of Weertman, the total time, t_{total}, that a dislocation spends deforming is partitioned into glide and climb components, such that

$$t_{\text{total}} = t_{\text{glide}} + t_{\text{climb}}$$

By (i) dividing through by the product $\rho b L$, where ρ is the dislocation density, b is the Burgers vector and L is the average distance that a dislocation moves parallel to the slip plane, and (ii) defining the total, glide and climb components of the strain rate in an appropriate way, demonstrate that

$$\frac{1}{\dot{\epsilon}_{\text{total}}} = \frac{1}{\dot{\epsilon}_{\text{glide}}} + \frac{1}{\dot{\epsilon}_{\text{climb}}}$$

so that one has a sequential process for the net strain rate.

2.37 Creep-rupture data for polycrystalline, nominally pure nickel are plotted on Figure 2.66. Use these data to plot a line for Ni on the Larson–Miller plot of Figure 1.16. Hence show that many intermetallics are superior to Ni in creep, but that they cannot compete at present with the single-crystal superalloys.

References

[1] C. T. Sims, N. S. Stoloff and W. C. Hagel, eds, *Superalloys II: High Temperature Materials for Aerospace and Industrial Power* (New York: John Wiley and Sons, 1987).

[2] M. Durand-Charre, *The Microstructure of Superalloys* (Amsterdam: Gordon & Breach Science Publishers, 1997).

[3]   P. Nash, ed., *Phase Diagrams of Binary Nickel Alloys* (Materials Park, OH: ASM International, 1991).

[4]   A.T. Dinsdale, SGTE data for pure elements, *Calphad*, **15** (1991), 317–425.

[5]   L. Kaufman and H. Bernstein, *Computer Calculation of Phase Diagrams* (New York: Academic Press, 1970).

[6]   J. O. Andersson, T. Helander, L. Hoglund, P. F. Shi and B. Sundman, Thermo-Calc and DICTRA: computational tools for materials science, *Calphad*, **26** (2001), 273–312.

[7]   N. Saunders, Phase diagram calculations for high-temperature materials, *Philosophical Transactions of the Royal Society of London A*, **351** (1995), 543–561.

[8]   K. J. Blobaum, D. Van Heerden, A. J. Gavens and T. P. Weihs, Al/Ni formation reactions: characterisation of the metastable Al_9Ni_2 phase and analysis of its formation, *Acta Materialia*, **51** (2003), 3871–3884.

[9]   S. Ochiai, Y. Oya and T. Suzuki, Alloying behaviour of Ni_3Al, *Acta Metallurgica*, **32** (1984), 289–298.

[10]   K. Hono, Private communication, The National Institute of Materials Science, Tsukuba Science City, Japan (2003).

[11]   R. W. Cahn, P. A. Siemers and J. E. Geiger, The order-disorder transformation in Ni_3Al and Ni_3Al − Fe alloys: 1. Determination of the transition temperatures and their relation to ductility, *Acta Metallurgica*, **35** (1987), 2737–2751.

[12]   F. J. Bremer, M. Beyss and H. Wenzl, The order-disorder transition of the intermetallic phase Ni_3Al, *Physica Status Solidi*, **110A** (1988), 77–82.

[13]   S. V. Prikhodko and A. J. Ardell, Coarsening of gamma prime in Ni-Al alloys aged under uniaxial compression: III. Characterisation of the morphology, *Acta Materialia*, **51** (2003), 5021–5036.

[14]   R. A. Ricks, A. J. Porter and R. C. Ecob, The growth of gamma prime precipitates in nickel-base superalloys, *Acta Metallurgica*, **31** (1983), 43–53.

[15]   E. E. Brown and D. R. Muzyka, Nickel-iron alloys, in C. T. Sins, N. S. Stoloff and W. C. Hagel, eds, *Superalloys II: High Temperature Materials for Aerospace and Industrial Power* (New York: John Wiley and Sons, 1987), pp. 165–188.

[16]   J. W. Brooks and P. J. Bridges, Metallurgical stability of Inconel 718, in D. N. Duhl, G. Maurer, S. Antolovich, C. Lund and S. Reichman, eds, *Superalloys 1988* (Warrendale, PA: The Metallurgical Society, 1988), pp. 33–42.

[17]   R. Cozar and A. Pineau, Morphology of gamma prime and gamma double prime precipitates and thermal stability of Inconel 718 type alloys, *Metallurgical Transactions*, **4** (1973), 47–59.

[18]   R. C. Reed, M. P. Jackson and Y. S. Na, Characterisation and modelling of the precipitation of the sigma phase in Udimet 720 and Udimet 720Li, *Metallurgical and Materials Transactions*, **30A** (1999), 521–533.

[19]   A. K. Sinha, Topologically close-packed structures of transition metal alloys, *Progress in Materials Science*, **15** (1972), 79–185.

[20]   S. Tin, T. M. Pollock and W. T. King, Carbon additions and grain boundary formation in high refractory nickel-base single crystal superalloys, in T. M. Pollock,

R. D. Kissinger, R. R. Bowman *et al.*, eds, *Superalloys 2000* (Warrendale, PA: The Minerals, Metals and Materials Society (TMS), 2000), pp. 201–210.

[21] Q. Z. Chen and D. M. Knowles, The microstructures of base/modified RR2072 SX superalloys and their effects on creep properties at elevated temperatures, *Acta Materialia*, **50** (2002), 1095–1112.

[22] A. Kelly, G. W. Groves and P. Kidd, *Crystallography and Crystal Defects*, revised edn (Chichester, UK: John Wiley & Sons, 2000).

[23] F. A. Mohamed and T. G. Langdon, The transition from dislocation climb to viscous glide in creep of solid solution alloys, *Acta Metallurgica*, **30** (1974), 779–788.

[24] P. C. J. Gallagher, The influence of alloying, temperature and related effects on the stacking fault energy, *Metallurgical Transactions*, **1** (1970), 2429–2461.

[25] L. Delehouzee and A. Deruyttere, The stacking fault density in solid solutions based on copper, silver, nickel, aluminium and lead, *Acta Metallurgica*, **15** (1967), 727–734.

[26] W. Wycisk and M. Feller-Kniepmeier, Quenching experiments in high purity Ni, *Journal of Nuclear Materials*, **69/70** (1978), 616–619.

[27] R. O. Simmons and R. W. Balluffi, Measurements of equilibrium vacancy concentrations in aluminum, *Physical Review*, **117** (1960), 52–61.

[28] B. Jonsson, Assessment of the mobilities of Cr, Fe and Ni in binary FCC Cr-Fe and Cr-Ni alloys, *Scandinavian Journal of Metallurgy*, **24** (1995), 21–27.

[29] M. S. A. Karunaratne, P. Carter and R. C. Reed, Interdiffusion in the FCC-A1 phase of the Ni-Re, Ni-Ta and Ni-W systems between 900 and 1300 deg C, *Materials Science and Engineering*, **A281** (2000), 229–233.

[30] M. S. A. Karunaratne and R. C. Reed, Interdiffusion of the platinum group metals in nickel at elevated temperatures, *Acta Materialia*, **51** (2003), 2905–2919.

[31] C. L. Fu, R. C. Reed, A. Janotti and M. Krcmar, On the diffusion of alloying elements in the nickel-base superalloys, in K. A. Green, T. M. Pollock, H. Harada, *et al.*, eds, *Superalloys 2004* (Warrendale, PA: The Minerals, Metals and Materials Society (TMS), 2004), pp. 867–876.

[32] A. Janotti, M. Krcmar, C. L. Fu and R. C. Reed, Solute diffusion in metals: larger atoms can move faster, *Physical Review Letters*, **92** (2004), 085901.

[33] P. A. Flinn, Theory of deformation in superlattices, *Transactions of the Metallurgical Society of AIME*, **218** (1960), 145–154.

[34] B. H. Kear, A. F. Giamei, J. M. Silcock and R. K. Ham, Slip and climb processes in gamma prime precipitation hardened nickel-base alloys, *Scripta Metallurgica*, **2** (1968), 287–294.

[35] B. H. Kear, A. F. Giamei, G. R. Leverant and J. M. Oblak, On intrinsic/extrinsic stacking fault pairs in the L1$_2$ lattice, *Scripta Metallurgica*, **3** (1969), 123–130.

[36] D. P. Pope and S. S. Ezz, Mechanical properties of Ni$_3$Al and nickel-base alloys with high volume fraction of γ', *International Metals Reviews*, **29** (1984), 136–167.

[37]   F. R. N. Nabarro and H. L. de Villiers, *The Physics of Creep* (London: Taylor and Francis, 1995).

[38]   B. H. Kear and H. G. F. Wilsdorf, Dislocation configurations in plastically deformed Cu_3Au alloys, *Transactions of the Metallurgical Society of AIME*, **224** (1962), 382–386.

[39]   K. Aoki and O. Izumi, Defect structures and long-range order parameters in off-stoichiometric Ni_3Al, *Physica Status Solidi*, **32A** (1975), 657–664.

[40]   H. Numakura, N. Kurita and M. Koiwa, On the origin of the anelastic relaxation effect in Ni_3Al, *Philosophical Magazine*, **79A** (1999), 943–953.

[41]   A. J. Bradley and A. Taylor, An x-ray analysis of the nickel-aluminium system, *Proceedings of the Royal Society of London*, **A159** (1937), 56–72.

[42]   A. Mourisco, N. Baluc, J. Bonneville and R. Schaller, Mechanical loss spectrum of Ni_3(Al, Ta) single crystals, *Materials Science and Engineering*, **A239-240** (1997), 281–286.

[43]   B. Reppich, Some new aspects concerning particle hardening mechanisms in γ' precipitating Ni-Base alloys – I Theoretical concept, *Acta Metallurgica*, **30** (1982), 87–94.

[44]   B. Reppich, P. Schepp and G. Wehner, Some new aspects concerning particle hardening mechanisms in γ' precipitating Ni-Base alloys – II Experiments, *Acta Metallurgica*, **30** (1982), 95–104.

[45]   E. Nembach, K. Suzuki, M. Ichihara and S. Takeuchi, In-situ deformation of the gamma prime hardened superalloy Nimonic PE16 in high-voltage electron microscopes, *Philosophical Magazine*, **51A** (1985), 607–618.

[46]   E. Nembach, *Particle Strengthening of Metals and Alloys* (New York: John Wiley and Sons, 1997).

[47]   T. Kruml, E. Conforto, B. Lo Piccolo, D. Caillard and J. L. Martin, From dislocation cores to strength and work hardening: a study of binary Ni_3Al, *Acta Materialia*, **50** (2002), 5091–5101.

[48]   A. J. Ardell, Precipitation hardening, *Metallurgical Transactions*, **16A** (1985), 2131–2165.

[49]   W. Huther and B. Reppich, Interaction of dislocations with coherent, stress-free, ordered precipitates, *Zeitschrift für Metallkunde*, **69** (1978), 628–634.

[50]   D. Hull and D. J. Bacon, *Introduction to Dislocations*, 4th edn (Sevenoaks: Butterworth-Heinemann, 2001).

[51]   M. P. Jackson and R. C. Reed, Heat treatment of Udimet 720Li: the effect of microstructure on properties, *Materials Science and Engineering*, **A259** (1999), 85–97.

[52]   B. Reppich, Particle strengthening, in R. W. Cahn, P. Haasen and E. T. Kramer, eds, *Materials Science and Technology, Volume 6: Plastic Deformation and Fracture* (Weinham, Germany: VCH, 1993), pp. 311–357.

[53]   R. G. Davies and N. S. Stoloff, On the yield stress of aged Ni-Al alloys, *Transactions of the Metallurgical Society of AIME*, **233** (1965), 714–719.

[54] B. J. Piearcey, B. H. Kear and R. W. Smashey, Correlation of structure with properties in a directionally solidified nickel-base superalloy, *Transactions of the American Society for Metals*, **60** (1967), 634–643.

[55] P. Beardmore, R. G. Davies and T. L. Johnston, On the temperature dependence of the flow stress of nickel-base alloys, *Transactions of the Metallurgical Society of AIME*, **245** (1969), 1537–1545.

[56] M. H. Yoo, On the theory of anomalous yield behaviour of Ni_3Al – effect of elastic anisotropy, *Scripta Metallurgica*, **20** (1986), 915–920.

[57] B. H. Kear and H. G. F. Wilsdorf, Dislocation configurations in plastically deformed polycrystalline Cu_3Au alloys, *Transactions of the Metallallurgical Society of AIME*, **224** (1962), 382–386.

[58] A. E. Staton-Bevan and R. D. Rawlings, The deformation behaviour of single crystal $Ni_3(Al, Ti)$, *Physica Status Solidi*, **29A** (1975), 613–622.

[59] P. Veyssiere, Yield stress anomalies in ordered alloys: a review of microstructural findings and related hypotheses, *Materials Science and Engineering*, **A309–310** (2001), 44–48.

[60] P. Veyssiere and G. Saada, Microscopy and plasticity of the $L1_2$ γ' phase, in F. R. N. Nabarro and M. S. Duesbery, eds, *Dislocations in Solids, Volume 10: $L1_2$ Ordered Alloys* (Amsterdam: Elsevier, 1996), pp. 252–441.

[61] R. A. Mulford and D. P. Pope, The yield stress of $Ni_3(Al, W)$, *Acta Metallurgica*, **21** (1973), 1375–1380.

[62] P. H. Thornton, R. G. Davies and T. L. Johnston, The temperature dependence of the flow stress of the γ' phase based upon Ni_3Al, *Metallurgical Transactions*, **1** (1970), 207–218.

[63] S. S. Ezz, D. P. Pope and V. Paidar, The tension compression flow-stress asymmetry in $Ni_3(Al, Nb)$ single crystals, *Acta Metallurgica*, **30** (1982), 921–926.

[64] D. P. Pope and S. S. Ezz, Mechanical properties of Ni_3Al and nickel-base alloys with high volume fraction of γ', *International Metals Reviews*, **29** (1984), 136–167.

[65] S. S. Ezz, D. P. Pope and V. Vitek, Asymmetry of plastic-flow in Ni_3Ga single crystals, *Acta Metallurgica*, **35** (1987), 1879–1885.

[66] D. C. Chrzan and M. J. Mills, Dynamics of dislocation motion in $L1_2$ compounds, in F. R. N. Nabarro and M. S. Duesbery, eds, *Dislocations in Solids, Volume 10: $L1_2$ Ordered Alloys* (Amsterdam: Elsevier, 1996), pp. 187–252.

[67] X. Shi, Doctoral Thesis, University Paris-Nord, France (1995).

[68] P. B. Hirsch, A model of the anomalous yield stress for (111) slip in $L1_2$ alloys, *Progress in Materials Science*, **36** (1992), 63–88.

[69] W. D. Jenkins, T. G. Digges and C. R. Johnson, Creep of high purity nickel, *Journal of Research of the National Bureau of Standards*, **53** (1954), 329–352.

[70] A. H. Clauer and B. A. Wilcox, Steady-state creep of dispersion-strengthened nickel, *Metal Science*, **1** (1967), 86–90.

[71] H. J. Frost and M. F. Ashby, *Deformation-Mechanism Maps: The Plasticity and Creep of Metals and Ceramics* (Oxford: Pergamon Press, 1982).

[72] A. K. Mukherjee, J. E. Bird and J. E. Dorn, Experimental correlations for high-temperature creep, *Transactions of the American Society for Metals*, **62** (1969), 155–179.

[73] J. P. Poirier, *Creep of Crystals: High Temperature Deformation Processes in Metals, Ceramics and Minerals* (Cambridge: Cambridge University Press, 1985).

[74] J. Weertman, Creep of indium, lead and some of their alloys, *Transactions of the Metallurgical Society of AIME*, **218** (1960), 207–218.

[75] J. Weertman and J. R. Weertman, Mechanical properties, strongly temperature dependent, in R. W. Cohn ed., *Physical Metallurgy* (New York: John Wiley & Sons, 1965), pp. 793–819.

[76] G. S. Daehn, H. Brehm, H. Lee and B-S. Lim, A model for creep based on microstructural length scale evolution, *Materials Science and Engineering*, **387** (2004), 576–584.

[77] H. Brehm and G. S. Daehn, A framework for modeling creep in pure metals, *Metallurgical and Materials Transactions*, **32A** (2001), 363–371.

[78] J. J. Gilman, *Micromechanics of Flow in Solids* (New York: McGraw-Hill, 1969).

[79] J. W. Martin and R. D. Doherty, *Stability of Microstructure in Metallic Systems* (Cambridge: Cambridge University Press, 1976).

[80] M. Winning, G. Gottstein and L. S. Shvindlerman, On the mechanisms of grain boundary migration, *Acta Materialia*, **50** (2002), 353–363.

[81] R. W. Evans and B. Wilshire, *Creep of Metals and Alloys* (London: The Institute of Metals, 1985).

[82] J. P. Dennison, R. J. Llewellyn and B. Wilshire, The creep and fracture behaviour of some dilute nickel alloys at 500 and 600 °C, *Journal of the Institute of Metals*, **94** (1966), 130–134.

[83] J. P. Dennison, R. J. Llewellyn and B. Wilshire, The creep and fracture properties of some nickel-chromium alloys at 600 °C, *Journal of the Institute of Metals*, **95** (1967), 115–118.

[84] O. D. Sherby and P. M. Burke, The mechanical behaviour of crystalline solids at elevated temperatures, *Progress in Materials Science*, **13** (1967), 325–390.

[85] C. M. Sellars and A. G. Quarrell, The high temperature creep of gold-nickel alloys, *Journal of the Institute of Metals*, **90** (1962), 329–336.

[86] W. R. Cannon and O. D. Sherby, High temperature creep behaviour of class I and class II solid solution alloys, *Metallurgical Transactions*, **1** (1970), 1030–1032.

[87] F. A. Mohamed and T. G. Langdon, The transition from dislocation climb to viscous glide in creep of solid solution alloys, *Acta Metallurgica*, **22** (1974), 779–788.

[88] T. B. Gibbons and B. E. Hopkins, Creep behaviour and microstructure of Ni-Cr base alloys, *Metal Science*, **18** (1984), 273–280.

[89] M. McLean, Nickel-based alloys: recent developments for the aero-gas turbine, in H. M. Flower, ed., *High Performance Materials for Aerospace* (London: Chapman & Hall, 1995), pp. 135–154.

[90] B. F. Dyson, M. S. Loveday and M. J. Rodgers, Grain boundary cavitation under various states of applied stress, *Proceedings of the Royal Society A*, **349** (1976), 245–259.

[91] J. F. Nye, *Physical Properties of Crystals: Their Representation by Tensors and Matrices* (Oxford: Clarendon Press, 1957).

[92] P. Caron, D. Cornu, T. Khan and J. M. de Monicault, Development of a hydrogen resistant superalloy for single crystal blade application in rocket engine turbopumps, in R. D. Kissinger, D. J. Daye, D. L. Anton *et al.*, eds, *Superalloys 1996* (Warrendale, PA: The Minerals, Metals and Materials Society (TMS), 1996), pp. 53–60.

[93] J. M. Oblak, D. F. Paulonis and D. S. Duvall, Coherency strengthening in Ni–base alloys hardened by $DO_{22}\gamma''$ precipitates, *Metallurgical Transactions*, **5** (1974), 143–153.

3 Single-crystal superalloys for blade applications

The gas turbine consists of many different pieces of turbomachinery, but the rows of turbine blading are of the greatest importance since many engine characteristics, for example the fuel economy and thrust, depend very strongly on the operating conditions which can be withstood by them. Thus very arduous temperatures and stresses are experienced by the materials employed, which are pushed near to the limits of their capability. This is particularly the case for the high-pressure blades, which are located nearest to the hot gases emerging from the combustion chamber. Their function is to extract work from the gas stream and to convert it to mechanical energy in the form of a rotating shaft, which drives the high-pressure compressor.

A consideration of the operating conditions experienced by the high-pressure turbine blades in a large civil turbofan engine, such as the Rolls-Royce Trent 800 or General Electric GE90, confirms this point. The temperature of the gas stream is about 1750 K, which is above the melting temperature of the superalloys from which the blades are made. The high-pressure shaft develops a power of about 50 MW – hence, with about 100 blades, each extracts about 500 kW, which is sufficient to satisfy the electricity requirement of about 500 homes. Each row of blades is expected to last at least 3 years, assuming they operate at 9 h/day. This is equivalent to about 5 million miles of flight, or ~500 circumferences of the world. The blades rotate at an angular speed, ω, of more than 10 000 revolutions per minute, so that the tip velocity is greater than 1200 km/h. The stresses experienced by the blading, which are largely centrifugal in nature due to the significant rotational speeds, are considerable; if the blade cross-section is assumed to be uniform, the stress, σ, at the blade root is approximately given by

$$\sigma = \int_{r_{\text{root}}}^{r_{\text{tip}}} \rho\omega^2 r \, dr = \frac{\rho\omega^2}{2}\left(r_{\text{tip}}^2 - r_{\text{root}}^2\right)$$

where ρ is the blade density. The radial positions of the blade tip and root are given by r_{tip} and r_{root}, respectively. Taking $\rho = 9000 \, \text{kg/m}^3$, the blade length as 10 cm and the mean blade radius to be 0.5 m, the stress at the blade root is given by

$$\sigma = 9000 \times (100 \times 2\pi)^2 \times (0.55^2 - 0.45^2)/2 = 180 \, \text{MPa}$$

Since the cross-sectional area is, in general, no greater than a few square centimetres, this is equivalent to the weight of a heavy truck hanging on each blade. The mass of a shroud and coatings will add to the value calculated. These considerations confirm that rather special materials are required for this application.

121

It has become common to process turbine blading by investment casting of the superalloys, very often into single-crystal form. Not surprisingly, considerable attention has been paid to the optimisation of the compositions of the alloys employed. The mechanical properties of the alloys are of great importance. Furthermore, the nature of the application means that great care needs to be taken in the design of the shape of the turbine blading. These topics are considered in this chapter.

3.1   Processing of turbine blading by solidification processing

The blading for the very first gas turbine engines was produced by extrusion and forging operations [1]. However, in the 1970s it became apparent that there were severe limitations with this approach. First, the blades tended to be heavy, since it proved impossible to make them in hollow form, without subsequent machining operations. Second, they were prone to cracking and incipient melting, since high working temperatures were required, particularly for alloys with high yield stresses. Thus, nowadays, superalloy turbine blading is always produced by investment casting [2]; it has been estimated recently that this industry alone is worth about $3 billion per annum. Hollow aerofoils with complicated shapes are produced routinely, incorporating internal cooling passages to allow cool air to flow inside the blade during operation, to restrict the metal temperatures to reasonable levels. A further advantage is that the grain boundaries can be eliminated, allowing the very best creep properties to be attained. This could not be achieved using thermomechanical processing.

3.1.1   The practice of investment casting: directional solidification

The investment casting or 'lost-wax' process [3,4] involves the following steps. First, a wax model of the casting is prepared by injecting molten wax into a metallic 'master' mould – if necessary (for hollow blades) by allowing wax to set around a ceramic core, which is a replica of the cooling passages required. These are arranged in clusters connected by wax replicas of runners and risers; this enables several blades to be produced in a single casting. Next, an investment shell is produced by dipping the model into ceramic slurries consisting of binding agents and mixtures of zircon ($ZrSiO_4$), alumina (Al_2O_3) and silica (SiO_2), followed by stuccoing with larger particles of these same materials. This operation is usually repeated three or four times until the shell thickness is adequate. Finally, the mould is baked to build up its strength. The first step involves a temperature just sufficient to melt out the wax – usually a steam autoclave is used. Further steps at higher temperatures are employed to fire the ceramic mould. After preheating and degassing, the mould is ready to receive the molten superalloy, which is poured under vacuum at a temperature of ~1550 °C. After solidification is complete, the investment shell is removed and the internal ceramic core leached out by chemical means, using a high-pressure autoclave. It is clear from this description that many steps are required. Fortunately, in most modern foundries considerable amounts of automation have been introduced. In Figure 3.1, the various steps in the investment casting process are summarised. A wax model, the resulting casting and the final machined high-pressure turbine blade are shown in Figure 3.2.

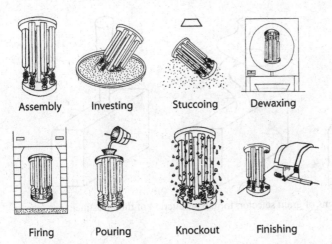

Assembly   Investing   Stuccoing   Dewaxing

Firing   Pouring   Knockout   Finishing

Fig. 3.1. Illustration of the various stages of the investment casting process. (Courtesy of Mr Frank Reed.)

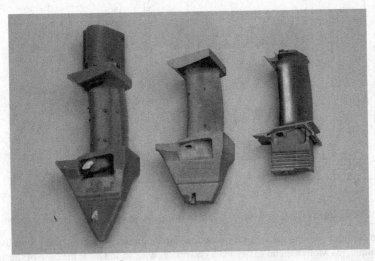

Fig. 3.2. The Trent 800 high-pressure turbine blade at different stages of its production by investment casting. Left: wax model containing ceramic core, ready to receive the investment shell. Centre: finished casting with pig-tail selector removed. Right: finished blade, after machining.

When investment casting was first applied to the production of turbine blading, equiaxed castings were produced by the 'power-down' method, which involved the switching off of the furnace after pouring of the molten metal [3]. However, it has been found that the creep properties are improved markedly if the process of *directional solidification* is used – these methods were pioneered by Versnyder and co-workers at the Pratt & Whitney company in the 1970s, leading to the introduction of directionally solidified blading cast from PWA1422 in the 1980s. After pouring, the casting is withdrawn at a controlled rate from the furnace – as in the conventional Bridgman crystal-growing method. A speed of a few inches per hour is typical – so that the solid/liquid interface progresses gradually along the casting, beginning at its base. This has the effect of producing large, columnar grains which are

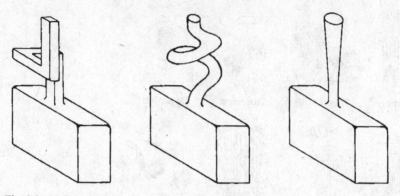

Fig. 3.3. Various designs of grain selector, for the production of the investment castings in single-crystal form.

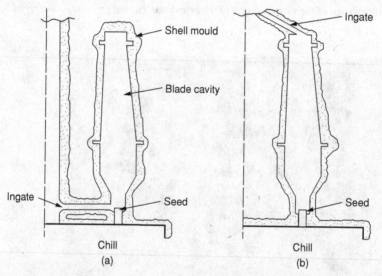

Fig. 3.4. Two configurations by which turbine blading can be grown in single-crystal form using the seeding technique. (a) Molten metal enters via a downpole, with an ingate at the base of the casting. (b) Downpole at the top in a so-called 'top-pouring' configuration.

elongated in the direction of withdrawal, so that transverse grain boundaries are absent [5]. In a variant of this process, the grain boundaries are removed entirely. Most typically, this is achieved [6] by adding a 'grain selector' to the very base of the wax mould, typically in the form of a pig-tail-shaped spiral; see Figure 3.3. Since this is not significantly larger in cross-section than the grain size, only a single grain enters the cavity of the casting, which is then in monocrystalline form. Alternatively, a seed can be introduced at the base of the casting (see Figure 3.4); provided the processing conditions are chosen such that this is not entirely remelted, growth occurs with an orientation consistent with that of the seed. In Figure 3.5, a high-pressure turbine blade casting is shown, with pig-tail grain selector still in place. Figure 3.6 shows a close-up of the base of the casting. Note, in particular, the columnar grain structure in the region below the grain selector.

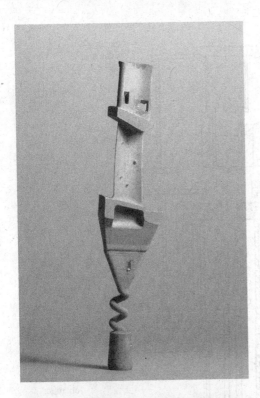

Fig. 3.5. Rolls-Royce Trent 800 high-pressure turbine blade casting, after removal of the investment shell.

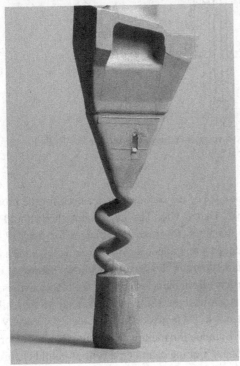

Fig. 3.6. Details of the starter block and pig-tail grain selector from the Trent 800 high-pressure turbine blade casting.

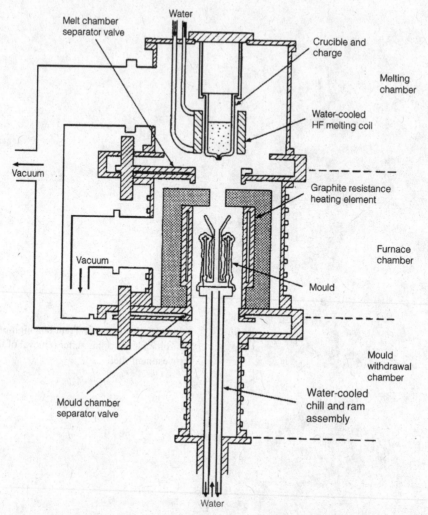

Fig. 3.7. Schematic illustration [6] of the investment casting furnace used by Rolls-Royce plc for the production of single-crystal turbine blading.

Figure 3.7 is a schematic illustration [6] of the casting furnace used for the production of turbine blading at Rolls-Royce plc, Derby, UK. The investment shell mould sits on a copper chill plate of about 14 cm diameter; this size is sufficient for five HP turbine blades for a civil jet engine to be produced simultaneously, or alternatively 20 for a smaller helicopter engine. A bottom-run mould system is chosen, which allows the use of integral ceramic filters to prevent ceramic particles from being swept into the blade cavities, which are then filled in a controlled, predictable fashion. Since the diameter of the chill plate is small, as little as 5 kg of charge needs to be melted. This is achieved, typically within 2 minutes, by using high-frequency (HF) induction melting in a disposable, refractory-fibre crucible. The furnace consists of an upper melting chamber, a central mould chamber and a lower withdrawal chamber, each of which are individually pumped and valved. Once a suitable vacuum is achieved in all chambers, the mould is raised into the central chamber, which is maintained at a temperature above the liquidus by a graphite resistance heater. Immediately after pouring, the mould is withdrawn at the programmed rate. In practice, a high thermal gradient of typically 4000 K/m is achieved, which is sufficient to produce cast microstructures of dendritic form (see Figure 3.8); the primary dendrite arm spacing is then in the range 100–500 μm [7]. The cooling time between the critical temperatures of 1400 °C and 1000 °C is typically about 20 minutes.

In this way, it has proven possible to cast turbine blade aerofoils with complicated patterns of cooling channels. Over time, significant improvements have been made to the cooling configurations. For example, the early extruded and cast equiaxed HP blades introduced in the 1970s on the Rolls-Royce RB211-22B and early RB211-524s had single-pass cooling configurations; see Figure 3.9(a). The air had a relatively short path inside the blade before being exhausted, which limited its potential to absorb heat through contact with the metal; this was so despite the incorporation of extensive numbers of film-cooling holes, typically drilled using Nd–YAG lasers. The latest generation RB211 and Trent blades use a multi-pass, or 'serpentine', configuration (see Figure 3.9(b)), in which the flow passes through longer passages before being exhausted. This maximises the heat pick-up of the cooling air, enabling increases to the main stream gas temperatures and reductions in the required cooling flow. These improvements in cooling technologies have allowed in part the turbine entry temperature of a typical large turbofan engine to be increased by about 250 °C, at a conservative estimate (see Figure 1.5), with associated improvements in engine efficiency.

The fact that the creep properties are improved if columnar structures are produced implies that the grain boundaries are a source of weakness. This has been demonstrated very clearly with elegant experiments on the CMSX-4 superalloy, which is designed to be used in single-crystal form, i.e. with no grain boundaries present. Castings were fabricated with the presence of high-angle grain boundaries, using two seeds of known orientation – this allowed the creep life to be measured of a specimen containing a grain boundary defect of known misorientation θ. The results are given in Figure 3.10. The data indicate that for an applied stress of 300 MPa at 850 °C, a perfect single crystal has a stress rupture life in excess of 10 000 h; however, a grain boundary of misorientation $\theta = 7°$ reduces the creep life to 100 h. When $\theta > 10°$, the creep life is only a few hours. Clearly, the creep rupture life depends very sensitively on θ. This is one reason why so-called 'grain-boundary-tolerant' single-crystal superalloys have recently become available – these contain the grain-boundary strengtheners such as carbon and boron present in the early superalloys which were used for equiaxed castings. The sensitivity of the creep life to the presence of grain boundaries is supported by early work [8] on Mar-M200, one of the first alloys to be used in single-crystal form; see Figure 3.11. Under creep loading of about 200 MPa and 980 °C, the removal of the transverse grain boundaries increases the creep ductility from below 5% to in excess of 25%. In single-crystal form, the creep life is improved to greater than 100 h from about 70 h for the directionally solidified, columnar-grained structure. When grain boundaries are present in these systems, creep damage occurs in the form of voids and cavitation; see Figure 3.12. If grain-boundary-strengthening elements such as carbon are present, cracking of the carbide precipitates and their decohesion from grain boundaries is commonplace.

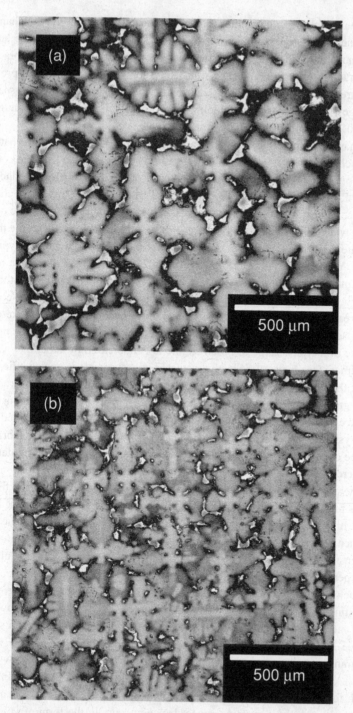

Fig. 3.8. Scanning electron micrographs [7] of the dendritic microstructure of a single-crystal
superalloy, in the as-cast state. (a) 2.5 mm/min, conventional Bridgman furnace; (b) 2.5 mm/min
with liquid tin as a metal coolant. (Courtesy of Andrew Elliott.)

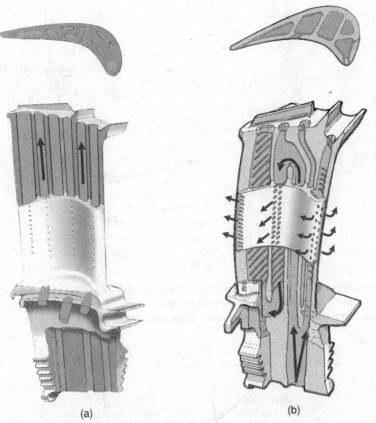

(a)   (b)

Fig. 3.9. Schematic illustration of the cooling configurations used for turbine blade aerofoils. (a) Single-pass cooling. (b) Multipass 'serpentine' cooling. In both cases, holes for film cooling are present. (Courtesy of Rolls-Royce.)

The removal of the grain boundaries using a grain selector, or else by seeding, has ramifications for the crystallographic orientation of the casting. The preferred growth direction for nickel and its alloys, in common with all known FCC alloys, is $\langle 001 \rangle$. Thus turbine blade aerofoils have a $\langle 001 \rangle$ direction aligned along, or close to, the axis of the casting. The texture arises as a consequence of the competitive growth of dendrites in the starter block of the casting [9]; see Figure 3.13. The solid nuclei which form adjacent to the chill block have an orientation which is random, but after only 1 mm of growth a strong $\langle 001 \rangle$-orientated texture is present; see Figure 3.14. This effect occurs because the dendrites grow at a rate which is largely controlled by solute diffusion, since the solid phase grows from the liquid with a very different composition from it; thus the local dendrite tip undercooling, ΔT_{tip}, scales monotonically with the velocity, v, of the dendrite, measured along the temperature gradient. It follows that dendrites which are misaligned by an angle θ with respect to perfectly aligned ones must grow at a greater undercooling and hence at the rear of the growth front; see Figure 3.15. This provides an opportunity for the secondary and tertiary arms of the well aligned dendrites to suppress the primary arms of those which are less so. The net

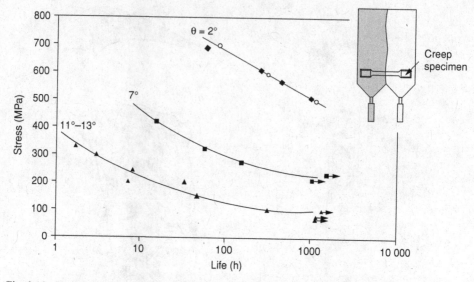

Fig. 3.10. Variation of the creep rupture properties of the superalloy CMSX-4 at 850 °C, measured on specimens with a single grain boundary of misorientation, θ, introduced. Note the very great reduction in performance caused by the presence of the grain boundaries. (Courtesy of Bob Broomfield.)

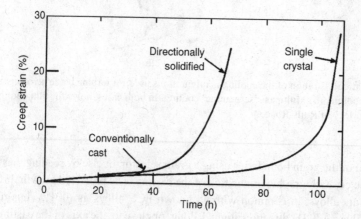

Fig. 3.11. Evolution of creep strain with time for the superalloy Mar-M200 under loading of 206 MPa and 982 °C, in the conventionally cast (equiaxed) state, the directionally solidified (columnar) state and the single-crystal form [8].

result is a reduction in the number of grain boundaries and the establishment of a strong $\langle 001 \rangle$-orientated texture.

3.1.2 Analysis of heat transfer during directional solidification

In practice, directional solidification requires careful control of the rate of heat removal from the casting. Mathematical analysis of the heat transfer processes is therefore useful, since it provides insight into the effects occurring.

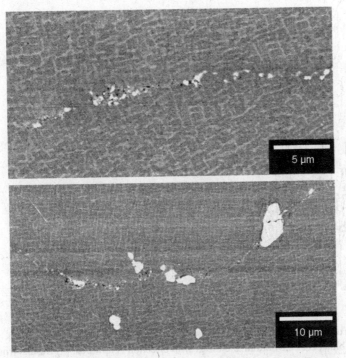

Fig. 3.12. Scanning electron micrographs of the grain boundaries in the directionally solidified superalloy RR2072, modified with 0.05 wt%C and 0.005 wt%B. (a) 18.8° misorientation, crept to 64 hours (80% of creep life); (b) 10.5° misorientation, crept to 100 hours (50% of creep life). (Courtesy of Quizi Chen.)

A   Treatment of infinite rod – estimation of withdrawal velocity

One can make a first estimate of the *maximum* speed at which the casting should be removed from the furnace in the following way. As a first approximation, one treats the casting as an infinite, cylindrical rod; this moves at a uniform rate, v, from a furnace of temperature T_1 into a medium of ambient temperature T_0; see Figure 3.16. The heat flow is taken to be unidirectional, i.e. it occurs only in the axial direction (denoted x). The temperature field, $T\{x\}$, comes to a steady-state, which satisfies the heat flow equation in the moving frame,

$$\kappa \frac{\partial^2 T}{\partial x^2} + vc\frac{\partial T}{\partial x} + \dot{Q} = 0 \tag{3.1}$$

where $\dot{Q}$ is the local volumetric rate of heat production or extraction, c is the volumetric specific heat capacity and κ is the thermal conductivity. At the melting temperature, when $T = T_m$ (note $T_1 > T_m > T_0$), the *production* of latent heat occurs; this requires that the following boundary condition is satisfied at the solid/liquid interface:

$$\kappa_s \frac{\partial T}{\partial x}\bigg|_s = \kappa_1 \frac{\partial T}{\partial x}\bigg|_1 + Lv \tag{3.2}$$

where L is the volumetric latent heat of fusion and the subscripts s and l refer to the solid and liquid, respectively. Conversely, heat is *extracted* at a rate $q = h_{eff}(T - T_0)$ by Newtonian

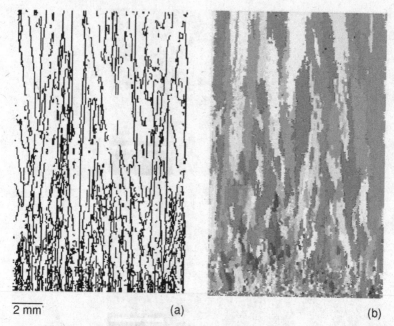

2 mm   (a)   (b)

Fig. 3.13. Grain structure in the starter block of a single-crystal turbine blade casting, cast from CMSX-4; the chill block is at the bottom and the growth direction is upwards. (a) Results from theoretical calculations made using the phase-transformation theory. (b) Experimental analysis using electron back-scattered diffraction patterns of the starter block in Figure 3.6, confirming the evolution of the (001) texture. Data taken from ref. [9].

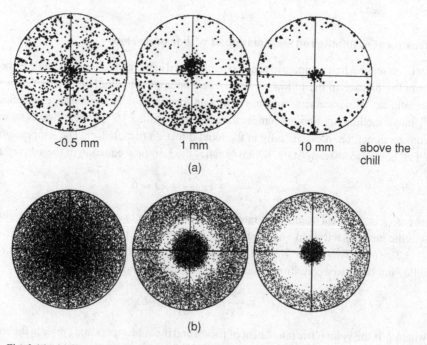

<0.5 mm   1 mm   10 mm   above the
                                                       chill
                     (a)

                     (b)

Fig. 3.14. (a) Measured and (b) predicted (001) pole figures taken from the starter block shown in Figure 3.6 at various heights above the chill block [9].

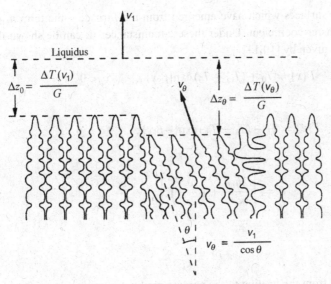

Fig. 3.15. Illustration of the competitive grain growth process, in which misaligned dendrites are suppressed by the secondary arms of well aligned ones.

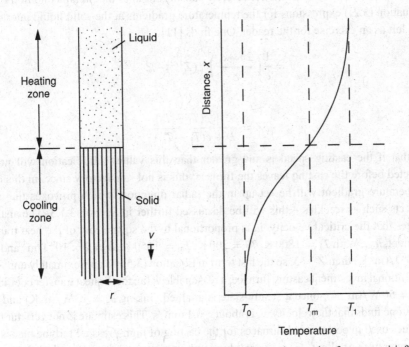

Fig. 3.16. The directional solidification of an infinite rod emerging from a semi-infinite furnace into a semi-infinite heat sink, at a velocity consistent with the temperature at the exit of the furnace equal to the melting temperature. Adapted from ref. [3].

heat transfer at the surfaces which have emerged from the furnace – the term h_{eff} is the associated heat-transfer coefficient. Under these circumstances, it can be shown that the temperature field is given by [10,11]

$$T\{x\} = T_0 + (T_m - T_0)\exp\{\beta_s x\}, \qquad x < 0 \qquad (3.3a)$$

and

$$T\{x\} = T_1 - (T_1 - T_m)\exp\{-\beta_1 x\}, \qquad x > 0 \qquad (3.3b)$$

where

$$\beta_s = \left[\frac{2h_{eff}}{\kappa_s r} + \left(\frac{vc}{2\kappa_s}\right)^2\right]^{1/2} - \frac{vc}{\kappa_s} \qquad (3.4a)$$

and

$$\beta_1 = \frac{vc}{\kappa_1} \qquad (3.4b)$$

where the transition from the heating to the cooling medium is assumed to occur abruptly at $x = 0$; see Figure 3.16.

With these assumptions, one can derive an expression for the velocity at which the casting can be removed from the furnace, by substituting into the Stefan condition (given by Equation (3.2)) expressions for the temperature gradients at the solid/liquid interface – this is left as an exercise for the reader. One finds [11]

$$v = \frac{1}{c}\left(\frac{2h_{eff}\kappa_s}{r}\right)^{1/2}(Z^{1/2} - Z^{-1/2}) \qquad (3.5a)$$

where

$$Z = \frac{L + c(T_1 - T_0)}{L + c(T_1 - T_m)} \qquad (3.5b)$$

Note that if the casting speed is any greater than this value, solidification will not be completed before the casting leaves the furnace; this is not satisfactory since, in this case, a temperature gradient will be set up in the radial direction, which provokes the onset of defects such as freckles – this will be discussed further in Section 3.1.3. The analysis indicates that the critical velocity, v, is proportional to the square root of the heat transfer coefficient, h_{eff}. With $T_1 = 1800\,K$, $T_0 = 300\,K$, $T_m = 1650\,K$, $L = 2 \times 10^9$ J/m³ and $c = 4 \times 10^6$ J/(m³ K) then $Z \sim 3$, so the last term in Equation (3.5a) is approximately unity. For a conventional investment casting furnace, a reasonable value for the heat transfer coefficient is $h_{eff} = 50$ W/(m² K), once a steady-state is attained. Taking $\kappa_s = 15$ W/(m K) and $r = 0.02$ m, one finds that the velocity, v, is about ~ 0.1 mm/s. This estimate is not very far from the values used in commercial furnaces for the casting of high-pressure turbine blades [9]. The temperature gradient, G, at the solidus can be determined from Equation (3.3a); it is $(T_m - T_0)\beta_s$, or about 75 K/cm, given the estimates that have been made.

In this analysis it has been assumed that there is no heat loss from the furnace to the cooling zone. This is impossible to achieve in practice, although much effort has been expended in designing optimum baffle arrangements by placing radiation shields at the base of the furnace, with openings just large enough to allow the castings to emerge from the furnace. These have been shown to be efficient at keeping the solid/liquid interface flat, and in the vicinity of the baffle.

B   Comparison of axial and transverse contributions to heat transfer

The above analysis indicates that, at steady-state, a balance exists between (i) the rate of heat extraction by conduction along the casting in the axial direction, and (ii) the rate of transverse heat loss by radiation through the curved, cylindrical surfaces. It follows that, as the casting first emerges from the furnace, this equilibrium has yet to be attained. The conditions pertaining to these early stages of directional solidification are important, particularly for single-crystal castings, since they govern the processes of competitive grain growth, texture development and grain selection.

At the beginning of directional solidification, the primary source of heat loss is due to conduction to the chill through the solidified portion of the ingot; this contribution diminishes as X, the length of the solidified ingot, increases. Thereafter, heat loss by radiation is expected to dominate, at least whilst the casting and ceramic investment shell remain in close contact and provided that the investment shell is relatively thin. The heat flux, q, should therefore be represented accurately by [3,12]

$$q = h_{eff}(T - T_0) = (h_C + h_R)(T - T_0) \tag{3.6}$$

where h_{eff} has now been partitioned into two terms, h_C and h_R, corresponding to the heat transfer coefficients for axial conduction and transverse radiation, respectively; these are given by

$$h_C = \frac{\kappa_s}{X} \tag{3.7}$$

and

$$h_R = \frac{\sigma(\epsilon T^4 - \alpha T_0^4)}{T - T_0} \tag{3.8}$$

where σ is the Stefan constant, ϵ is the emissivity of the ingot and α is the absorbtivity of the surroundings.

It is helpful to determine the length of the ingot which solidifies before the steady-state is attained. If $\epsilon = \alpha = 0.45$ and $\sigma = 5.67 \times 10^{-8}$ J/(m^2 s K^4), $T = 1500$ K and $T_0 = 400$ K, then

$$h_R = 120 \text{ W}/(\text{m}^2 \text{ K}) \tag{3.9}$$

and, with $\kappa_s = 16$ W/(m K),

$$h_C = 1600 \text{ W}/(\text{m}^2 \text{ K}) \tag{3.10}$$

when $X = 10$ mm. Thus, initially, $h_C > h_R$; however, at the point when $X = 130$ mm the two contributions are equal, and soon thereafter cooling is controlled by radiative heat transfer. The dominance of the radiative mode at later times is aided also by the relatively small cross-sectional area available for conduction through the copper chill, the presence of a grain selector and the substantial surface area of the mould, particularly for turbine blades which are of complex shape. It is clear therefore that, in practice, only the first part of the casting is cooled effectively by conduction of heat through the water chill; for the majority of the casting one is relying upon radiation to the water-cooled walls of the vacuum chamber.

C   Effects of quenching medium – liquid metal cooling

Equation (3.5a) indicates that the velocity of withdrawal, v, scales inversely with $\sqrt{r}$, the square root of the radius of the casting; thus, for castings of significant size, the time necessary for solidification to be completed can become prohibitively long. A further difficulty is that the economics of the single-crystal casting process depend also upon the yield, which is decreased if defects such as freckles, spurious grains and low-angle boundaries are formed; this is most likely to occur when the temperature gradient is low since the width of the semi-solid 'mushy' region is then considerable. The temperature gradient, G, is proportional to the parameter β_s in Equation (3.3a) – and Equation (3.4a) confirms that this decreases as r increases. Thus, the tendency for grain defect formation is enhanced for components of larger size. An additional challenge is that it becomes increasingly difficult to keep the profile of the solid/liquid interface flat as the casting becomes large; this can further increase the probability of defect formation, particularly freckles. These considerations are particularly pertinent to the casting of the blades for an industrial gas turbine (IGT) used for electricity generation – these might be about 500 mm long with a mass in excess of 5 kg.

Under these circumstances, more efficient methods are being sought for cooling the mould as it emerges from the furnace [13]. The rationale for this is as follows: a larger h_{eff} has the effect of increasing v and G, as indicated by Equations (3.3), (3.4) and (3.5). Enhanced cooling has been achieved in the laboratory by various means, including water sprays, fluidised bed cooling, gas cooling using jets of Ar/He mixtures, and liquid metal cooling; all depend upon a close thermal contact of the mould and the cooling medium. Of the different techniques, the liquid metal cooling (LMC) technique appears to be the most likely to make the transition into the commercial foundry – indeed it has been used already in Russia. The mould is immersed into a bath of molten metal to increase the heat transfer rate (see Figure 3.17). The choice of liquid metal coolant is important – the requirements are a low melting temperature, low vapour pressure, high thermal conductivity, low viscosity, lack of toxicity and low cost, on account of the quantities of metal required. For these reasons, aluminium and tin have emerged as the prime candidates for the LMC process. The important physical properties of the two metals are listed in Table 3.1. Although the vapour pressures of Al and Sn are more or less equivalent, the lower melting temperature of Sn is advantageous since it allows a lower bath temperature to be used. The thermal conductivity of Al is considerably higher than for Sn, which might indicate that heat transfer by conduction through the molten metal will be particularly effective.

In order to compare the advantages of Sn and Al as coolants for the LMC process, one must consider in more detail the various resistances to heat flow arising as the ingot and mould are immersed into the liquid metal. The conventional Bridgman process is considered first. The conduction and radiation coefficients, h_C and h_R, must be considered as before; the estimates of 1600 W/(m² K) and 120 W/(m² K) of Equations (3.9) and (3.10), respectively, are used again. However, for large IGT castings one must include the coefficient h_{gap}, to

Table 3.1. *The physical properties of two candidate metals for the liquid metal cooling (LMC) process – aluminium and tin [13]*

Property	Unit	Aluminium	Tin
Melting point	°C	660	232
Density	kg/m³	2382	6980
Specific heat	kJ/(m³ K)	2597	1857
Thermal conductivity	W/(m K)	104	33.5
Thermal diffusivity	m²/s	40×10^{-6}	18×10^{-6}
Latent heat	kJ/m³	922 000	419 000
Kinematic viscosity	m²/s	6.3×10^{-7}	2.58×10^{-7}

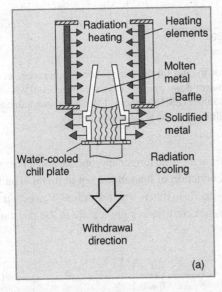
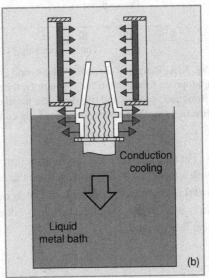

Fig. 3.17. Schematic illustration of two of the solidification processes used for casting. (a) Conventional high-rate-solidification (HRS) Bridgman process; (b) liquid metal cooling (LMC) process. (Courtesy of Robert Singer.)

acknowledge the need for radiation across the gap at the metal/shell interface, and h_{shell} for conduction through the investment shell, since it is relatively thick. A reasonable estimate for h_{gap} is 250 W/(m² K). The conductivity of the porous ceramic shell is about 1.5 W/(m K), and a typical thickness for IGT applications is 12 mm; this yields $h_{shell} = 125$ W/(m²/K). The effective heat transfer coefficient, h_{eff}, is given by summing the four contributions, since the resistances to heat flow are in series; hence [14]

$$h_{eff}^{Bridgman} = \left[\frac{1}{h_C} + \frac{1}{h_{gap}} + \frac{1}{h_{shell}} + \frac{1}{h_R} \right]^{-1} \tag{3.11}$$

so that $h_{eff} = 50$ W/(m² K) for the conventional Bridgman process. With $T_1 = 1700$ K and $T_0 = 400$ K, the net rate of heat transfer is $50 \times (1700 - 400)$ or 65 kW/m².

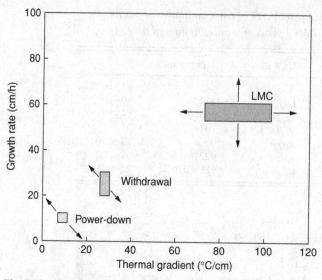

Fig. 3.18. Ranges of thermal gradients and growth rates for the three solidification processes appropriate for turbine blade castings: (i) the power-down process, (ii) the conventional withdrawal (Bridgman) process and (iii) the liquid metal cooling (LMC) process [12]. The arrows indicate directions in which changes in processing conditions can be accomplished easily.

For the LMC process, the contribution from h_R is no longer appropriate and it is replaced with a term, h_{LMC}, which describes the conduction of heat in the liquid metal. The liquid metal is an efficient heat sink so that $h_{LMC} \gg h_R$ – this is true regardless of whether Al or Sn is used as the coolant. Thus, h_{LMC} does not contribute significantly to the resistance to heat flow; hence,

$$h_{eff}^{LMC} = \left[\frac{1}{h_C} + \frac{1}{h_{gap}} + \frac{1}{h_{shell}} \right]^{-1} \tag{3.12}$$

so that $h_{eff} = 80 \, \text{W}/(\text{m}^2 \, \text{K})$. The net rate of heat transfer is then given by $h_{eff}(T - T_{bath})$, where T_{bath} is the bath temperature; appropriate values are 950 K and 550 K for Al and Sn, respectively. With $T = 1700$ K, the net rates of heat transfer are then $80 \times (1700 - 950)$ or $60 \, \text{kW}/\text{m}^2$ for Al and $80 \times (1700 - 550)$ or $90 \, \text{kW}/\text{m}^2$ for Sn. These calculations confirm that the net rate of heat extraction can be improved markedly if molten Sn is used as a cooling medium. Liquid Al, on the other hand, does not improve matters appreciably – in fact, these calculations indicate that the net rate of heat transfer is lower than for the conventional Bridgman technique. This is a consequence of the higher bath temperature required by the higher melting temperature of Al.

In Figure 3.18, the growth rates and withdrawal velocities typical of the LMC process are compared with those which can be achieved using the conventional Bridgman (withdrawal) and power-down methods [12]. Liquid Sn is used as the cooling medium. One can see that the LMC process affords an increased thermal gradient of at least a factor of 3, and a growth velocity of at least a factor or 2, when compared to the conventional Bridgman method.

Fault	Six-month average		Fault	Six-month average
Orientation	8.0%		Inclusions	4.7%
High-angle boundaries	5.8%		High-angle boundaries	4.6%
Blocked core	3.1%		Core breakthrough	3.5%
			Root wall	3.4%
Recrystallisation	3.0%		Freckles	2.2%
Inclusions	2.3%		Aerofoil shape	2.1%
Aerofoil shape	2.3%		Furnace malfunction	2.0%
Aerofoil wall	2.2%		Blocked core	1.4%
Furnace malfunction	1.1%		Recrystallisation	1.1%
Overall yield	69%		**Overall yield**	67%

(a)   (b)

Fig. 3.19. Non-conformance statistics for two single-crystal castings processed in a commercial foundry: (a) seeded blade cast from second-generation alloy, and (b) blade cast from third-generation alloy, with a grain selector.

3.1.3 Formation of defects during directional solidification

A decision to cast a turbine blade aerofoil in single-crystal form represents a considerable challenge to the foundry. This is because very stringent requirements are then placed on the quality of the casting that can be accepted. Defects, such as high-angle grain boundaries, which arise when the grain selector fails to function properly or else due to nucleation of spurious grains around re-entrant features such as platforms and shrouds, cannot be tolerated. The crystallographic orientation must be such that the $\langle 001 \rangle$ direction is closely aligned to the axis of the casting – usually within 15° or so; in practice, the use of a grain selector or seed fails to ensure that this happens completely routinely. Recrystallisation during subsequent solutioning heat treatment, driven by plastic strains introduced by the shrinkage of the metal around the ceramic core, introduces high-angle grain boundaries, and therefore necessitates rejection [15]. Additional requirements, common to columnar or equiaxed castings used for lower-temperature applications such as low-pressure (LP) blading, must also be respected. The design of the component requires that tight dimensional tolerances are placed on the shape of the aerofoil and, if the blading is hollow, the thickness of its walls. The presence of entrained ceramic particles swept into the molten metal from the shell cannot be tolerated since they impair the fatigue properties; these are detected during the non-destructive inspection using radiographic methods. Given these considerations, it is clear that the probability of component rejection casting is likely to be reasonably high. The data in Figure 3.19, which are taken from a typical commercial foundry specialising in

 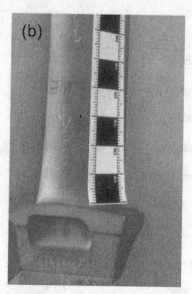

Fig. 3.20. Photographs of sections of blade castings illustrating the presence of defects: (a) high-angle grain boundaries; (b) freckle chains.

the casting of blading in single-crystal form, confirm this. For the seeded casting, failure to control the orientation of the blading is the major cause of non-conformance; this problem disappears when a grain selector is employed, but in both cases the overall yield is rather low at around 70%. Images of castings with grain defects are shown in Figure 3.20.

Case study: The freckle defect

The above data indicate that there are many possible defects which can arise when casting a turbine blade in single-crystal form. Of these, the phenomenon of freckling [16] is worthy of further consideration since the probability of its occurrence has been found to be strongly dependent upon the composition of the alloy being cast – the third- or fourth-generation alloys, which are rich in elements such as Re, are particularly prone to this effect. Freckles, also known as channel segregates, are thin chains of equiaxed grains associated with *localised* heterogeneities in chemical composition; see Figure 3.20(b). Single-crystal castings, both for aeroengine and IGT applications, can exhibit the freckling phenomenon. The *localisation* is always found to be on the scale of the primary dendrite arm spacing, with the freckles being found in the interdendritic regions. Chemical analysis of the freckle chains confirms that they are enriched with regard to elements which partition to the interdendritic liquid, for example Al, Ti, Ta and Nb, where these are present.

There are other pieces of evidence which shed light on the mechanism by which freckles form. Consider Figure 3.21, which summarises the results from a considerable programme of experimental work carried out by Tin, Pollock and co-workers aimed at identifying the factors which promote freckling [17]. More than 16 different alloys were cast into cylindrical

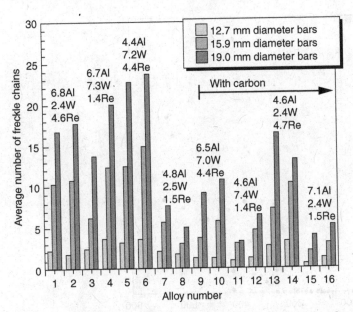

Fig. 3.21. Statistics for the formation of freckle chains in cast bars of varying diameters, in superalloys of various compositions [17]. For all alloys, Ta ~ 6.1, Hf ~ 0.14. All numbers refer to concentrations in wt%.

bars of three different diameters: 12.7 mm, 15.9 mm and 19.0 mm, at otherwise identical conditions. Some of the alloys were deliberately doped with carbon. The average number of freckle chains appearing on the surface of the castings was then characterised. A number of interesting observations are seen. First, the extent of freckling becomes more pronounced as the bar diameter increases. Second, there is a greater tendency for freckling when the combined W and Re content, i.e. W + Re, is significant. Third, additions of carbon are found to be effective at reducing the degree of freckling. Finally, and significantly, when casting is carried out at different casting conditions (see Figure 3.22) it is found that there is a critical minimum primary dendrite arm spacing (corresponding to a lower limit for the temperature gradient) below which freckles are not found [18].

These observations support the following mechanism for the freckling effect. On thermodynamic grounds, the alloying additions prefer to partition to either the interdendritic regions or alternatively to the dendrites themselves. Experiments using electron probe microanalysis (EPMA) indicate that Re and W segregate to the dendrite core, and that Al, Ti and Ta partition to the interdendritic regions [19]; see Figure 3.23. Other elements such as Cr, Co and Ru are found to be rather neutral in partitioning behaviour, with a partitioning coefficient close to unity. Particularly when the Re and W contents are large, liquid in the interdendritic region becomes depleted in these heavy elements, causing it to become lighter than the liquid at a temperature closer to the liquidus. Thus, a so-called 'density inversion' becomes possible [20]. For freckling to occur, the enhanced buoyancy of the interdendritic liquid (due to the loss of heavy elements) must overcome frictional effects which prevent the lighter liquid from rising – if this occurs, then thermosolutal convection currents, or

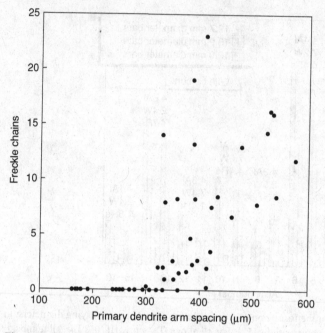

Fig. 3.22. Dependence of the number of freckle chains on the primary dendrite arm spacing. Note that the critical spacing for the formation of freckles is ~300 μm [18].

'freckle plumes', are set up. When casting is of significant cross-sectional area, the width of the mushy zone is wide, and freckling is exacerbated. Similarly, additions of elements such as Re and W, which exacerbate the density inversion on account of their partitioning behaviour, promote freckling; conversely, elements such as Ta restrict it. Additions of carbon cause the formation of carbides at temperatures around the liquidus, and these appear to be responsible for its beneficial effect [17]. When the dendrite arm spacing is low, freckling does not occur since any possible density inversion provides insufficient driving force to overcome the resistance to interdendritic fluid flow, since the permeability of the mushy zone is then low.

That the single-crystal superalloys are prone to the freckle defect can be seen by making calculations for the density of the liquid phase as the function of temperature; see Figure 3.24. Experimental measurements for the density of liquid CMSX-4 above the melting temperature (about 1650 K) have been reported in ref. [21]; these data are plotted on the figure, along with an extrapolation to lower temperatures (the dotted line), which represents the density of the *unpartitioned* liquid in the mushy zone. Estimates of the density of the *partitioned* liquid can be made assuming equilibrium is attained, by accounting for the mass loss due to the partitioning of elements such as Re and W into the solid phase. The density of the partitioned liquid is given by the curved solid line. The driving force for freckling is proportional to the *difference* between the solid and dotted lines, denoted $\Delta\rho$; for CMSX-4 it is a maximum at about 20 K below the melting temperature. These calculations can be carried out for single-crystal superalloys of different compositions, for example

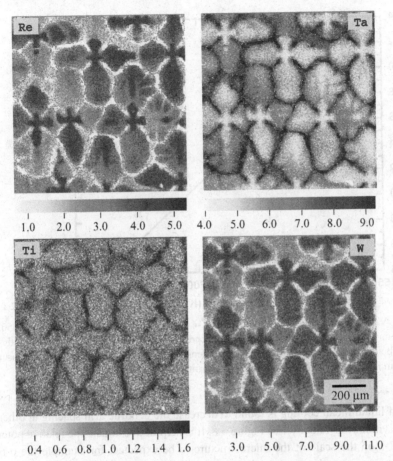

Fig. 3.23. Electron probe microanalysis (EPMA) maps of the transverse section of the as-cast structure of the CMSX-4 superalloy, which confirms the partitioning of W and Re to the dendrite core, and Ti and Ta to the interdendritic regions. (Courtesy of Mudith Karunaratne.)

for the experimental alloys reported in ref. [18] for which freckling data are available; see Figure 3.25. When the density difference, $\Delta\rho$, is evaluated at a supercooling of 20 K below the liquidus, a correlation is found with the average number of freckle chains found per blade sample. This provides strong evidence that the density inversion is responsible for the driving force for the freckling phenomenon.

3.1.4 The influence of processing conditions on the scale of the dendritic structure

During processing by directional solidification, the temperature gradient, G, and the withdrawal velocity, v, influence the scale of the dendritic structure which is formed. For example,

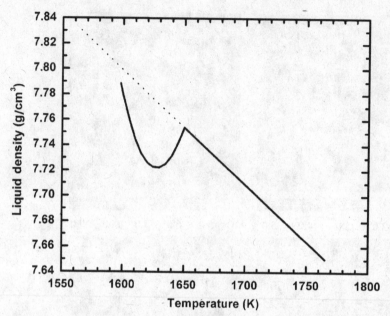

Fig. 3.24. Variation of the density of liquid CMSX-4 with temperature. The data beyond 1650 K are those recommended by ref. [21] – the dotted line represents an extrapolation below the liquidus. The solid line below 1650 K represents the calculated density of the liquid in the mushy zone, assuming equilibrium is attained; thus, the effect of solute partitioning is accounted for.

the use of liquid metal cooling is found to decrease the dendrite arm spacing at otherwise identical withdrawal velocity; see Figure 3.8. It is helpful to estimate the relative potencies of G and v on the scale of the microstructure to be expected, for example the primary dendrite arm spacing, L_p.

In order to do this, consider first the temperature interval, ΔT_0, required for solidification to be completed. If the phase diagram is taken to be a binary one with a liquidus of slope m and a solid/liquid partitioning coefficient of k, then it can be demonstrated that ΔT_0 is given by $mc_0(k - 1)/k$, where c_0 is the bulk composition of the alloy under consideration. The scale of the dendritic structure is then as depicted in Figure 3.26. The dendrites are assumed to possess a radius of curvature R at their tips, which are spaced a distance L_p apart. The mean height of the dendrite trunks is then $\Delta T_0/G$, where G is the temperature gradient; the thickness of the trunks at their roots is given the symbol λ, which is approximately one half of L_p. A steady-state is assumed – thus, the dendritic structure translates at a velocity v, and its shape is invariant as it does so. A first estimate of the radius of curvature of the tip of each dendrite is given by fitting a parabola to its shape [22]; from elementary geometry, it is then equal to the ratio of the square of the minor axis of the parabola (which is equal to λ) to the major axis ($\Delta T_0/G$) – so that $R = \lambda^2 G/\Delta T_0$.

For the solidification conditions pertinent to the superalloys, there are two contributions to the total undercooling ΔT. The first is due to the solutal undercooling, $\Delta T_{solutal}$, arising from the partitioning of solute; the second, $\Delta T_{curvature}$, is a consequence of the Gibbs–Thomson

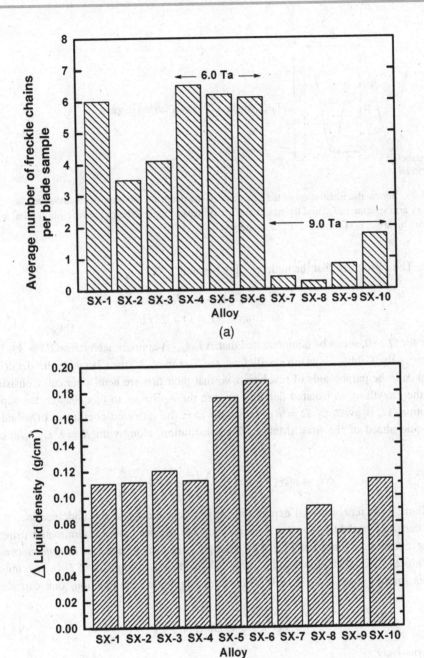

Fig. 3.25. Data for the freckling propensity of a number of experimental single-crystal superalloys [18]. (a) Average number of freckle chains per blade sample; (b) corresponding differences between the densities of the unpartitioned and partitioned liquid at an interval of 20 K below the liquid of the alloy concerned. Note the correlation between the data in (a) and (b).

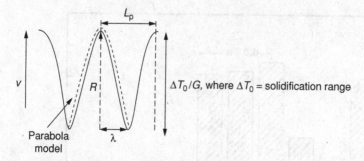

Fig. 3.26. Schematic illustration of the geometry of the dendritic structure at steady-state: the primary arm spacing is L_p, and the major and minor axes of the paraboloid of revolution fitted to the dendrite tip are $\Delta T_0/G$ and $\lambda = L_p/2$, respectively.

effect. Thus, provided that the undercooling is small,

$$\Delta T = \Delta T_{\text{solutal}} + \Delta T_{\text{curvature}}$$
$$= mc_0\Omega(1 - k) + 2\theta/R \tag{3.13}$$

since, for $\Omega \sim 0$, it can be demonstrated that $\Delta T_{\text{solutal}}$ is approximately $mc_0\Omega(1 - k)$. The term θ is the Gibbs–Thomson coefficient; the factor of 2 arises because the dendrites are taken to be paraboloids of revolution, so that their tips are hemispherical. Consistent with the growth of an isolated dendrite tip, see the Appendix to this chapter, the supersaturation, Ω, is given by $\Omega = Rv/D$, where D is the diffusion coefficient of solute in the liquid ahead of the front. Making this substitution, along with $R = \lambda^2 G/\Delta T_0$, one has

$$\Delta T = mc_0(1 - k)\left(\frac{v}{2D}\right)\left(\frac{\lambda^2 G}{\Delta T_0}\right) + \frac{2\theta\Delta T_0}{\lambda^2 G} \tag{3.14}$$

Equation (3.14) represents an expression for the dependence of the undercooling, ΔT, upon the three variables $\lambda = L_p/2$, G and v. One of these can be eliminated using a further equation or criterion [22]. For example, one can invoke the *extremum condition*, which states that growth will occur at whatever λ minimises ΔT or maximises v. This implies that $\partial \Delta T/\partial \lambda = 0$; after some algebraic manipulation, one can show that

$$vG^2\lambda^4 = 4\theta Dmc_0(1 - k) \tag{3.15}$$

or alternatively

$$vG^2 L_p^4 = 64\theta Dmc_0(1 - k) \tag{3.16}$$

Note that the terms on the right-hand side of Equation (3.16) are expected to be invariant for any given alloy system. Thus, the primary arm spacing, L_p, is predicted to be proportional to $G^{-1/2}v^{-1/4}$. It follows that G is expected to have a more profound influence on the spacing than v – as found, for example, in the liquid metal cooling process. The experimental results

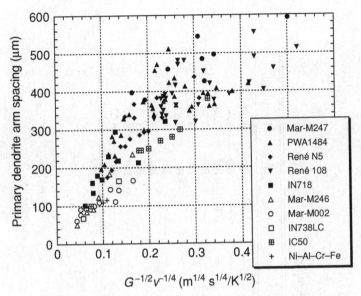

Fig. 3.27. Data [23] for the variation of the primary dendrite arm spacing with the combination $G^{-1/2}v^{-1/4}$, where G is the temperature gradient and v is the withdrawal velocity, for a number of commercial superalloys.

confirm this (see Figure 3.27), which provides data [23] for the primary dendrite arm spacing for a number of commercial superalloys solidified under a range of G and v conditions. The agreement between the experimental data and the theory gives some credence to the use of the extremum condition in deriving it, despite the lack of a firm physical basis for its application.

3.2 Optimisation of the chemistry of single-crystal superalloys

Over a period of about 20 years, the chemistry of the single-crystal superalloys has been refined in order to improve the properties displayed. Although much of this progress has been incremental, a clear picture has emerged of the roles played by the various elements in conferring the best properties. Whilst it should be remembered that the various alloy producers and engine manufacturers have settled on proprietary chemistries to maintain their competitive advantage, the compositions of the single-crystal superalloys have evolved significantly in the period since 1980 when the first ones emerged. The compositions of important first-, second-, third- and fourth-generation single-crystal superalloys are given in Tables 3.2, 3.3, 3.4 and 3.5. Of note is the introduction of significant additions of Re, and the reduction of Ti and Mo to low concentrations. Some interesting insights into the early alloy development efforts are given in refs. [24] and [25]. With the benefit of hindsight, the improvements in performance have been won by respecting the four guidelines set out on pp. 152–170.

Table 3.2. *The compositions (in weight%) of selected first-generation single-crystal superalloys*

Alloy	Cr	Co	Mo	W	Al	Ti	Ta	Nb	V	Hf	Ni	Density (g/cm³)
Nasair 100	9	—	1	10.5	5.75	1.2	3.3	—	—	—	Bal	8.54
CMSX-2	8	4.6	0.6	8	5.6	1	6	—	—	—	Bal	8.60
CMSX-6	9.8	5	3	—	4.8	4.7	2	—	—	0.1	Bal	7.98
PWA1480	10	5	—	4	5	1.5	12	—	—	—	Bal	8.70
SRR99	8	5	—	10	5.5	2.2	3	—	—	—	Bal	8.56
RR2000	10	15	3	—	5.5	4	—	—	1	—	Bal	7.87
Rene N4	9	8	2	6	3.7	4.2	4	0.5	—	—	Bal	8.56
AM1	7.8	6.5	2	5.7	5.2	1.1	7.9	—	—	—	Bal	8.60
AM3	8	5.5	2.25	5	6	2	3.5	—	—	—	Bal	8.25
TMS-6	9.2	—	—	8.7	5.3	—	10.4	—	—	—	Bal	8.90
TMS-12	6.6	—	—	12.8	5.2	—	7.7	—	—	—	Bal	9.07

Table 3.3. *The compositions (in weight%) of selected second-generation single-crystal superalloys*

Alloy	Cr	Co	Mo	Re	W	Al	Ti	Ta	Nb	Hf	Ni	Density (g/cm^3)
CMSX-4	6.5	9	0.6	3	6	5.6	1	6.5	—	0.1	Bal	8.70
PWA1484	5	10	2	3	6	5.6	—	8.7	—	0.1	Bal	8.95
Rene N5	7	8	2	3	5	6.2	—	7	—	0.2	Bal	8.70
MC2	8	5	2	—	8	5	1.5	6	—	—	Bal	8.63
TMS-82+	4.9	7.8	1.9	2.4	8.7	5.3	0.5	6	—	0.1	Bal	8.93

Table 3.4. *The compositions (in weight%) of selected third-generation single-crystal superalloys*

Alloy	Cr	Co	Mo	Re	W	Al	Ti	Ta	Nb	Hf	Others	Ni	Density (g/cm^3)
CMSX-10	2	3	0.4	6	5	5.7	0.2	8	0.1	0.03	—	Bal	9.05
Rene N6	4.2	12.5	1.4	5.4	6	5.75	—	7.2	—	0.15	0.05C 0.004B 0.01Y	Bal	8.97
TMS-75	3	12	2	5	6	6	—	6	—	0.1	—	Bal	8.89

Table 3.5. *The compositions (in weight%) of selected fourth-generation single-crystal superalloys*

Alloy	Cr	Co	Mo	Re	Ru	W	Al	Ti	Ta	Hf	Others	Ni	Density (g/cm³)
MC-NG	4	<0.2	1	4	4	5	6.0	0.5	5	0.10	–	Bal	8.75
MX4 / PW1497	2	16.5	2.0	5.95	3	6.0	5.55	–	8.25	0.15	0.03C 0.004B	Bal	9.20
TMS-138	2.8	5.8	2.9	5.1	1.9	6.1	5.8	–	5.6	0.05	–	Bal	8.95
TMS-162	2.9	5.8	3.9	4.9	6	5.8	5.8	–	5.6	0.09	–	Bal	9.04

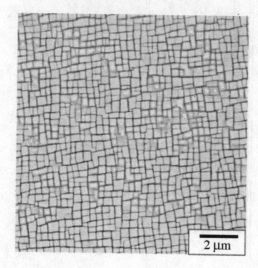

Fig. 3.28. Scanning electron micrograph of the γ/γ' microstructure of the CMSX-4 single-crystal superalloy. (Courtesy of Nirundorn Matan.)

3.2.1 Guideline 1

> *Proportions of γ'-forming elements such as Al, Ti and Ta should be high, such that the γ' fraction is $\sim$70%.*

Why is 70% an appropriate figure? To answer this question, one must have an appreciation of the micromechanisms of creep deformation in the single-crystal superalloys. This is discussed in more detail in Section 3.3, but here it is sufficient to note that, across a wide range of temperature and levels of applied stress, creep deformation on the microscale is restricted to the γ channels which lie between the precipitates of γ'. Thus, creep dislocations do not penetrate the γ' precipitates. It follows then that a reduced fraction of γ yields improved properties. The optimum microstructure then consists of many fine γ' particles with very thin layers of the γ matrix separating them; see Figure 3.28. This can be rationalised by making a rough estimate of the Orowan stress, τ_{loop}, following ref. [26]. This is approximately Gb/l, where G is the shear modulus, b is the Burgers vector and l is the width of the γ' channel. With $G = 50\,\text{GPa}$, $b = 0.25\,\text{nm}$ and $l = 100\,\text{nm}$, this yields 125 MPa. For uniaxial loading along $\langle 100 \rangle$, the Schmid factor assuming $\{111\}\langle 1\bar{1}0\rangle$ slip is $1/\sqrt{6}$; the applied stress for bowing is then 306 MPa. This is significant, and is one of the major reasons why the single-crystal superalloys perform so well at elevated temperatures.

A second point is that the creep performance does not increase monotonically as the γ' fraction is increased – thus, an alloy with 70% γ' performs better than one with 100% γ'. This implies a strong strengthening effect from the γ/γ' interfaces, which impart resistance to creep deformation. This effect has been confirmed by measuring the creep rupture lives of a series of alloys containing various fractions of γ' *of identical composition* [27]. The set of tie-lines characterising a multicomponent alloy of interest was first determined, and a family of related alloys cast with chemistries chosen such that the compositions of both γ and γ' were identical in all cases. The results are given in Figure 3.29. Note that the

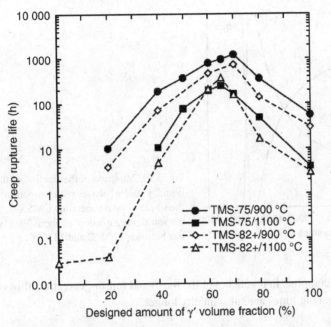

Fig. 3.29. Variation of the creep rupture lives of the single-crystal superalloys TMS-75 and TMS-82+, as a function of the amount of γ' phase [27]. The creep rupture life is largest when the γ' fraction is about 70%. (Courtesy of Hiroshi Harada.)

creep strength of γ' is somewhat superior to that of γ, but that microstructures consisting of mixtures of γ and γ' perform best. The optimum γ' fraction is ~70%.

With respect to the optimisation of the chemistry, it is useful to consider the Ni–Al binary diagram. At 700 °C, the solubility of Al in Ni is about 10 at%. The Ni$_3$Al has an Al content of 25 at%. Therefore, in the binary system one needs a composition, $\bar{c}$, satisfying $(\bar{c} - 10)/(25 - 10) = 0.75$. Hence $\bar{c} = 21.25$ at%. This corresponds to about 11 wt%. A modern single-crystal superalloy would have about 6 wt% Al. Clearly, significant reliance is being placed on the γ'-forming elements such as Ta and Ti.

Thus the balance of γ- and γ'-forming elements must be chosen such that the γ' fraction is optimal. But there is another consideration. There should exist a sufficient solutioning window, ΔT, below the melting temperature, across which γ is the only stable phase. This is necessary because, after solidification, heat treatment is required to remove residual microsegregation and eutectic mixtures rich in γ'. This puts the γ' precipitates into solution; during subsequent exposure to lower temperatures, the γ' develops a uniform microstructure with an optimum precipitate size. The heat treatment for a typical single-crystal superalloy such as CMSX-4 would involve a solutioning of 8 h at 1314 °C, a primary age of 4 h at 1140 °C followed by a secondary age of 16 h at 870 °C. Figure 3.30 shows a prediction of the phase stability of the CMSX-4 superalloy around the melting temperature. The solutioning window is about 50 °C, which is rather small; this emphasises the need for precise temperature control during heat treatment. Since the solutioning window is expected to decrease as the γ' fraction is increased, the heat treatment of the higher-strength alloys

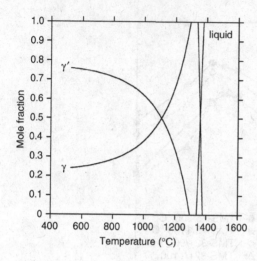

Fig. 3.30. Variation of the fraction of the liquid, γ and γ' phases with temperature for the single-crystal superalloy CMSX-4. Note the solutioning window of about $50\,°C$ which lies between $1300\,°C$ and $1350\,°C$.

can be difficult to accomplish in practice. For the third- and fourth-generation alloys which are rich in Re, solutioning times are substantially longer.

3.2.2   Guideline 2

> *The composition of the alloy must be chosen such that the γ/γ' lattice misfit is small; this minimises the γ/γ' interfacial energy so that γ' coarsening is restricted.*

The lattice misfit, δ, is defined according to

$$\delta = 2 \times \left[\frac{a_{\gamma'} - a_{\gamma}}{a_{\gamma'} + a_{\gamma}} \right] \tag{3.17}$$

where the lattice parameters of the γ and γ' phases, a_{γ} and $a_{\gamma'}$, respectively, depend *via* Vegard's law on the elemental compositions *in each phase*, and consequently on the extent of γ/γ' partitioning. For the γ phase, one has, to a good approximation (refer to Figure 2.18),

$$a_{\gamma} = 3.524 + 0.110x_{Cr}^{\gamma} + 0.478x_{Mo}^{\gamma} + 0.444x_{W}^{\gamma} + 0.441x_{Re}^{\gamma}$$
$$+ 0.179x_{Al}^{\gamma} + 0.422x_{Ti}^{\gamma} + 0.700x_{Ta}^{\gamma}\,Å \tag{3.18}$$

where the x_i^{γ} are the mole fractions of i in γ, and the various constants are taken from correlations such as those given in Chapter 2. For the γ' phase, one finds

$$a_{\gamma'} = 3.570 - 0.004x_{Cr}^{\gamma'} + 0.208x_{Mo}^{\gamma'} + 0.194x_{W}^{\gamma'} + 0.262x_{Re}^{\gamma'} + 0.258x_{Ti}^{\gamma'} + 0.500x_{Ta}^{\gamma'}\,Å$$
$$\tag{3.18b}$$

where the mole fractions are now evaluated in that phase. The value of the misfit has been found to exhibit a significant effect on the morphology of the γ' precipitates [28]; see Figure 3.31. When the misfit is small, less than about 0.5%, the γ' particles are cuboids

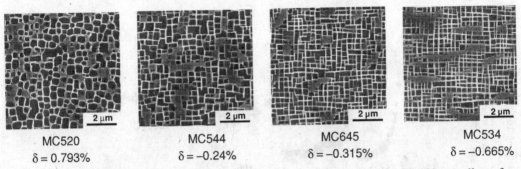

MC520	MC544	MC645	MC534
$\delta = 0.793\%$	$\delta = -0.24\%$	$\delta = -0.315\%$	$\delta = -0.665\%$

Fig. 3.31. The γ/γ' microstructures displayed by four experimental single-crystal superalloys of varying chemistries [28]. Note the correlation between the γ' morphology and the estimated values of the lattice misfit, δ. (Courtesy of Pierre Caron.)

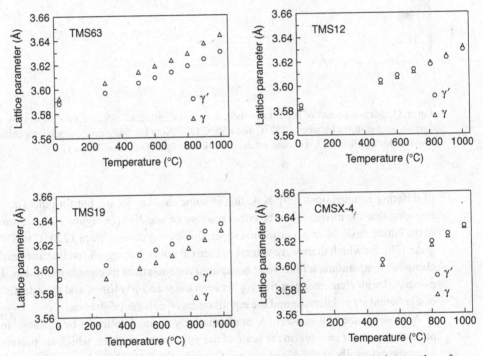

Fig. 3.32. Variation of the lattice parameter of the γ matrix and γ' precipitates in a number of nickel-based single-crystal superalloys, determined using X-ray diffraction. Data from ref. [29]. (Courtesy of Hiroshi Harada.)

with sharp corners characteristic of elastic coherency. The precipitates become spherical as the magnitude of the misfit increases, since coherency is lost.

When designing alloys, an additional consideration is the strong temperature dependence of the lattice misfit. Experiments using high-temperature diffractometry indicate that the expansion coefficient of γ' is considerably less than that of γ; thus, for all known single-crystal superalloys the lattice misfit becomes *more negative* as the temperature increases. This is demonstrated in Figure 3.32, in which results are given for a_γ and $a_{\gamma'}$ for four alloys

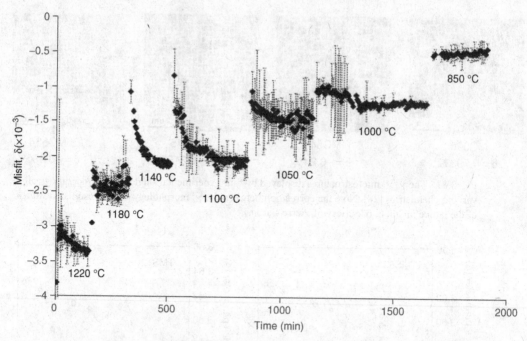

Fig. 3.33. Measurements of the lattice misfit, δ, in an experimental single-crystal superalloy at various temperatures between 1220 °C and 850 °C [30]. Note the finite time at each temperature required for equilibrium to become established. (Courtesy of Giovanni Bruno.)

of differing compositions [29]. Note that in some cases $a_{\gamma'} > a_\gamma$, but for others $a_{\gamma'} < a_\gamma$, implying that the misfit, δ, can be either positive or negative. In Figure 3.33, the variation of the lattice misfit of an experimental superalloy during cooling from 1220 °C to 850 °C is given [30], for which there were several isothermal holds. It is significant that after each step change in temperature, a finite time is required for δ to attain its equilibrium value. This is associated with elemental partitioning between the γ and γ' phases, and the establishment of a coherent γ/γ' interface and the equilibration of coherency stresses.

A consequence of a non-zero δ, or alternatively a zero δ followed by a change in temperature, are *misfit stresses* on the scale of the γ/γ' microstructure, which are present even when the externally applied load is zero. Consider the SRR99 single-crystal superalloy. It has been reported [31] that $a_{\gamma'} = 0.358\,37$ nm and $a_\gamma = 0.358\,87$ nm at room temperature, which yields a negative misfit, δ, of -1.4×10^{-3}. Assuming that the γ/γ' interfaces are forced into elastic coherence, it follows that the γ' particles are constrained in a state of hydrostatic tension, and that the γ channels separating them are in a state of biaxial compression; see Figure 3.34. The effect of elastic coherency is thus to *increase* $a_{\gamma'}$ from its value that would be measured if the particles are removed, which is possible by means of wet chemical electrolysis; conversely, a_γ is decreased. One should then refer to δ as the *constrained* lattice misfit to distinguish it from an *unconstrained* value, which is expected to be smaller in magnitude.

In practice, it has proven possible to design alloys with a constrained lattice misfit, δ, measured at room temperature, within ±1%. Alloys such as CMSX-4, Rene N5 and Rene

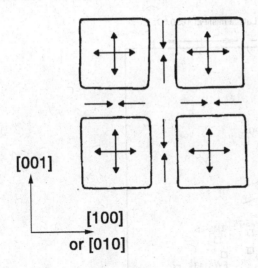

[001]

[100]
or [010]

Fig. 3.34. Schematic illustration of the coherency stresses to be expected in a nickel-based superalloy when the misfit, δ, is negative. (Courtesy of Hael Mughrabi.)

N6 have values in the range 0 to +0.4%. There is some evidence, see Figure 3.35, that alloys of negative misfit possess greater creep resistance than those with positive misfit [29]. This is because creep deformation is controlled by dislocation activity in the γ channels promoted by externally applied tensile stresses, which are superimposed on the misfit stresses. As a consideration of Figure 3.34 confirms, the net stresses seen in the γ channels will be lower in a negatively misfitting alloy than in a positively misfitting one, for loading along the normal to the unit cell faces, $\langle 001 \rangle$. The influence of the lattice misfit on the creep performance is discussed further in Section 3.3.

3.2.3 Guideline 3

> *Concentrations of creep-strengthening elements, particularly Re, W, Ta, Mo and Ru, must be significant but not so great that precipitation of topologically close-packed (TCP) phases is promoted.*

The success of this strategy can be judged by considering the creep performance of the first-, second- and third-generation single-crystal superalloys. The data in Figure 1.16 indicate that creep rupture lives vary from about 250 h at 850 °C/500 MPa for a first-generation alloy such as SRR99, to about 2500 h for a third-generation alloy such as CMSX-10/RR3000. Under more demanding conditions, for example 1050 °C/150 MPa, rupture life improves four-fold from 250 h to 1000 h. As a consideration of Table 1.3 shows, the major differences between the chemistries of these first-, second- and third-generation alloys lies in the addition of Re. Although absent in the first-generation alloys, the second-generation alloys contain about 3 wt%Re, and the third-generation alloys contain between 5 and 6 wt%. Thus, a strong creep-strengthening effect can be attributed to Re. The elements W and Ta appear to be less potent, with Cr and Co having little effect. Experience has shown that improvements

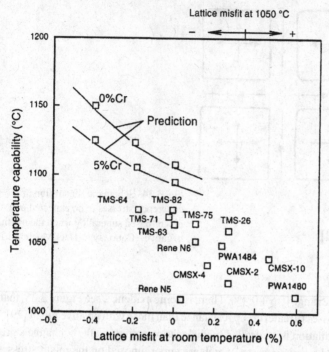

Fig. 3.35. Variation with lattice misfit of the creep capability, characterised in terms of the temperature for 1000 h creep life at a stress of 137 MPa. (Courtesy of Hiroshi Harada.)

in creep strengthening occur in the following order:

$$Co \quad \rightarrow \quad Cr \quad \rightarrow \quad Ta \quad \rightarrow \quad W \quad \rightarrow \quad Re$$

The best strengtheners thus appear to be those which diffuse the slowest in nickel – this is unsurprising, given the importance of diffusion in the dislocation climb processes, when mass transport at dislocation cores is required.

In practice, there is a limit to the concentrations of the refractory elements which can be added. Excessive quantities of Cr, Mo, W and Re promote the precipitation of intermetallic phases which are rich in these elements, as discussed in Section 2.1.3. Precipitation of these TCP phases implies a lack of stability, which is detrimental to creep behaviour. Examine the Larson–Miller diagram given in Figure 3.36, on which the creep performance of a number of prototype second-generation alloys containing 3 wt%Re is plotted. At low temperatures of about 800 °C and stresses of 500 MPa, the behaviour is found to vary only weakly with alloy composition. However, at higher temperatures of ~1000 °C and stresses of ~150 MPa, the alloys fall into two groups of contrasting behaviour. Creep performance is very much inferior when TCP formation occurs, with the rupture life being reduced accordingly. Figure 3.37 illustrates the type of extensive TCP formation characteristic of the unstable alloys. The conclusion that can be drawn from these experiments is that the TCP precipitation is detrimental to properties. The depletion of the remaining γ/γ' structure of the elements which improve creep strength is responsible for this effect.

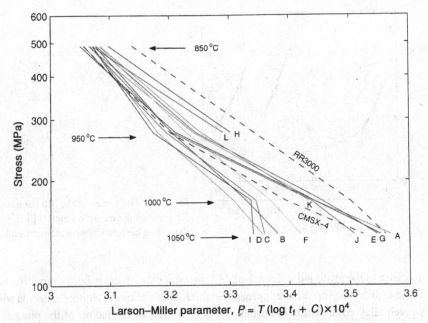

Fig. 3.36. Variation of the stress for rupture with the Larson–Miller parameter, P, for various second-generation single-crystal superalloys. The alloys B, C, D, F and I were found to be prone to TCP formation; thus the creep properties of the stable alloys are very much superior. (Courtesy of Martin Rist.)

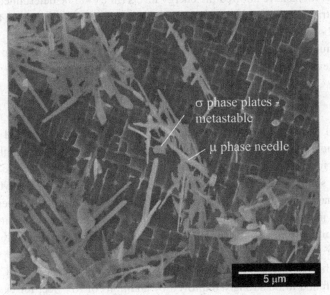

Fig. 3.37. Scanning electron micrograph of the experimental second-generation superalloy RR2071, aged at 900 °C for 500 h, showing extensive precipitation of the TCP phases σ and P. (Courtesy of Cathie Rae.)

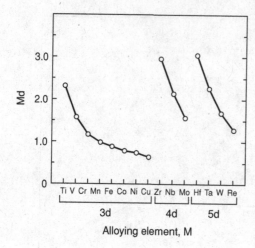

Fig. 3.38. The values of the Md parameters for some of the transition metals [32]; note the strong correlation with atomic number.

Because of the profound influence of TCP precipitation on creep behaviour, efforts have been made to develop predictive approaches which enable alloy developers to decide whether any given alloy composition is unstable or not. In view of the nature of the phases, band structure calculations are the most appropriate. For example, for the first row of transition metals, the 4p fraction of the hybrids can be ignored and the outer electrons regarded as existing in two distinct, overlapping bands derived from the 3d band (providing up to ten electrons) and the 4s band (providing two). Similar arguments apply to the second and third rows. These considerations have allowed the energy levels (in eV) to be determined using band theory for the transition metals at, or around, the Fermi level. These have been termed metal-d levels (symbol Md), and have been found to correlate well with the electronegativity and the metallic radius. In the absence of more rigorous electron theory, one then works out the effective $\overline{Md}_{eff}$ level from [32]

$$\overline{Md}_{eff} = \sum_i \overline{x}_i Md_i \qquad (3.19)$$

where the values of Md_i for the various i values are given in Figure 3.38, and the $\overline{x}_i$ correspond to the mole fraction of i in the alloy. The *higher* value of $\overline{Md}_{eff}$ the higher the probability of TCP formation. The critical value appears to around 0.99. The procedure described here is a simplified version of the procedure known as PHACOMP, which is an acronym for 'phase computation'. In a variant of the procedure described here, the concentration terms in the summation can be taken to be the mole fractions in γ, since the TCPs usually precipitate from this phase. However, there is no strong evidence that this provides a greater predictive capability.

In recent years, the stability of the single-crystal superalloys with respect to TCP precipitation has been found to be strongly improved by additions of Ru. See, for example, Figure 3.39, in which the time–temperature-transformation (TTT) curves for the formation of TCP phases is given for two experimental alloys, the second of which is doped with 2 wt%Ru [33]. A consequence of this is that the creep properties are found to be significantly improved (see Figure 3.40), particularly at temperatures beyond about 950 °C;

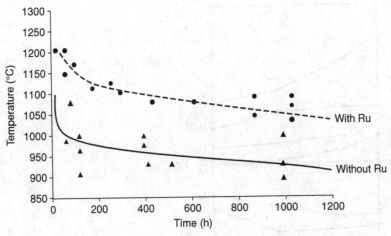

Fig. 3.39. Time–temperature-transformation (TTT) diagrams for the formation of TCP phases in two single-crystal superalloys, one with and the other without Ru-doping – the base composition being Ni–12Co–2.5Cr–9W–6.4Re–6Al–5.5Ta–0.15Hf (wt%). (Courtesy of An-Chou Yeh.)

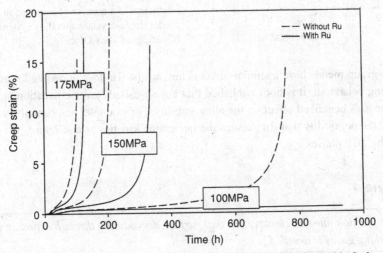

Fig. 3.40. Creep curves for two single-crystal superalloys tested at 1100 °C, the second doped with 2 wt%Ru [33] – the base composition being Ni–12Co–2.5Cr–9W–6.4Re–6Al–5.5Ta–0.15Hf (wt%). (Courtesy of An-Chou Yeh.)

for this reason, Ru additions are now commonly added to the latest 'fourth-generation' single-crystal superalloys, although, at the time of writing, these are not yet being used in commercial engines. The reasons for the so-called 'Ru effect' are controversial; some researchers have noted that Ru, which partitions strongly to the γ phase, has a strong influence on the γ/γ' partitioning behaviour of elements such as Al, Re and Cr [34,35]. Significant additions of Ru have the effect of reducing the partitioning coefficients of many of the alloying additions (see Figure 3.41); this effect, however, is not unique to Ru, since other

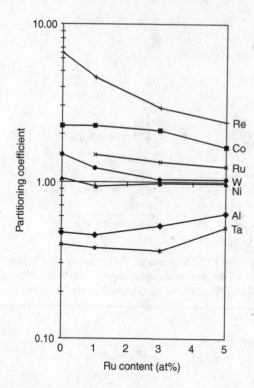

Fig. 3.41. Measurements of the influence of Ru-doping on the γ/γ' partitioning coefficient of a number of elements in a model single-crystal superalloy system [35]. (Courtesy of Tony Ofori.)

platinum-group metals have a similar effect. Thus, whilst Ru does undoubtedly influence partitioning behaviour, it is not established that any so-called 'reverse partitioning' is the reason for Ru's beneficial effect on the alloy stability. For example, one cannot rule out at this stage the possibility that Ru poisons the nucleation kinetics or else retards the growth rates of the TCP phases.

3.2.4   Guideline 4

> *The composition must be chosen such that surface degradation through exposure to the hot, working gases is avoided.*

Attack by oxidation is a possibility. When this occurs, wall thickness and thus loading bearing capacity are lost, but more importantly it can be the precursor to mechanical failure by fatigue. One should examine first the thermodynamics of formation of the important oxides NiO, Cr_2O_3 and Al_2O_3. The Gibbs energies of formation indicate that Al_2O_3 is the most stable of the three, NiO the least, with Cr_2O_3 between the two – as demonstrated by a consideration of the Ellingham diagram. On this basis alone, one might expect Al_2O_3 to be the oxide of prevalence in superalloy systems. However, of equal relevance in practice are the kinetics of oxidation. NiO has a high concentration of nickel vacancies and thus forms very rapidly with voids and microcracking prevalent – thus it is friable, which promotes spallation and hence sustained attack. Chromia, Cr_2O_3, is less so, but at elevated temperatures it is

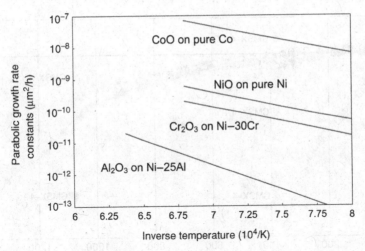

Fig. 3.42. Variation of the parabolic thickening constants for oxide formation with temperature, for various species on different substrates. Note that the formation of Al_2O_3 is slow.

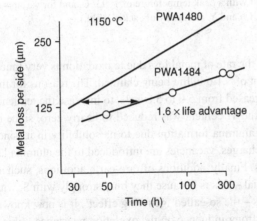

Fig. 3.43. Data for the rate of metal loss due to oxidation of the PWA1480 and PWA1484 single-crystal superalloys, confirming the improved oxidation behaviour of the second-generation alloy. Data taken from ref. [36].

oxidised to the gaseous CrO_3, with a significant rate of oxidation being determined by transport through a gaseous boundary layer, which is greatly increased in rapidly moving gas streams. This effect limits the temperature at which a chromia-forming alloy can be used to about 900 °C. In binary Ni–Cr alloys, Cr_2O_3 will not form if the Cr content falls below about ∼12 wt%, although alloying with Al reduces this value. When growth of an Al_2O_3-based scale occurs, the rate of oxidation is always found to be slow. Typical parabolic rate constants for the various oxides are given in Figure 3.42. For these reasons, the first single-crystal superalloys – for example the first- and second-generation ones – were designed to be alumina formers. To ensure this, the aluminium content was chosen to be around ∼6 wt%.

It is instructive to consider the improvements in oxidation resistance which arose as the second-generation single-crystal alloys were developed. Consider Pratt & Whitney's success with PWA1484, which is a derivative of the first-generation alloy PWA1480 [36]. Figure 3.43, which compares the oxidation performance of the two alloys during isothermal

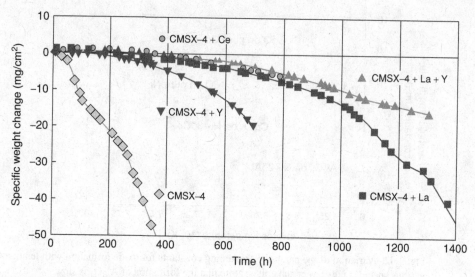

Fig. 3.44. Weight change data [38] due to cyclic oxidation for the CMSX-4 single-crystal superalloy during a cyclic oxidation test with a peak temperature of 1093 °C, and for variants of it doped with the rare-earth elements La, Ce and Y. (Courtesy of Jackie Wahl.)

exposure to 1000 °C, indicates that the rate of metal loss due to oxidation is very much less for PWA1484, with an improvement of ×1.6 in life being claimed. The reasons for this are as follows. The Al content was increased from 5.0 to 5.6 wt%, to increase the tendency for Al_2O_3-scale formation. Secondly, the Ti content was reduced to nearly zero, since it was found that Ti increases the rate of alumina formation due to its solubility in it; since the Ti^{4+} and Al^{3+} ions have different charges, vacancies are introduced to the alumina lattice, thus increasing the ionic mobilities. Finally, additions of rare-earth additions, such as Hf, La and Y, were found to be beneficial – this is because they bind strongly with S, which is inevitably present at impurity levels – the so-called 'gettering effect'. It is now known that S present in the superalloys has a strong influence on the oxidation resistance – due to its solubility in the oxide scale and its likely segregation to the metal/scale interface – it has been suggested, for example, that sulphur destroys the strong Van der Waals bond between the alumina scale and the metal substrate [37]. This explains the considerable emphasis placed by the alloy manufacturers on reducing the S levels of superalloy melts; sulphur should be restricted to 10–20 ppm, although recent melt desulphurisation techniques have allowed residual levels to reach less than 1 ppm on a routine basis [38]. Results from dynamic cyclic oxidation tests for bare CMSX-4 with and without rare-earth doping (see Figure 3.44) provide very convincing evidence that the performance of the second-generation alloy CMSX-4 is improved by additions of the rare-earth elements. Recent experiments on the TMS-138 superalloy doped with 4, 40, 100 and 1000 ppm by weight of yttrium indicate that the oxidation resistance is optimal at about 100 ppm; field emission electron microprobe analysis (FE-EPMA) confirms that sulphur is present primarily as the YS compound at this yttrium concentration, but as Y_2S_3 when it falls to 40 ppm; see Figure 3.45. Moreover, sulphur is present in the oxide scale for the lower yttrium contents considered. For the very

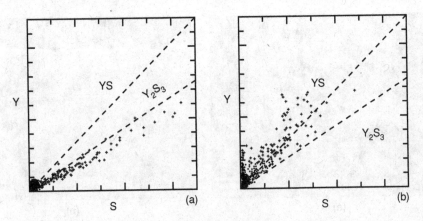

Fig. 3.45. Field emission electron microprobe analysis (FE-EPMA) data from secondary particles in the TMS-138 superalloy doped with (a) 40 and (b) 100 ppm (by weight) of yttrium, confirming the presence of Y_2S_3 and YS, respectively. (Courtesy of Atsushi Sato.)

best oxidation performance in single-crystal superalloys, it also appears to be necessary to tie up residual carbon, which is often present at the 400 ppm level; for this reason, carbide formers such as Hf are often added at greater than 400 ppm concentrations.

In practice, the oxidation behaviour of the single-crystal superalloys is complicated by the many different elements present. Consider, for example, a third-generation alloy such as CMSX-10, which does not form an alumina scale [39]; see Figure 3.46. During oxidation, an external scale of NiO forms very rapidly and spallation occurs readily. The underlying metal, being depleted in Ni, forms β–NiAl, which grows into the metal by a process of internal oxidation. Degradation is further complicated by the formation of various Cr-, W- and Ta-rich oxides beneath the NiO scale [39]. These effects are due to the choice of alloy composition: the Re content is high at 6 wt% to confer creep strengthening with the Cr content reduced to 2 wt% in an attempt to prevent TCP formation. The Cr content is too low to prevent either the formation of a thick external Ni scale or to resist attack by internal oxidation. For these reasons, for adequate environmental performance it is necessary to coat CMSX-10 by the process of pack or gas-phase aluminisation. A recent comparative study of the cyclic oxidation performance of typical first-, second-, third- and fourth-generation single-crystal superalloys has confirmed the superior performance of the earlier variants; see Figure 3.47. These considerations indicate that, during the development of the latest generations of single-crystal superalloys, less emphasis has been given to imparting oxidation resistance by appropriate choice of alloy chemistry; instead, one is increasingly reliant on coating technologies for imparting resistance to environmental degradation. Chapter 5 deals with these technologies in detail.

Case study – hot corrosion

Attack by hot corrosion is also a likelihood which should be accounted for when designing single-crystal superalloys [40,41]. Whilst a number of different corrosion degradation reactions can arise, the pervasive feature of them is chemical attack by gaseous, oxidising

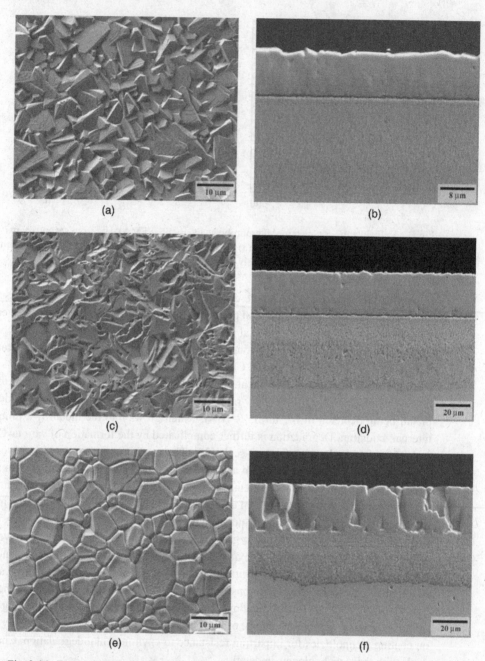

Fig. 3.46. External NiO scale and internal oxidation in the third-generation superalloy CMSX-10 following isothermal heat treatment for 100 h at (a), (b) 800 °C, (c), (d) 900 °C and (e), (f) 1000 °C, viewed from above and in cross-section. (Courtesy of Matthew Hook.)

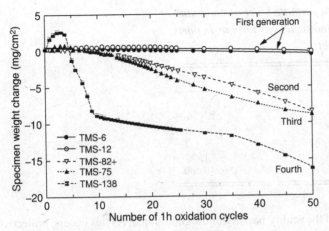

Fig. 3.47. Results from cyclic oxidation testing at 1100 °C (1 h hold at temperature) for various single-crystal superalloys. Note the superior performance of the earlier generations. (Courtesy of Kyoko Kawagishi.)

environments rich in sulphur and consequently SO_3 – the stable oxide at typical TET temperatures [42]. The sulphur is introduced since it is present as an impurity in the fuel burned in the combustor. Thus, sulphidisation – which causes the formation of nickel, chromium, cobalt and other metal sulphides, and which is thus a direct analogue of oxidation – is a possibility. However, in practice, the corrosiveness of the environment is exacerbated by its contamination with compounds (most notably NaCl, but also KCl, $MgCl_2$ and $CaCl_2$) ingested from the incoming air, which occurs particularly in marine environments; in the presence of sulphur- and oxygen-rich gases, these are converted to salts such as Na_2SO_4, which are then deposited on the blading. Attack by deposits produced in this way is known as hot corrosion. The combination of NaCl and Na_2SO_4 is particularly pernicious, as a eutectic mixture forms when the two compounds are mixed in equimolar proportions which melts at ~620 °C, which is considerably lower than the melting temperatures of the pure constituents (800 °C and 890 °C for NaCl and Na_2SO_4, respectively). Molten salts so produced can dissolve oxides such as Al_2O_3 and Cr_2O_3, and thus are able to destroy any protective scale present by a fluxing mechanism. The hot-corrosion effect is exacerbated by the presence of other metallic salts containing Li, Mg, Ca or K as airborne matter and/or other impurities in the fuel such as V or Pb. For this reason, most turbine manufacturers place tolerances on the levels of fuel contaminants; see Table 3.6.

The reaction conditions characteristic of hot-corrosion attack may be either basic or acidic in nature. Basic conditions arise because of reactions of the type

$$SO_4^{2-} \rightleftharpoons O^{2-} + SO_3 \rightleftharpoons O^{2-} + SO_2 + \tfrac{1}{2}O_2 \tag{3.20}$$

or, if the prevailing ion is sodium,

$$Na_2SO_4 \rightleftharpoons Na_2O + SO_3 \rightleftharpoons Na_2O + SO_2 + \tfrac{1}{2}O_2 \tag{3.21}$$

Hence, the liquid becomes more basic with respect to its initial composition on account of an increase in the oxide ion concentration (the equilibria between SO_4^{2-}, O^{2-} and SO_3

Table 3.6. *Maximum permitted fuel contaminant levels for a typical industrial gas turbine, in parts per million by weight*

Contaminant	Level
Na+K	1
V	0.5
Pb	1
F_2	1
Cl_2	1500
S	10 000

being an indicator of the acidity/basicity of the conditions). As this occurs, protective oxide scales may react with the deposit via a fluxing process according to

$$Al_2O_3 + O^{2-} \rightleftharpoons 2AlO_2^- \tag{3.22}$$

or, alternatively,

$$Na_2O + Al_2O_3 \rightleftharpoons 2NaAlO_2 \tag{3.23}$$

with the species formed being soluble in the liquid deposit. Such basic fluxing reactions have a number of distinctive features. First, this form of attack is usually restricted to high temperatures (above 900 °C) since the chemical reactions which produce oxide ions are otherwise kinetically constrained – thus it has become known as 'high-temperature' or 'type I' hot corrosion [41]. Second, a porous outer oxide scale is produced with internal sulphides in the alloy substrate near the scale–alloy interface, as a consequence of sulphur removal from the Na_2SO_4 to produce oxide ions. Finally, the amount of attack depends upon the production of oxide ions by the melt; hence, a supply of Na_2SO_4 is necessary for the attack to continue. Alternatively, acidic conditions might prevail. For example, 'gas-phase-induced' acidic fluxing processes can occur, for example

$$Ni + SO_3 + \tfrac{1}{2}O_2 \rightarrow Ni^{2+} + SO_4^{2-} \tag{3.24}$$

such that the acidic component is supplied to the liquid deposit by the gaseous atmosphere (for example SO_3 or, alternatively, V_2O_5), with the removal of nickel from the alloy preventing the development of a continuous, protective layer of either Al_2O_3 or Cr_2O_3 over the alloy surface. In this case, the attack – usually giving heavy pitting but no internal sulphides – is known as 'low-temperature' or 'type II' hot corrosion [41] since it occurs between 650 °C and 800 °C. The low-temperature restriction results from the necessity to form sulphates such as $CoSO_4$, Ni_2SO_4 or $Al_2(SO_4)_3$, which require prohibitively high SO_3 partial pressures, which are not available in typical combustion environments as the temperature is increased. So-called 'alloy-induced' acidic fluxing modes have also been proposed, in which the deposit becomes more acidic by reacting with certain oxides in the corrosion product, for example

$$Mo + \tfrac{3}{2}O_2 \rightarrow MoO_3 \tag{3.25}$$

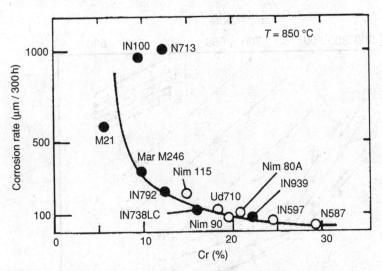

Fig. 3.48. Variation of the corrosion rate of various superalloys with Cr content, confirming the significant attack in Cr-lean systems. (Courtesy of Malcolm McLean.)

and then

$$MoO_3 + SO_4^{2-} \rightleftharpoons MoO_4^{2-} + SO_3 \qquad (3.26)$$

implying an equilibration with SO_2 gas; subsequently reactions occur between the protective oxide barriers and the deposit according to

$$Al_2O_3 + 3SO_3 \rightleftharpoons 2Al^{3+} + 3SO_4^{2-} \qquad (3.27)$$

Alloys containing high concentrations of the refractory elements Mo, W and V are prone to this effect since they form oxides which cause Na_2SO_4 to become more acidic, with only one application of the salt being catastrophic on account of the self-sustaining nature of the reaction.

In view of such accelerated attack, which can rapidly destroy rows of turbine blades should conditions prove favourable to it, attempts have been made to design superalloys with compositions that provide resistance to hot corrosion. Alloys with high chromium content are found to resist degradation by hot corrosion very strongly (see Figure 3.48), and thus chromia- rather than alumina-forming alloys are to be preferred when corrosive conditions are prevalent. Although the degradation mechanisms of hot corrosion and sulphidisation should be distinguished, this finding is consistent with studies carried out by Mrowec *et al.* [43] on binary Ni–Cr alloys exposed to sulphur vapour at 1 atm in the range 600 to 900 °C. Below ~2 wt%Cr, the parabolic sulphidisation rate was found to be marginally greater than that of pure nickel, the scale being identified as nickel sulphide with ~0.4 wt%Cr dissolved in it; however, for greater chromium concentrations, the rate of sulphidisation was substantially less than that of pure Ni, and indeed, for Cr concentrations above 20 wt%Cr, the resistance was found to be greater than for pure Cr provided that the temperature is below 800 °C; see Figure 3.49. When the chromium content reached 50 wt%, chromium sulphide replaced nickel sulphide as the majority phase in the scale, the latter now being dispersed

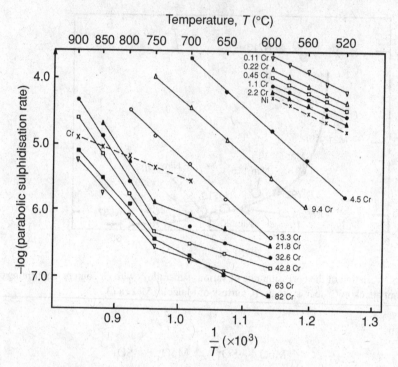

Fig. 3.49. Dependence of the parabolic sulphidisation rate for binary Ni–Cr alloys as a function of the reciprocal of the absolute reaction temperature and the chromium content [43].

in the former. Thus, resistance to both sulphidisation as well as hot corrosion is imparted by progressive additions of chromium to the superalloys. Unfortunately, the preference for high Cr content is at odds with the very best oxidation resistance, which demands an alumina former and hence a high aluminium content. A further significant factor is that Cr, being a significant γ former, is incompatible with alloys which have the very greatest creep resistance – since this requires a high γ' fraction. Furthermore, excess Cr content impairs alloy stability by promoting TCP formation. These considerations emphasise that trade-offs must be made when designing alloys for turbine applications, to achieve an optimal combination of resistances to oxidation, corrosion and creep.

3.3 Mechanical behaviour of the single-crystal superalloys

Our considerations so far have emphasised that significant stresses are set up in the turbine blading during operation. The mechanical performance of the single-crystal superalloys is then quite critical to the successful operation of the turbine engine. In this section, the response to the two most important modes of loading are considered: creep and fatigue. Since it is of the greatest practical significance, attention is restricted to the performance under uniaxial loading with the stress axis aligned along the $\langle 001 \rangle$ crystallographic direction.

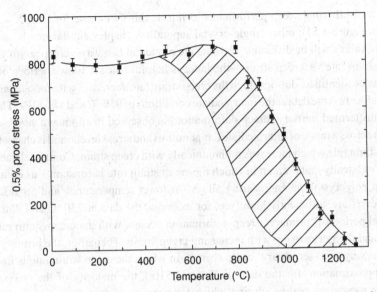

Fig. 3.50. Variation of the uniaxial 0.5% yield stress of ⟨001⟩-orientated single-crystal superalloy CMSX-4, showing (hatched region) the conditions under which significant inelastic creep deformation is observed.

3.3.1 Performance in creep

In Chapter 2, it was seen that the yield stress of a typical ⟨001⟩-oriented single-crystal superalloy remains rather constant from ambient conditions to approximately 700 °C. It falls to substantially lower values if the temperature is raised any further; however, in practice, of greater significance is the possibility of *time-dependent* inelastic deformation at a stress substantially lower than that required for yielding. This process is known as *creep deformation*. It is instructive to consider some numbers, for example for the second-generation single-crystal superalloy CMSX-4. The yield stress at 750 °C is about 900 MPa, yet a stress of 750 MPa is sufficient to cause a creep strain of 5% in about 5 h. The yield stress at 950 °C is about 600 MPa, but a stress of 200 MPa is sufficient for a creep strain of 5% after 1800 h. Beyond 1100 °C the yield stress falls to very low values, and in practice it is difficult to measure with accuracy because creep deformation occurs rapidly such that a substantial strain-rate dependence is displayed; however, at a temperature of 1150 °C, a stress of only 100 MPa causes 1% strain within 200 h. In Figure 3.50, yield stress data for CMSX-4 are plotted along with the values of stress at which substantial creep deformation is observed. It is clear that temperatures and stresses must be restricted if excessive creep deformation is not to occur. This is of practical significance, since components such as turbine blades are machined such that blade tip/seal clearances satisfy tight tolerances, to prevent engine performance being compromised. Typically, a set of blades must be removed from service if creep strains of greater than a few per cent are accumulated.

The mechanism of creep deformation in ⟨001⟩-oriented single-crystal superalloys is very sensitive to the conditions of temperature and stress that are acting. A number of regimes can be identified, and in each a distinct mode of microstructural degradation is predominant.

Data [44] for isothermal creep performance confirms this (see those for CMSX-4, which are given in Figure 3.51); other single-crystal superalloys display similar trends, although the absolute values will be different. The data are plotted in two ways: creep strain vs. time and creep strain rate vs. creep strain – the latter is helpful since it allows periods of creep hardening to be identified, during which the creep strain rate *decreases* with *increasing* creep strain. Consider first the data at the intermediate conditions of 950 °C and 185 MPa. This type of behaviour (termed 'tertiary' creep deformation) is observed in all known single-crystal superalloys across a range of intermediate temperatures and stress levels, and is characterised by a creep strain rate which increases monotonically with creep strain. Completely absent is any period of 'steady-state' creep in which the creep strain rate is constant – as commonly observed in polycrystalline engineering alloys. At lower temperatures, and provided that the stress levels are sufficiently high (see, for example, the data at 750 °C and 750 MPa), a substantial period of 'primary' creep deformation occurs, with the creep strain rate first increasing and then decreasing with increasing creep strain. Thereafter, the primary creep regime gives way to a 'secondary' creep regime, in which the creep strain rate is invariant to a first approximation; for the data in Figure 3.51(a), the majority of the creep life is spent in the secondary regime, during which time the secondary creep rate is of order 10^{-7}/s. It is emphasised that such 'secondary' creep occurs only after a substantial period of primary strain, and that there have been no observed cases of single-crystal alloys which proceed straight to the 'secondary' regime. Finally, at temperatures beyond about 1050 °C, the creep curve displays a distinct plateau before the creep strain increases catastophically, with rupture occurring almost immediately afterwards. There is a creep-hardening effect at first, before the plateau, during which the creep strain does not vary strongly with time. At these high temperatures, the creep behaviour is affected by a degradation of the γ/γ' microstructure, which gives rise to a 'rafted' morphology; one refers to a 'rafting regime' to distinguish this distinct behaviour from the 'tertiary' and 'primary' regimes arising at lower temperatures and greater stresses. The conditions of stress and temperature required for the operation of the primary, tertiary and rafting regimes in CMSX-4 are summarised in Figure 3.52, which is taken from ref. [45].

The micromechanisms of deformation in the different regimes are now discussed in detail.

A   Tertiary creep regime

In the intermediate, 'tertiary', creep regime, evidence from transmission electron microscopy [26,46–49] indicates that dislocation activity is of the $a/2\langle 1\bar{1}0\rangle\{111\}$ type, and is restricted to the γ channels between the γ' particles (see Figure 3.53). The density of dislocations is found to increase with increasing creep strain, while other forms of microstructural degradation (γ' coarsening, creep cavitation) are absent. The γ' particles remain intact, i.e. they are not sheared by the creep dislocations; therefore, substantial cross-slip is required for substantial deformation to occur, since the γ' fraction is high and consequently the γ channels too thin to allow the dislocations to percolate the γ/γ' structure if this were not the case. The leading segments of the dislocation loops are screw in character, consistent with the requirement for cross-slip; as they expand through γ matrix channels, segments

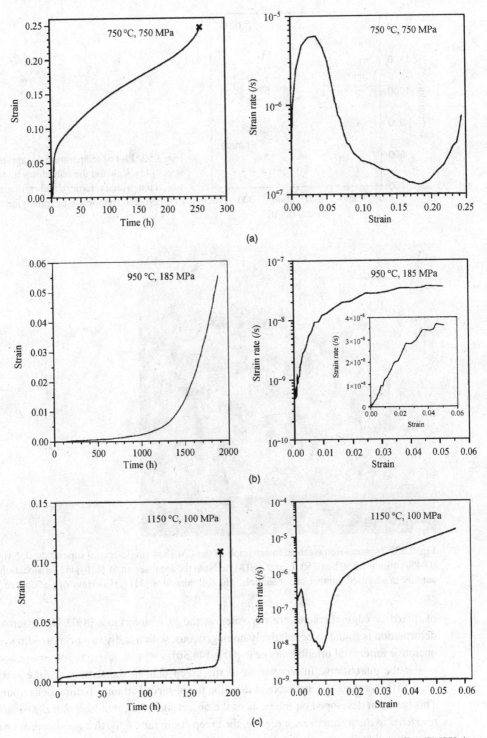

Fig. 3.51. Constant load creep data [45,46] for ⟨001⟩-oriented single-crystal superalloy CMSX-4 at various temperatures and stresses: (a) 750 °C and 750 MPa, (b) 950 °C and 185 MPa, (c) 1150 °C and 100 MPa.

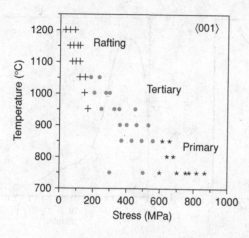

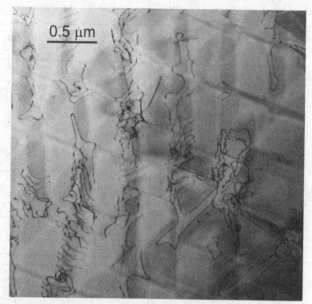

Fig. 3.52. Plot of temperature vs. applied stress [45], showing the conditions under which the primary, tertiary and rafting modes of creep deformation are operative for CMSX-4.

Fig. 3.53. Transmission electron micrograph of the CMSX-4 single-crystal superalloy, deformed to 0.04% strain in 1890 h at 750 °C and 450 MPa. Note the localisation of $\langle 1\bar{1}0 \rangle \{111\}$ dislocation activity in a limited number of γ channels. The foil normal is $\{111\}$. (Courtesy of Cathie Rae.)

of mixed or edge character are deposited at the γ/γ' interfaces [49]. The macroscopic deformation is found to be relatively homogeneous, with usually two or more slip systems operative for crystal orientations near $\langle 001 \rangle$ [46,50].

For the quantitative interpretation of the creep data in the tertiary regime (see Figure 3.51(b)), a theory is available to describe the observed strain softening behaviour [51]. This has been developed on the basis of the observation that, *provided that deformation is restricted to the tertiary creep regime*, the creep strain rate, $\dot{\epsilon}$, is, to a good approximation, proportional to the accumulated macroscopic creep strain, ϵ; see Figure 3.51(b). It is reasonable to assume that the mobile dislocation density, ρ_m, which is taken to be the variable characterising the state of damage, varies with ϵ according to

$$\rho_{\mathrm{m}} = \rho_0 + M\epsilon \qquad (3.28)$$

where ρ_0 is the initial dislocation density and M is a constant describing the dislocation multiplication rate. The macroscopic creep strain rate, $\dot{\epsilon}$, is then given by combining Equation (3.28) with Orowan's equation

$$\dot{\epsilon} = \rho_{\mathrm{m}} b v \qquad (3.29)$$

where b is the Burgers vector and v is the dislocation velocity. Provided v is constant, the proportionality of $\dot{\epsilon}$ and ϵ follows from the substitution of Equation (3.28) into Equation (3.29), yielding

$$\dot{\epsilon} = \dot{\Gamma}_{\mathrm{i}} + \Omega\epsilon \qquad (3.30)$$

where $\dot{\Gamma}_{\mathrm{i}}$ is the initial strain rate. Calculations indicate that each of Ω and $\dot{\Gamma}_{\mathrm{i}}$ shows an Arrhenius-type dependence upon temperature, consistent with

$$\Omega = a \exp\left\{ b\sigma - \frac{Q_1}{RT} \right\} \qquad (3.31)$$

and

$$\Gamma_{\mathrm{i}} = c \exp\left\{ d\sigma - \frac{Q_2}{RT} \right\} \qquad (3.32)$$

where a, b, c and d and Q_1 and Q_2 are constants and R is the gas constant. In the tertiary creep regime, Equations (3.28)–(3.32) have been found to model the macroscopic creep curves of CMSX-4 with reasonable accuracy (see Figure 3.54), although it is seen that the creep performance at the highest temperatures is underpredicted; more will be said about this later in the context of the rafting effect. It follows from the model that, at constant temperature and stress, the creep strain is proportional to $\exp\{kt\}$, where k is a constant and t is time; therefore it increases exponentially with time t.

The model ignores one important aspect of the deformation behaviour in the tertiary regime: the incubation period during which $a/2\langle 1\bar{1}0\rangle\{111\}$ dislocations multiply from dislocations sources, typically low-angle grain boundaries, into regions which are initially dislocation-free [26]. Particularly when the temperature is low, the creep strain rate is very small and the creep strain is less great than predicted by an interpolation from the behaviour at longer times, using Equations (3.28)–(3.32). During the incubation period, the TEM evidence indicates that the glide of dislocations occurs preferentially through the horizontal γ channels, i.e. the channels which lie perpendicular to the direction of the applied stress. This can be rationalised in the following way: the superposition of the (negative) misfit stress for a negatively misfitting superalloy and the externally applied (positive) tensile stress along $\langle 001\rangle$ is such as to restrict dislocation activity in the vertical channels; conversely, the stress in the horizontal channels is, to a first approximation, unaffected. Detailed calculations which take account of the Poisson effect and a determination of the misfit stress indicate that the horizontal channels experience a shear stress at least twice as great as that in the vertical channels [26,52]. However, the incubation period is short under most circumstances,

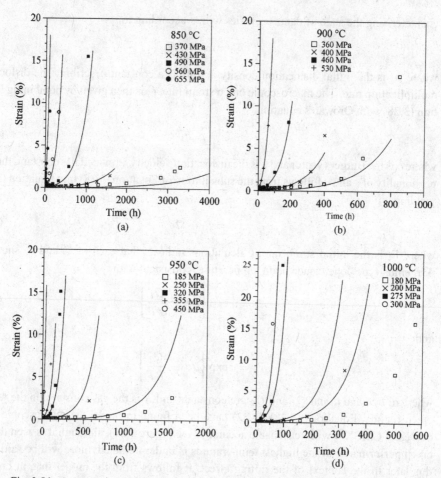

Fig. 3.54. A comparison of the creep data for the CMSX-4 single-crystal superalloy and the tertiary creep model described by Equations (3.28)–(3.32). Note that the agreement between model and experiment is generally good, except at high temperatures and low stresses. Data from refs. [45] and [46].

and the equations given above can be used to describe the creep strain evolution in the tertiary regime with reasonable accuracy.

B   Primary creep regime

At low temperatures, and provided the applied stress is sufficiently high, ⟨001⟩-oriented single-crystal superalloys undergo significant primary creep deformation. In Figure 3.55, creep curves for CMSX-4 are plotted at temperatures of 750 °C and 850 °C for specimens of identical orientation and within a few degrees of ⟨001⟩. A number of observations can be made. First, the primary creep strain increases with applied stress, with primary creep eventually giving way to secondary creep soon after the knee of each curve; thus, significant primary creep strain occurs only when the applied stress is greater than a critical threshold

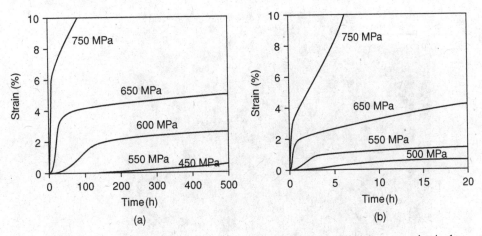

Fig. 3.55. Constant load creep data for CMSX-4 at low temperatures and high stresses, i.e. in the primary creep regime. (a) 750 °C; (b) 850 °C.

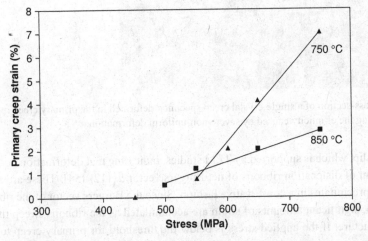

Fig. 3.56. Plot of percentage primary creep strain against applied stress, confirming a threshold stress of ∼500 MPa for operation of the primary creep mechanism.

value of about 500 ± 50 MPa; see Figure 3.56. Second, the transition from primary to secondary creep occurs at a time which decreases as the applied stress increases; moreover, further analysis reveals that the secondary creep rate scales in proportion to the primary creep strain [46].

A characteristic of primary creep is a significant heterogeneity of slip. If the creep specimen is cylindrical in shape, as is most usually the case, then after creep deformation the cross-section is found to be elliptical (see Figure 3.57), with the degree of ellipticity increasing with the primary creep strain incurred. More significantly, careful characterisation of the shape change and correlation of it with the specimen orientation indicates that the macroscopic shape change is consistent with $\langle 11\bar{2}\rangle\{111\}$ deformation [46,53]. This observation was first made by Kear and co-workers in the late 1960s; they proposed a model for

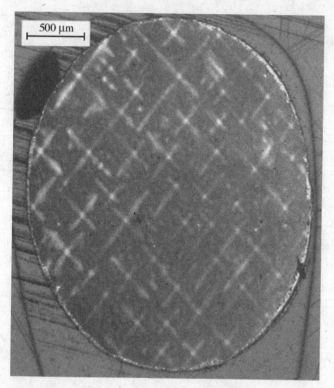

Fig. 3.57. Cross-section of a single-crystal creep specimen deformed in the primary creep regime [53], illustrating the ellipticity caused by severe non-uniform deformation.

$\langle 11\bar{2}\rangle\{111\}$ slip, which is supported by TEM studies, indicating that deformation occurs by the movement of dislocation ribbons of net Burgers vector $a\langle 11\bar{2}\rangle$ [54]. These are able to shear the γ' precipitates in a conservative fashion. Since the Burgers vector of the ribbon is large at $a\sqrt{6}$, significant amounts of strain are accumulated as the ribbons sweep through the γ/γ' structure. If the applied stress is below the threshold for primary creep to operate, then the microscopic evidence shows that only dislocations of form $a/2\langle 1\bar{1}0\rangle\{111\}$ are present, and that these reside in γ. This is very strong evidence that the $a\langle 11\bar{2}\rangle$ ribbons are responsible for primary creep. Figure 3.58 shows a transmission electron micrograph of an $a\langle 11\bar{2}\rangle$ ribbon in a creep-deformed TMS-82 single-crystal superalloy.

There has been considerable debate about the mechanism of formation of the $a\langle 11\bar{2}\rangle$ dislocation ribbons, but it is now known to involve the reaction of $a/2\langle 1\bar{1}0\rangle\{111\}$ dislocations in γ, which are dissociated into their Shockley partials. A typical reaction might then be given by [55]

$$a/2[011] + a/2[\bar{1}01] \rightarrow a/3[\bar{1}12] + a/6[\bar{1}12] \tag{3.33}$$

If the applied stress is sufficient, the $a/3[\bar{1}12]$ dislocation is able to enter the γ', leaving a superlattice intrinsic stacking fault (SISF) behind it (see Figure 3.59(a)) and the remaining $a/6[\bar{1}12]$ at the γ/γ' interface. Formation of the ribbon in γ' then requires further reactions between the $a/2\langle 1\bar{1}0\rangle\{111\}$-type dislocations, such that the original $a/6[\bar{1}12]$, a further $a/6[\bar{1}12]$ and eventually a second $a/3[\bar{1}12]$, enter the γ'; see Figure 3.59(b). An anti-phase

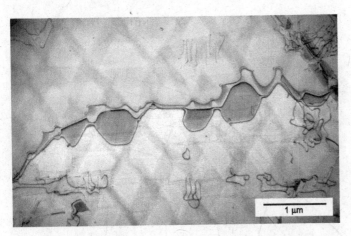

Fig. 3.58. TEM micrograph of an $a\langle 11\overline{2}\rangle$ ribbon in the single-crystal superalloy TMS-82, deformed in creep at 750 °C and 750 MPa to 11% strain. The foil normal is {111}. (Courtesy of Cathie Rae.)

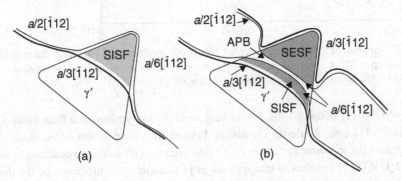

Fig. 3.59. Illustration of the mechanism of shearing of the γ' by an $a\langle 11\overline{2}\rangle$ ribbon during primary creep. An $a/3[\overline{1}12]$ dislocation enters the γ', leaving a superlattice intrinsic stacking fault (SISF) behind it, (a), leaving an $a/6[\overline{1}12]$ partial at the γ/γ' interface. In (b), the original $a/6[\overline{1}12]$ and a further $a/6[\overline{1}12]$ have entered the γ', leaving a $a/3[\overline{1}12]$ at the interface.

boundary (APB) then separates the two $a/6[\overline{1}12]$ dislocations, and a superlattice extrinsic stacking fault (SESF) the second $a/6[\overline{1}12]$ and the final $a/3[\overline{1}12]$. Once formed, the ribbon is able to glide through the γ' precipitates in a glissile fashion. Ribbons intersecting a γ' precipitate cause it to be populated with stacking faults. The cessation of primary creep occurs when the $a\langle 11\overline{2}\rangle$ dislocation ribbon is unable to glide conservatively through the γ/γ' structure, primarily due to the adsorption of $a/2\langle 1\overline{1}0\rangle\{111\}$ dislocation segments at the γ/γ' interfaces.

A consequence of the mechanism by which primary creep occurs in single-crystal superalloys is a considerable creep anisotropy. By this one means that the creep life is affected very strongly by small changes in the orientation of the tensile axis with respect to the $\langle 001\rangle$ crystallographic axis. This was noted by MacKay and Maier [56], who plotted the variation of the creep life of Mar-M247 single-crystals deformed at 760 °C on a $\langle 001\rangle/\langle 011\rangle/\langle 111\rangle$ standard triangle of the stereographic projection; see Figure 3.60. Misorientations of only 10° from $\langle 001\rangle$ were found to have a profound effect on creep life, with orientations near

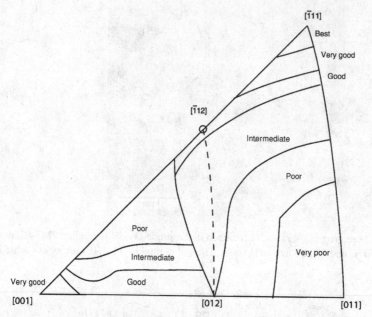

Fig. 3.60. Variation of the performance of the Mar-M247 single-crystal superalloy deformed in primary creep at 760 °C. Orientations near ⟨011⟩ have the shortest rupture lives, orientations near ⟨111⟩ the best, with ⟨001⟩ somewhere between. After ref. [56].

to the ⟨001⟩/⟨011⟩ boundary displaying very much better performance than those near to the ⟨001⟩/⟨111⟩ one. While the details are beyond the scope of this book, these effects arise because the movement of the $a\langle11\bar{2}\rangle$ ribbon requires (i) sufficient densities of at least two $a/2\langle1\bar{1}0\rangle\{111\}$ families of dislocations in γ, to enable the nucleation of the dislocation ribbons in accordance with the reaction given above, and (ii) sufficient resolved shear stress for the propagation of the $a\langle11\bar{2}\rangle$ ribbon. Thus conditions for both the nucleation and the propagation of the ribbon must be satisfied if appreciable primary creep is to occur [57]. Given this remarkably complex mechanism of primary creep deformation in single-crystal superalloys, it is unsurprising that its extent is remarkably sensitive to the chemical composition, with cobalt and rhenium in particular promoting primary creep and tantalum suppressing it [55,57,58]; it seems plausible that stacking fault shear of γ' is promoted by HCP elements such as Co and Re since they lower the stacking fault energy in γ. This suggestion is consistent with observations that primary creep in Co-containing single-crystal superalloys is highly sensitive to orientation, with creep anisotropy being suppressed as the Co content is reduced [58].

C   Rafting regime

In the rafting regime, the γ/γ' microstructure degrades rather quickly because thermally activated processes are favoured strongly since the temperature is high. Two effects occur at a very early stage of deformation: (i) equilibrium interfacial dislocation networks form at the γ/γ' interfaces [59], and (ii) the γ' particles coalesce by a process of directional coarsening

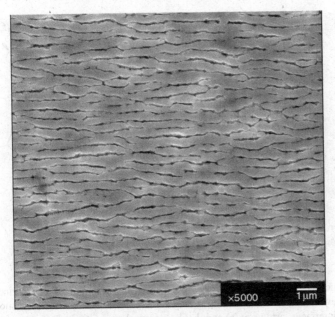

Fig. 3.61. Rafted γ/γ' structure in the CMSX-4 superalloy, deformed at 1150 °C and 100 MPa to 0.39% strain in 10 h [62].

known as the 'rafting' effect [49,60–62]. Figure 3.61 illustrates the rafted morphology in the CMSX-4 superalloy after tensile creep loading to 0.39% strain in 10 h at 1150 °C and 100 MPa [62]. The axis of loading lies in the plane of the micrograph in the vertical direction; thus the normals to the broad faces of the rafts – which are plate-shaped in form – lie perpendicular to the axis of loading. This is always found to be the case for negatively misfitting alloys such as CMSX-4, into which category most commercially developed alloys fall. When the misfit is positive, needle-shaped rafts are found whose long axes lie parallel to the direction of loading. Changing the sign of the stress, i.e. compressive rather than tensile loading, has an effect on the raft morphology which is equivalent to changing the sign of the misfit.

The interfacial dislocation networks arise from $a/2\langle1\bar{1}0\rangle\{111\}$ creep dislocations originating in the γ phase; these become captured by the γ/γ' interfaces, and the networks are formed by a series of dislocation reactions. The dislocations are primarily of edge character with the extra half-planes in the γ' phase if $\delta < 0$. Thus by this mechanism the perfect elastic coherency of the interfaces is lost and misfit stresses are relieved. Figure 3.62 is taken from the work of Field, Pollock and co-workers [63], who have analysed the networks in detail using TEM microscopy – this specimen was interrupted after 20 h with 0.5% creep strain having been accumulated at 1093 °C. Note in particular the square configurations of four dislocations of Burgers vector of $a/2\langle1\bar{1}0\rangle$ type; they are of edge character and are thus efficient at relieving the misfit, but these segments have no shear stress resolved on them – thus they must have been formed by reactions driven by reductions in misfit and dislocation line energies. A model has been proposed by these authors which accounts for their formation; see Figure 3.63. Initially, on a γ/γ' interface which has [001] as its

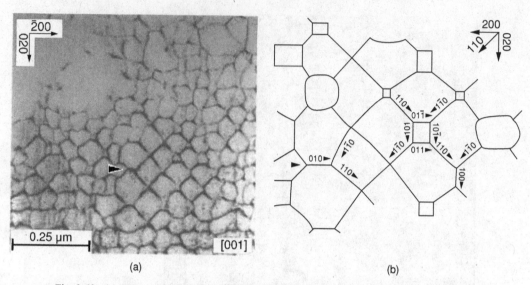

Fig. 3.62. (a) TEM micrograph and (b) accompanying schematic interpretation of an experimental single-crystal superalloy deformed to 0.5% creep strain in 20 h at 1093 °C, showing dislocation networks at the γ/γ' interfaces; arrows in (a) and (b) show equivalent positions. Foil normal {001}, equivalent to the direction of loading. After ref. [63].

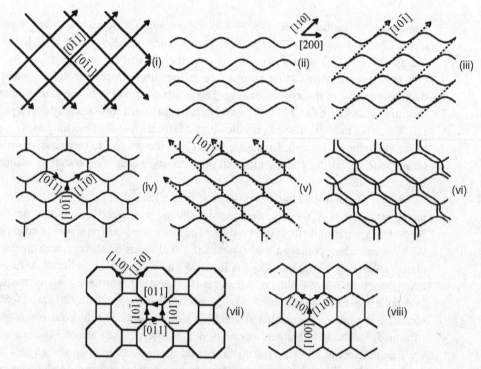

Fig. 3.63. Schematic representation of dislocation reactions leading to the formation of mismatch accommodating nets on the (001) γ/γ' interfaces during creep. After ref. [63].

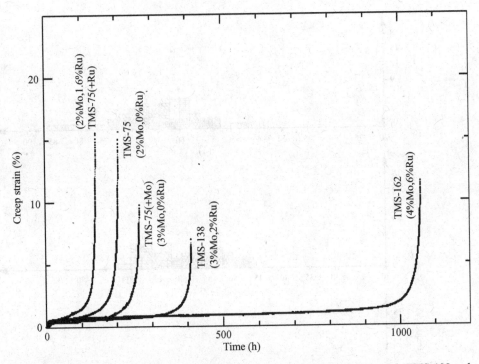

Fig. 3.64. Creep curves of the superalloys TMS-75, TMS-75(+Ru), TMS-75(+Mo), TMS-138 and TMS-162 tested at 1100 °C and 137 MPa [64,65].

normal, slip-deposited dislocations form two perpendicular sets, for example of $a/2[0\bar{1}1]$ dislocations, one resulting from slip on (111) and the other on $(\bar{1}11)$. These can react to form a single set with average line vector [100]. Further $a/2\langle1\bar{1}0\rangle$ dislocations resulting from slip on the other two octahedral planes, $(1\bar{1}1)$ and $(\bar{1}\bar{1}1)$, can then be knitted into the net, consistent with reactions of the type [63]

$$a/2[101] - a/2[0\bar{1}1] \rightarrow a/2[110] \tag{3.34a}$$

and

$$a/2[101] - a/2[1\bar{1}0] \rightarrow a/2[011] \tag{3.34b}$$

or, alternatively,

$$a/2[101] + a/2[10\bar{1}] \rightarrow a[100] \tag{3.34c}$$

Thus it is clear that the equilibrium network can arise only when sufficient creep dislocations have formed on all four of the {111}-type planes. Harada and co-workers have isolated a strong correlation between the dislocation density of the equilibrium network in single-crystal superalloys of varying compositions and their creep performance under loading at 1100 °C and 137 MPa (see Figures 3.64 and 3.65), with the dislocation spacing being found to be proportional to the logarithm of the minimum creep rate [64,65]. Use of this effect has been made in designing TMS-162, which is one of the strongest single-crystal superalloys

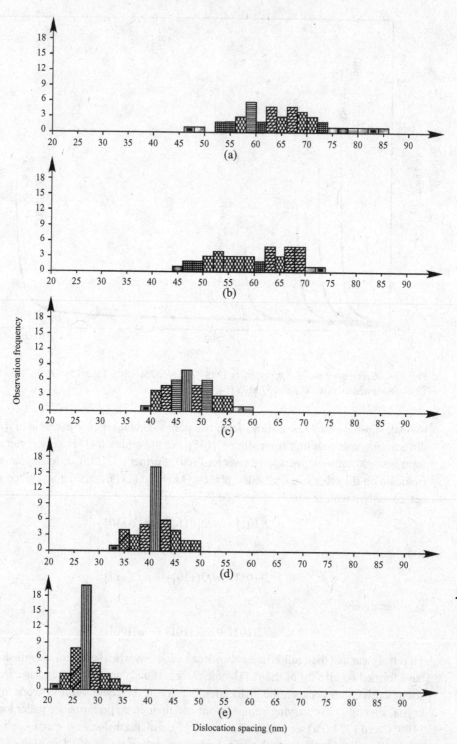

Fig. 3.65. Measurements of the equilibrium γ/γ' dislocation spacings in (a) TMS-75(+Ru), (b) TMS-75, (c) TMS-75(+Mo), (d) TMS-138 and (e) TMS-162 tested at 1100 °C and 137 MPa [64,65]; taken with the data in Figure 3.64, a strong creep strengthening effect is confirmed when the dislocation spacing is small.

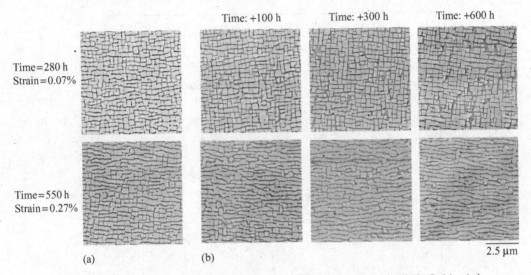

Fig. 3.66. SEM micrographs [62] of the CMSX-4 superalloy deformed at 950 °C. (a) γ/γ' microstructures after 280 h (0.07% strain) and 550 h (0.27% strain), with an applied stress of 185 MPa. (b) Microstructures after further heat treatment of 100, 300 and 600 h, with zero applied stress.

to date (see Figure 1.17). Additions of Mo are critical to the superior creep performance displayed, since the lattice misfit is reduced as the concentration of this element increases.

The relaxation of interfacial misfit by the above mechanism is important since it is now recognised that rafting occurs only once this has occurred [60–62]. Consider a negatively misfitting alloy such as CMSX-4. The micromechanism of creep deformation in the tertiary creep regime involves the multiplication of dislocations in γ, initially in the horizontal channels for reasons discussed earlier; the relaxation of interfacial misfit thus occurs preferentially on the γ/γ' interfaces whose normals lie parallel to the direction of loading. Rafting requires diffusion of γ-partitioning elements such as Cr, Re out of the vertical channels and γ'-partitioning elements such as Al, Ta into them; the evidence indicates that such mass transport can occur only once the lattice misfit has become relaxed, probably because diffusion occurs near or within the relaxed interfaces. The possibility of pipe diffusion of solutes within the dislocation networks should not be dismissed, although no direct evidence for this has been found as yet. The role of interfacial misfit relaxation in the rafting phenomenon has been confirmed with experiments which have demonstrated that partially crept single crystals continue to raft *even after the externally applied load is removed* [60,62]. Moreover, the kinetics of rafting are not slowed in any way by the removal of the stress field. Figure 3.66 shows the results of experiments on CMSX-4 [62]. For deformation at 950 °C and 185 MPa, an accumulated strain of about 0.1% is sufficient to relax the interfacial misfit and, thereafter, rafting without the externally applied load continues at the same rate as with the load present.

A controversial question concerns the influence of the rafting effect on the creep capability of the single-crystal superalloys. Note (see Figure 3.61) that as many as ten or more γ' precipitates coalesce in the transverse direction; thus a fully developed raft is formed from 100 or more precipitates of total cross-sectional area equal to several hundreds of square

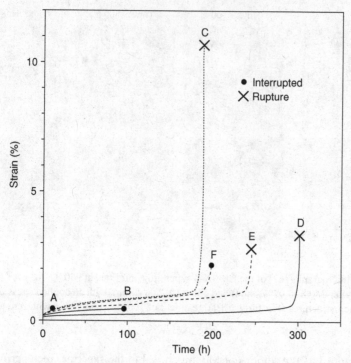

Fig. 3.67. Constant load creep data for the CMSX-4 superalloy, deformed at 1150 °C and 100 MPa.

micrometres, since the rafts have a thickness equal to one γ' particle. Extensive rafting leads eventually to the inversion of the γ/γ' structure such that the continuous, matrix phase is γ' rather than γ. Under these circumstances, rafting reduces the activity of the $a/2\langle 1\bar{1}0\rangle\{111\}$ dislocations in γ. This is supported by various pieces of evidence [66]. First, extrapolation of the tertiary creep law to temperatures where rafting occurs leads to an underestimate of the creep performance, particularly at small applied stresses when creep life is long; see Figure 3.54(d). Second, in the rafting regime, fully developed rafts are found to have formed at times equivalent to the onset of, or even before, the plateau observed in the creep curves at these very high temperatures; see Figure 3.51(c). Thus rafting is at least partially responsible for the creep-hardening effect which occurs initially and the creep plateau that occurs thereafter. Third, there is considerable scatter in the creep rupture lives at these very high temperatures (see Figure 3.67); however, the creep plateau is very reproducible. Final rupture at these temperatures is associated with extensive creep cavitation at porosity intro- duced either during the casting process into the interdendritic regions [67], or alternatively at pores forming around TCP phases which precipitate at these elevated temperatures [66]. Cavitation is, however, restricted to those pores present in the vicinity of fracture surface (see Figure 3.68), and it is probable therefore that their formation is strain-controlled. Hot isostatic pressing has, however, been found to increase the mean value of the rupture life somewhat. However, more research is needed to clarify the reasons for the onset of the rapid acceleration in creep strain rate leading to rupture, and the micromechanical deformation modes which lead to the cutting of the rafted γ' microstructure.

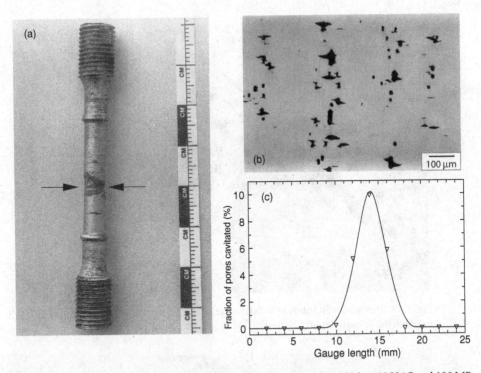

Fig. 3.68. CMSX-4 creep testpiece interrupted after 2.2% strain after 200 h at 1150 °C and 100 MPa (see the creep curve marked 'F' in Figure 3.67). (a) Necked creep specimen at point of fracture. (b) Micrograph of creep-cavitated casting porosity in necked region. (c) Variation of fraction of pores cavitated as a function of distance along specimen, providing confirmation that cavitation is restricted to the necked region.

3.3.2 Performance in fatigue

As for all metallic materials, the single-crystal superalloys are prone to fatigue failure due to oscillating loads, at stresses which are nominally elastic. This is because plastic deformation can occur in a very localised sense, typically near sites of stress concentration; this fatigue damage is most often in the form of intense, localised slip with the dislocation activity restricted to a small number of lattice planes [68]. This phenomenon is often referred to as 'cyclic slip localisation'; particularly at temperatures of around 700 °C and when stresses are high this encourages the formation of 'persistent slip bands' (PSBs) which shear the γ' particles [69]; see Figure 3.69. Fatigue initiation can always be traced to sites of stress concentration, for example casting pores or machining marks. At higher temperatures and when stresses are low, dislocation activity is restricted to the γ channels and cyclic slip localisation is less intense; thus deformation is more homogeneous, but nonetheless limited to the most highly stressed {111} planes [70]; see Figure 3.70. Under these conditions, fatigue initiation occurs at surface pits or cracks in the oxide films – and the life is improved markedly if protective coatings (for example the aluminides) are used.

In practice, the environment experienced by the blading in service is very difficult to replicate exactly in the laboratory. Consequently, reliance is placed on the characterisation of

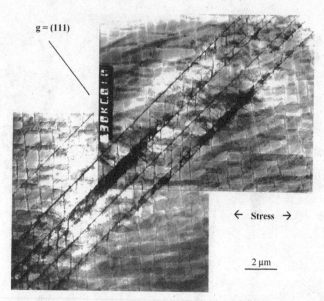

g = (111)

← Stress →

2 μm

Fig. 3.69. Persistent slip bands in a ⟨001⟩-oriented single-crystal superalloy cycled at 700 °C [69], in which the γ' particles are sheared.

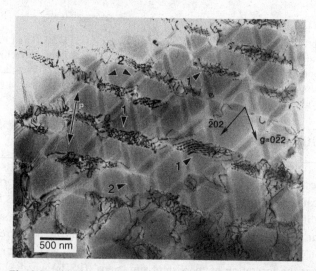

Fig. 3.70. Dislocation microstructure in the AM1 single-crystal superalloy deformed at 950 °C with $\Delta\epsilon_{total} = 1.27\%$ and with $R = 0$ [70]. P is the projection of the loading axis. Dislocations are localised at {001} interfaces, see arrows marked 1; {010} and {100} channels, see arrows marked 2, are free of dislocations.

fatigue performance under conditions which can be easily controlled, for example isothermal tests under uniaxial load or strain control. The cyclic stress range, $\Delta\sigma$, is defined as $\sigma_{max} - \sigma_{min}$, where σ_{max} and σ_{min} are the maximum and minimum stresses seen during the cycle, and the mean stress, σ_{mean}, is $\frac{1}{2}(\sigma_{max} + \sigma_{min})$. The R-ratio is defined as $\sigma_{min}/\sigma_{max}$, so that $R = -1$ corresponds to fully reversed loading with $\sigma_{max} = -\sigma_{min}$ and $R = 0$ corresponds

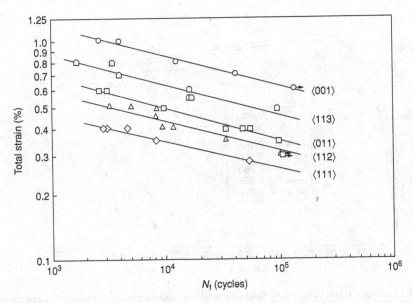

Fig. 3.71. The orientation dependence of the strain-controlled LCF life of an experimental single-crystal superalloy at 980 °C, for various values of the total strain range, $\Delta\epsilon_{total} = 0.6$ [71]. Axial load: $K_T = 1.0$; frequency $= 0.33$ Hz.

to pure cyclic conditions with $\sigma_{min} = 0$. Two distinct forms of fatigue failure are identified. Low-cycle fatigue (LCF) arises when the stress amplitude places the nominal stresses at or beyond the elastic limit – failure occurs typically within 10^5 cycles, with a majority of the fatigue life being spent in the propagation (rather than initiation) stage. From a practical perspective, the driving force for LCF damage originates from abrupt changes in loading due to engine start-up and stopping. Conversely, high-cycle fatigue (HCF) arises from smaller stress amplitudes, typical of those imposed on the blading by vibrations or resonance. In this case, the nominal stresses are always in the elastic regime, and the majority of the fatigue life is spent in the initiation stage; failure occurs in excess of 10^5 cycles.

A Low-cycle fatigue

Under LCF testing conditions, which are often carried out under total strain control, it is found that the fatigue performance depends very strongly on the orientation of the loading axis with respect to the crystallographic orientation, i.e. considerable anisotropy is displayed. Figure 3.71 shows data due to Dalal, Thomas and Dardi [71], who measured the LCF response of smooth, polished specimens of an experimental alloy (composition in wt%: Ni–6.8Al–13.8Mo–6W) at 980 °C, under total strain control and with $R = 0$. The logarithm of the total strain range, $\Delta\epsilon_{total}$, is found to be inversely proportional to the logarithm of the fatigue life, N_f, for any given orientation. Of the three orientations at the corners of the standard stereographic triangle, i.e. $\langle 001 \rangle$, $\langle 011 \rangle$ and $\langle \bar{1}11 \rangle$, the $\langle 001 \rangle$ orientation displayed the best performance and $\langle 111 \rangle$ the worst; all other orientations were between these two

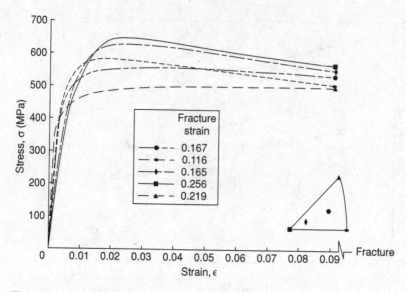

Fig. 3.72. Stress–strain curves for PWA 1480 single-crystal testpieces, for various orientations taken from the standard stereographic triangle [72].

limits. Thus the fatigue strength correlates strongly with the elastic modulus in the loading direction – so that a direction such as ⟨001⟩, which is elastically soft, displays very much better properties than one which is stiff, such as ⟨111⟩. In fact, it can be shown that if one plots the product $\Delta\epsilon_{total}E_{hkl}$ rather than $\Delta\epsilon_{total}$ against N_f, then the fatigue data collapse onto a straight line. The term E_{hkl} is the plane-specific elastic modulus along the loading axis.

These observations can be rationalised in the following way. Assuming for one moment that the yield stresses in the ⟨001⟩ and ⟨111⟩ directions are equal, the elastic strain at yield is smaller for a specimen tested along ⟨111⟩ on account of its greater stiffness; thus, under total strain control, the plastic strain available to drive the fatigue process is correspondingly greater, so that poorer properties are displayed. In fact, the yield stress shows a fair degree of anisotropy; see, for example, the data of ref. [72] for the PWA 1480 single-crystal superalloy (Figure 3.72). Such experiments indicate that the yield stress along ⟨001⟩ is greater than that along ⟨111⟩; the increment $\Delta\sigma_y$ can be in excess of 200 MPa; under total strain control, the plastic strain along ⟨001⟩ is reduced by a factor $\Delta\sigma_y/E_{001}$ due to this effect, giving further justification to the superior properties along ⟨001⟩. Under load control, the situation is more complicated. The low modulus along ⟨001⟩ might at first be expected to result in large strains and poor fatigue life, but this is not observed in practice. Clearly, under load-controlled conditions one must be relying on the larger flow stress along that direction, or, alternatively, the life is limited by stress concentrations which behave in a strain-controlled manner.

Under LCF conditions, the cyclic stress–strain hysteresis loops show a number of interesting characteristics. First, the width of the loops is larger when high inelastic strains are imposed, which occurs for the stiffest orientations. See, for example, Figure 3.73 [71], for the case of $R = 0$. Second, a considerable degree of tension–compression yield stress

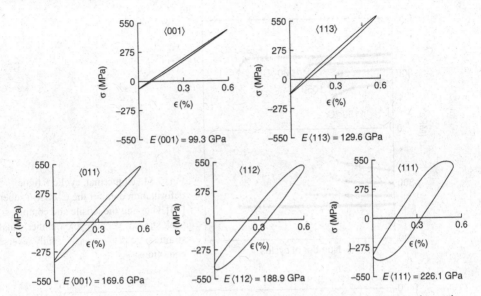

Fig. 3.73. Effect of crystallographic orientation on fatigue hysteresis loops for an experimental single-crystal superalloy at 980 °C and with $\Delta\epsilon_{total} = 0.6$, $R = 0$ [71].

asymmetry is observed for $\langle 001 \rangle$ orientations, with the yield stress in tension exceeding that in compression; Gabb et al. [73] report for Rene N4 values of 956 MPa and 818 MPa for tension and compression, respectively, at 760 °C. Interestingly, along $\langle 011 \rangle$ the yield stress is greater in compression than in tension – the values reported are 748 MPa and 905 MPa for tension and compression, respectively; along $\langle 111 \rangle$ little anisotropy was observed – 817 MPa and 842 MPa, respectively. Finally, the cyclic stress–strain curves show a considerable strain-softening response [74] – the values of σ_{max} and σ_{min} decline during the test consistent with a creep contribution inducing damage into the material. The experiments by Mughrabi et al. on the CMSX-6 superalloy (see Figure 3.74) indicate that the strain softening is most pronounced at higher temperatures, when the γ/γ' structure evolves via the rafting effect. These same workers have demonstrated that prestraining in tension, which induces rafts aligned normal to the stress axis, weakens the material in fatigue; conversely, prestraining in compression strengthens it.

As expected, the LCF life decreases quickly with increasing temperature. Figure 3.75 shows data [75] for the CMSX-4 single-crystal superalloy, measured under load control, with $R = 0$. One can see that no distinct endurance limit is displayed. Extensive shear banding and PSBs are found in the vicinity of the fracture surface; see Figure 3.76.

B   High-cycle fatigue

The natural resonant vibratory frequencies of single-crystal turbine blading range from a few to several thousand hertz. Consequently, if the excitation frequencies due to turbulent flow around the aerofoil are in this range, then high-cycle fatigue (HCF) failure will occur. HCF loads are superimposed on the steady-state mean stresses arising from centrifugal

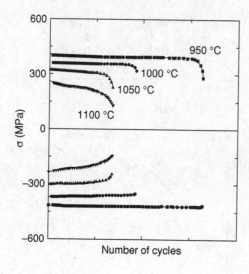

Fig. 3.74. Isothermal, cyclic fatigue deformation data for the CMSX-6 superalloy [74], showing the tensile and compressive peak stresses versus cycle number. Total strain range $\Delta\epsilon_{\text{total}} = 10^{-2}$; fully reversed conditions.

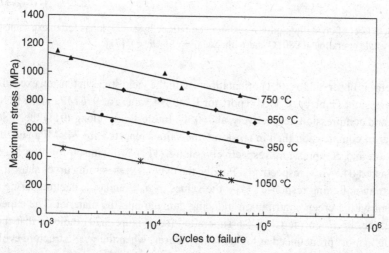

Fig. 3.75. Load control, LCF fatigue data [75] for the uncoated, second-generation superalloy CMSX-4 at various temperatures, with $R = 0$.

loading of the blade and thermal loads from thermal gradients set up by the action of the hot gas and the cooling air passing through the blading. Clearly, one then expects the HCF performance to depend on the R-ratio, and a considerable interaction of fatigue and creep damage mechanisms might be anticipated when the temperature is high and as $R \to 1$.

Recently, Wright and co-workers [76] have completed elegant and careful work which sheds light on the effects occurring during HCF loading. Load-controlled testing was carried out at $1038\,^{\circ}\text{C}$ on $\langle 001 \rangle$-oriented single-crystal specimens of PWA 1484 protected with a platinum aluminide coating. A first set of tests were carried out at 59 Hz for R-ratios of -1.0, -0.33, 0.1, 0.5, 0.8 and 1.0; obviously, when $R = 1.0$ there is no alternating component so that, in that limiting case, one is executing a constant load creep test. A second set of tests

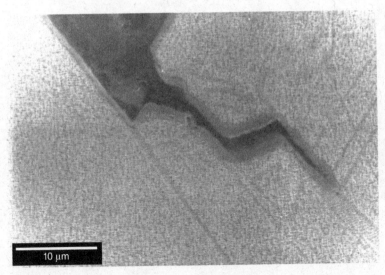

Fig. 3.76. Intense shear banding and persistent slip bands near the fracture surface of a CMSX-4 testpiece, cycled under constant load control at 850 °C, $R = 0$ and with $\sigma_{max} = 760$ MPa [75].

were carried out at $R = -1.0$ and $R = 0.1$ at a range of frequencies from 59 Hz to 900 Hz. The results for the fatigue life, N_f, are plotted against the stress range, $\Delta\sigma$, on a conventional σ–N diagram (Figure 3.77(a)); for each R-ratio the Coffin–Manson relationship is obeyed. No evidence of an endurance limit was found. Moreover, the fatigue capability is seen to decrease, and the slope of the curve is steeper as $R \rightarrow 1$. In Figure 3.77(b), the data are plotted instead as σ_{mean} against time to failure. Here, the data for the higher R-ratios (0.5, 0.8 and 1.0) collapse along a single line defined by the creep rupture capability. This implies that life is independent of the alternating stress $\Delta\sigma$; thus the dominant mode of damage is creep. The data for lower R-ratios fall at lower times, implying that a fatigue mode is operative in that regime. The data at varying frequencies allow the fatigue and creep components to be further distinguished. When $R = -1$ the results were found to be independent of frequency for the range examined (see Figure 3.78(a)); this implies that a time-independent fatigue process is occurring at low mean stresses. At higher mean stresses, corresponding to $R = 0.1$, life is larger at the higher frequencies for any given $\Delta\sigma$; see Figure 3.78(b). This is because the failure depends more heavily upon the creep component which is time- rather than cycle-dependent. The characteristics of the fracture surfaces were consistent with this interpretation. At low mean stresses, corresponding to $R = -1$, failure was initiated either internally at pores, carbides or eutectic γ', and crystallographic facets could be identified consistent with a fatigue mode of damage accumulation. At higher R-ratios, the appearance of the fracture surface resembled that of a creep rupture failure; it was rough with multiple surface cracking and a degree of necking was evident.

   Wright and co-workers [76] displayed their results on the so-called Goodman diagram, on which the stress range, $\Delta\sigma$, is plotted against the mean stress, σ_{mean}, for a constant HCF capability. Figure 3.79 shows a typical diagram for 10^7 cycles to failure; the numbers correspond to the percentage creep damage determined from a creep–fatigue damage

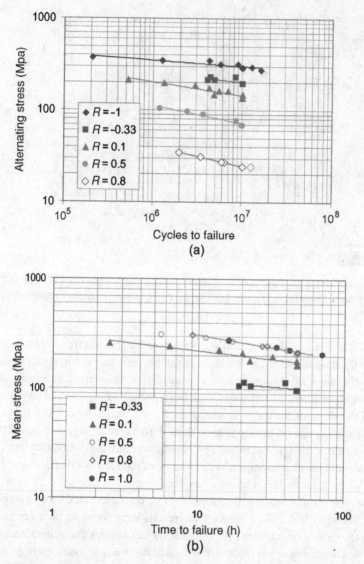

Fig. 3.77. HCF data for coated PW1484 at 1038 °C at 59 Hz, showing the effect of *R*-ratio: (a) conventional $\sigma-N$ curve, and (b) plot of mean stress versus time to failure [76].

mechanics model to which the data were fitted. For $R > 0$ one can see that creep mode becomes significant.

3.4   Turbine blading: design of its size and shape

The previous sections in this chapter have been concerned with the processing of turbine blade aerofoils by investment casting, with the metallurgical principles controlling the optimum superalloy compositions and with the mechanical behaviour observed during

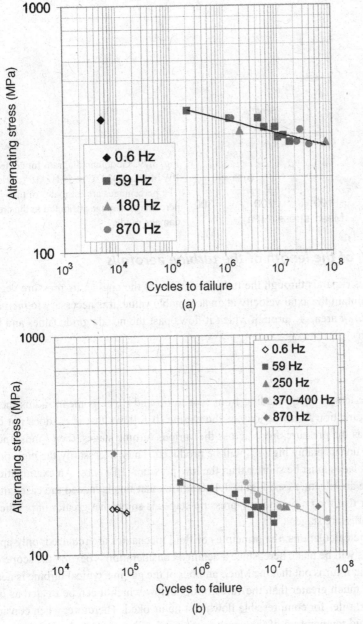

Fig. 3.78. HCF data for coated PW1484 at 1038 °C showing the effect of frequency: (a) $R = -1$ and (b) $R = 0.1$ [76].

service. Until now, the factors influencing the shape and geometry of the aerofoil have been largely ignored; these are dealt with in this section. It turns out that the relevant considerations depend largely on the laws of aerodynamics, fluid mechanics and thermodynamics [77,78].

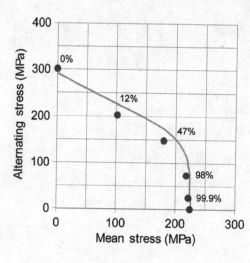

Fig. 3.79. Goodman diagram for coated PW1484 at 1038 °C at 59 Hz, 10^7 cycles [76]. The percentages are estimates of the percentage damage attributed to the creep damage mode.

3.4.1   Estimation of the length of the turbine aerofoils

As the hot gas expands through the turbine sections of the engine, its pressure decreases. Hence, to maintain the axial velocity at an acceptable value, it is necessary to *increase* the annulus or *throat* area, A, through which it flows past the nozzle guide vanes and blades; this is given by

$$A = 2\pi r_{\mathrm{m}} l \tag{3.35}$$

where l is the length of the blading from root to tip and r_{m} is its mean radius measured in the radial direction from the axis of the engine. It follows that if r_{m} does not change substantially as the pressure drops across the various turbine stages, then l must increase; otherwise, an unacceptably high v_{jet} will be produced. Thus, necessarily the high-pressure (HP) turbine blades must be shorter than the low-pressure (LP) ones. An examination of a typical engine architecture (see Figure 1.1) confirms that this is indeed the case, although it is also clear that the r_{m} for the low-pressure stages is somewhat greater in practice than for the high-pressure stages.

For accurate calculations, the principles of fluid mechanics are required; only approximate estimates will be made here, since a detailed consideration is beyond the scope of this book. However, it turns out that the Mach number of the gas in a typical turbine is near 1.0, which is very much greater than the value of 0.3 below which it can be treated as incompressible; thus, rules for compressible flow must be invoked. Therefore, when considering the pressure and temperature of the gas stream, it is important to distinguish between the so-called static quantities, p and T, and the stagnation values, p_0 and T_0 (those which would be attained if the gas were brought to rest without work and heat transfer), via [77]

$$T_0/T = 1 + \left(\frac{\gamma - 1}{2}\right) M^2 \tag{3.36}$$

$$p_0/p = \left[1 + \left(\frac{\gamma - 1}{2}\right) M^2\right]^{\gamma/(\gamma-1)} \tag{3.37}$$

where M is the local Mach number, the ratio of the velocity, V, to the speed of sound given by $\sqrt{\gamma RT}$. As in Chapter 1, γ is the ratio of the specific heats at constant pressure and constant temperature, R is the gas constant and T is absolute temperature.

In order to determine the rate of mass flow, $\dot{m}$, through a section of area A, one must consider the combination

$$\dot{m}/A = \rho V \tag{3.38}$$

where ρ is the density of the gas stream and V is its velocity. Conservation of energy requires

$$\dot{m}c_p T + \dot{m}V^2/2 = \dot{m}c_p T_0 \tag{3.39}$$

where, upon rearranging, one has $V = \sqrt{2c_p(T_0 - T)}$. Assuming isentropic flow such that $\rho = CT^{1/(\gamma-1)}$, where C is a constant given by $C = \rho_0^{\gamma-1}/T_0^{\gamma-1}$, one has

$$\dot{m}/A = CT^{1/(1-\gamma)}\sqrt{2c_p(T_0 - T)} \tag{3.40}$$

After some algebraic manipulation, one can show that the dimensionless mass flow rate per unit area, Q, defined by the quantity $\dot{m}\sqrt{c_p T_0}/Ap_0$, is given by [77]

$$Q = \frac{\dot{m}\sqrt{c_p T_0}}{Ap_0} = \frac{\gamma}{\sqrt{\gamma-1}}M\left[1 + \left(\frac{\gamma-1}{2}\right)M^2\right]^{-(\gamma+1)/(2(\gamma-1))} \tag{3.41}$$

One can see from the right-hand side that Q is constant for any given γ and M. Thus, this equation describes how the throat area, A, must change as p_0 and T_0 alter. The variation of Q with Mach number, M, is given in Figure 3.80, for the case $\gamma = 1.4$. It is found that the flow rate is a maximum at a Mach number of 1, when $Q = 1.281$; at this point, the cross-section, or nozzle, is said to be 'choked'. For the turbine sections in a jet engine, this is usually a good assumption. It can be shown that, close to sonic velocity, very small changes in A produce large variations in the Mach number and pressure. Note that it has been assumed that the flow is one-dimensional and adiabatic. Provided that these are reasonable assumptions, Figure 3.80 can be used to determine A if estimates of the other quantities in the expression for Q are known. To a first approximation, one can take estimates of the pressure and temperature of the gas stream from the ideal thermodynamic cycle described in Chapter 1; this turns out to be not a bad assumption.

These considerations demonstrate that as p_0 decreases, A must increase – so the blades must get longer. The effect of the corresponding temperature drop is found to be rather less important in influencing A. This explains why the low-pressure blades of a gas turbine engine are very much longer than the high-pressure ones.

Example calculation

A high-pressure (HP) turbine stage passes a mass flow of 200 kg/s. If the stagnation pressure and temperature of the gas stream are taken as 1800 kPa and 1450 K, respectively, choose an appropriate annulus area, A. Assume that the flow is occurring close to Mach 1, so that the cross-section can be assumed to be choked. Hence, if the mean radius of the stage is to be 0.5 m, determine the length of the turbine blading. Take $c_p = 1000$ J/(kg K).

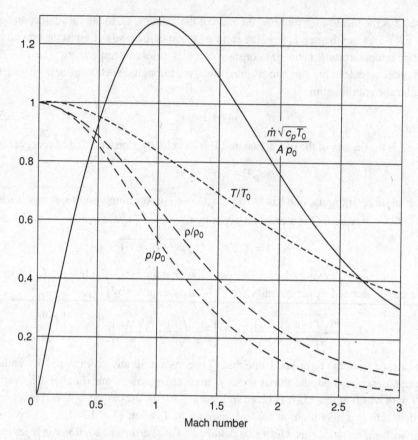

Fig. 3.80. Variation of the non-dimensional flow rate, $\dot{m}\sqrt{c_p T_0}/Ap_0$, with Mach number, M, for a perfect gas with $\gamma = 1.4$, flowing through a uniform orifice. Also shown are the ratios of static to stagnation values of pressure, density and temperature. Adapted from ref. [77].

Answer   At choking conditions, it follows that

$$Q = \frac{\dot{m}\sqrt{c_p T_0}}{Ap_0} = 1.281$$

so that

$$A = \frac{200\sqrt{1000 \times 1450}}{1.281 \times 1800 \times 10^3} = 0.1044\,\mathrm{m^2} = 2\pi r_m l$$

If $r_m = 0.5\,\mathrm{m}$, it follows that $l = 33\,\mathrm{mm}$.

3.4.2   Choice of mean radius for turbine blading

The function of the turbine blading is to extract mechanical work from the gas stream, and to do this as efficiently as possible. To a good approximation, the gas flow after exit from the nozzle guide vanes can be assumed to be axial; but, after interaction with the turbine blading, there is a significant tangential component.

Suppose that the flow enters at radius r (which remains unchanged) with tangential velocity $V_{\theta 1}$ and that it leaves with a value equal to $V_{\theta 2}$. The torque, T, required for this to occur is equal to the rate of change of angular momentum, and the power extracted, $\dot{W}$, is the torque times the angular velocity, Ω; hence,

$$\dot{W} = T\Omega = \dot{m}\Omega r_m(V_{\theta 2} - V_{\theta 1}) = \dot{m}U_m(V_{\theta 2} - V_{\theta 1}) \tag{3.42}$$

where U_m is the speed at the mean blade radius and the product $U_m(V_{\theta 2} - V_{\theta 1})$ is known as the stagnation enthalpy; it is given the symbol Δh_0.

Equation (3.42) is one form of the Euler equation. It can be written in non-dimensional form as

$$\frac{\Delta h_0}{U_m^2} = \frac{V_{\theta 2}}{U_m} - \frac{V_{\theta 1}}{U_m} = \frac{c_p \Delta T}{U_m^2} \tag{3.43}$$

where ΔT is the temperature drop across the stage. Extensive experimentation by aerodynamicists on turbine systems has shown that the non-dimensional work coefficient, $c_p \Delta T / U_m^2$, must not exceed ~ 2, for satisfactory performance of the aerofoils.

Example calculation

A row of turbine blades is spinning at 6000 rev/min, and a temperature of 250 K is dropped across the stage. Estimate the mean radius, r_m, at which the row should be placed, for optimum efficiency. Take $c_p = 1000$ J/(kg K) as before.

Answer From the data given, $\Omega = 100$ rev/s and the core turbine work, $\Delta h_0 \sim c_p \Delta T$, is 250 kJ/kg. Now for an efficient turbine

$$\frac{c_p \Delta T}{U_m^2} \sim 2$$

hence $U_m = 350$ m/s and therefore $r_m = U_m/(2\pi\Omega)$ is approximately 0.56 m. It follows that if ω drops to half of its value, to 50 revs/s, r_m must be doubled to 1.1 m. Conversely, if the temperature drop is reduced by a factor of 2 to 125 K, then r_m is reduced by a factor of $\sqrt{2}$ to 0.40 m.

This result has some severe consequences for the design of the low-pressure (LP) turbine for a civil turbofan engine such as the Rolls-Royce Trent 800 or General Electric GE90. Since there is a need to restrict the tip speed of the fan blade to values not greatly exceeding the speed of sound, the LP shaft speed is relatively low – about 3000 rev/min – and therefore approximately one-third of the speed of the HP spool. Thus, given the considerations so far, r_m should ideally be made very large to return U_m and the work coefficient, $c_p \Delta T / U_m^2$, to acceptable values for optimum aerodynamic performance; however, calculations will confirm that this becomes difficult as the required r_m is then large enough to cause problems with the flow of the bypass air past the engine core.

In practice, this problem is solved by building a multistage LP turbine. Across each stage, a relatively small ΔT is dropped, thus restricting the work extracted. The penalty incurred, unfortunately, is greater weight – since the amount of turbomachinery required is then significantly increased – and greater manufacturing and maintenance costs. All large civil turbofans suffer from this difficulty. For example, the three-spool Trent 800 has single-stage HP and intermediate pressure (IP) turbines, but a five-stage LP turbine. The two-spool GE90 is built with a two-stage HP turbine and a six-stage LP turbine. For Pratt & Whitney's two-spool PW4084, two HP stages are used, but a seven-stage LP turbine is required.

3.4.3 *Estimation of exit angle from blade cross-section*

The turbine consists of alternating rows of stationary guide vanes (stators) and blades (rotors); see Figure 3.81. The blades move in the tangential direction at a speed U_m, which, to a good approximation, is not very different from the local speed of sound. The function of the guide vanes is to remove the tangential component of the gas flow; the blades introduce this as a consequence of extracting work from the gas stream. Since it is not effective to allow the high-pressure gas exiting from the combustion chamber to undergo a single expansion across a single stage, the alternating rows of stationary vanes and moving blades facilitate a series of smaller expansions, which can be used efficiently [79].

In practice, the flow around the turbine blading is unsteady, and a full mathematical treatment is very complicated. However, it is helpful to consider the *relative* frame of reference local to the moving blades and the *absolute* reference frame pertinent to the stationary guide vanes. In the relative frame of reference, it has been found that the flow is steady enough to make calculations of sufficient accuracy for practical purposes. The situation is then as depicted in Figure 3.82. The gas is assumed to have an axial component of velocity given by V_x. The gas approaches the first row of stators with absolute velocity V_1^{abs} at an angle α_1^{abs} to the axial; its tangential component is thus $V_x \tan \alpha_1^{abs}$. After exit from the guide vane, the absolute velocity, V_2^{abs}, has increased, as has the tangential component, which is now given by $V_x \tan \alpha_2^{abs}$. Note that the absolute velocity of the gas exiting the guide vane is greater in the absolute frame than in the relative one; the converse is true for the gas exiting the blading – the angle α_3^{rel} is somewhat bigger than α_3^{abs}.

These considerations of so-called *velocity triangles* allow a first estimate of the shape of the trailing edge of the turbine blading, as the following example will demonstrate.

Example calculation

A first-row HP turbine stage is operating in the gas stream of a gas turbine engine. Estimate the flow direction which the gas should take out of the blading if the mean velocity, U_m, is set at twice the axial velocity of the gas stream.

Answer   The stagnation enthalpy, Δh_0, is given by

$$\Delta h_0 = U_m V_x \left(\tan\{\alpha_3^{rel}\} - \tan\{\alpha_2^{rel}\} \right) = U_m V_x \left(\tan\{\alpha_2^{abs}\} - \tan\{\alpha_3^{abs}\} \right)$$

Assuming that the velocity of the gas relative to the blading is perfectly axial on exit from the nozzle guide vane, $\tan\{\alpha_2^{rel}\} = 0$. With $\Delta h_0 / U_m^2 = 2$ and $U_m = 2V_x$ as given, then $\tan\{\alpha_3^{rel}\} = 4$ so that $\alpha_3^{rel} = 76°$.

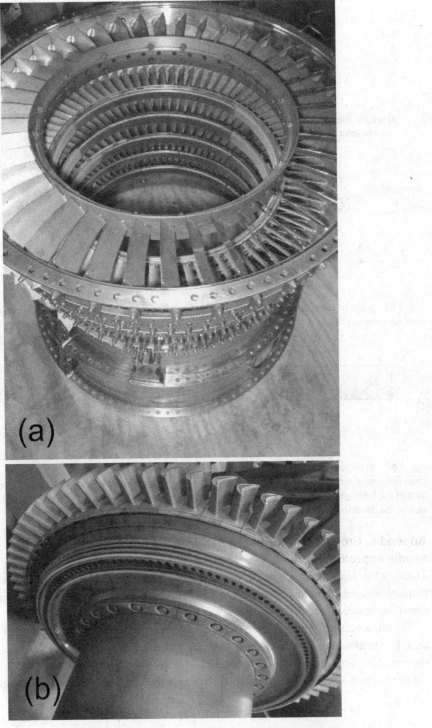

Fig. 3.81. Photographs of turbine components for the CFM-56 engine: (a) guide vanes (stators) in low-pressure turbine section, and (b) spool of high-pressure turbine blading.

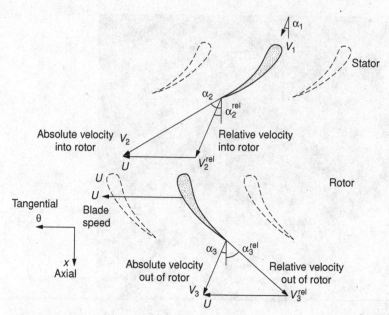

Fig. 3.82. Stage of an axial flow turbine, showing nozzle guide vanes (stators) and blading (rotors) in cross-section [77].

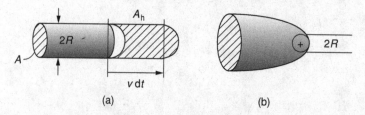

Fig. 3.83. Approximation of the dendrite tip as a cylinder with a hemispherical tip. (a) The cross-section of the cylinder, $A = \pi R^2$, determines the volume which grows in time dt. The surface area of the hemispherical cap, $A_h = 2\pi R^2$, determines the amount of radial solute diffusion. (b) The size of the dendrite in relation to the cylinder. After ref. [22].

Appendix. Growth of an isolated dendrite, using hemispherical needle approximation

The tip of a dendrite moving at a velocity v can be approximated as a cylinder with a hemispherical cap of radius R; see Figure 3.83. This allows some important factors affecting growth to be accounted for.

The cross-section of the cylinder, $A = \pi R^2$, determines the volume which grows in time dt; this is responsible for the rejection of solute into the melt. Likewise, the surface area of the cap, $A_h = 2\pi R^2$, determines the amount of radial solute diffusion. The flux, J_1, due to solute rejection and the flux, J_2, due to diffusion in the liquid ahead of the tip are given by

$$J_1 = Av(c_l^* - c_s^*) \qquad J_2 = -DA_h\frac{dc}{dr}\bigg|_{r=R} \tag{A1}$$

and, under steady-state conditions, these two fluxes must be equal.

The concentration gradient at the tip can be approximated by the value appropriate for a growing sphere, evolving under steady-state conditions:

$$\frac{dc}{dr}\bigg|_{r=R} = -\frac{c_1^* - c_0}{R} \tag{A2}$$

This can be proved by finding the solution to Laplace's equation in spherical coordinates:

$$\frac{d}{dr}\left(r^2 \frac{dc}{dr}\right) = 0 \tag{A3}$$

by satisfying the boundary conditions and by evaluating the first derivative at $r = R$. Combining the various expressions, one has

$$\frac{vR}{2D} = \frac{c_1^* - c_0}{c_1^* - c_s^*} \tag{A4}$$

where the term on the right-hand side is known as the supersaturation, Ω. At constant Ω, the result predicts that the velocity, v, is inversely proportional to the dendrite tip radius, R.

Note that the expression for the flux at the dendrite tip is only approximate, since Laplace's equation was used. A rigorous analysis yields the Ivantsov solution, see ref. [22], but it can be shown that the approximation given is accurate in the limit $\Omega \sim 0$.

Questions

3.1 The materials used for high-pressure turbine blade aerofoils are often referred to as *single-crystal superalloys*. Explain why the use of the term 'single-crystal' is disingenuous.

3.2 How might the (a) axial and (b) transverse temperature gradients compare during the investment casting of a typical high-pressure blade for (i) an aeroengine and (ii) an industrial gas turbine?

3.3 During investment casting of hollow turbine blading, platinum pins are sometimes used to keep the internal ceramic and shell from moving relative to each other. These are expensive: why does one bother with them?

3.4 During solidification of a single-crystal alloy, the variation of fraction solid with temperature is an important consideration. Sketch the variation which might be expected assuming (i) equilibrium freezing conditions and (ii) Scheil freezing with no back-diffusion.

In order to assess whether back-diffusion is important, it has been suggested that one consider the parameter

$$\alpha = \frac{D_s t_f}{L^2}$$

where D_s is the diffusion coefficient in the solidifying phase, L is a characteristic distance equivalent to the dendrite arm spacing and t_f is the solidification time $(G/(v\Delta T))$, where ΔT is the solidification range. Using appropriate values of the G, v and ΔT for directional solidification, estimate the solidification time, t_f.

In order to determine whether back-diffusion is important, one should consider the Clyne–Kurz parameter, α', where

$$\alpha' = \alpha \left(1 - \exp\left(-\frac{1}{\alpha}\right)\right) - 0.5 \exp\left(-\frac{1}{2\alpha}\right)$$

Note that $\alpha' = 0$ for Scheil solidification and $\alpha' = 0.5$ for equilibrium solidification given by the lever rule. Using this expression, and making estimates for the various parameters, determine whether back-diffusion is likely to be an important consideration for these alloys.

3.5   Investment cast components often exhibit dramatic changes in cross-section. How would one expect the spacing of isotherms to change when the cross-section of the casting increases in this way? Rationalise the observation that freckle chains are often found near re-entrant features, such as shrouds and platforms.

3.6   The preferred growth direction of a single-crystal superalloy is $\langle 100 \rangle$. Why?

3.7   Consider a typical Bridgeman arrangement for the casting of directionally solidified components. What would be the effect of increasing the withdrawal velocity on the temperature gradient at the liquidus? What is the disadvantage of a more rapid withdrawal rate?

3.8   The following defects can occur during the casting of single-crystal components: (i) high-angle grain boundaries, (ii) freckles and (iii) spurious grains. What is meant by these terms? Give a brief explanation of the origin of each effect.

3.9   Various empirical relationships have been derived to describe the susceptibility of a single-crystal superalloy to freckling. Expressions which have been proposed for a freckling index, F, include

$$\frac{W + Re}{Ta} \quad \text{or} \quad \frac{W + 1.2Re}{Ta + 1.5Hf + 0.5Mo - 0.5Ti}$$

where the alloy compositions are input in wt%. Is there any physical basis underlying their use? On the basis of the known mechanism of freckling, is the value of F required to be small or large if defect-free castings are to be produced?

Using typical compositions for the superalloys (wt%), rank the susceptibility of the first-, second- and third-generation alloys to freckle formation. Are your findings consistent with the observation that the later generations are more freckle prone?

Many alloy developers are considering additions of Ru to single-crystal superalloys. How should you modify the expression for F to account for the presence of this element?

3.10   Find a good estimate of the thermal expansion coefficient of a typical single-crystal superalloy used for a turbine aerofoil. Use this to estimate the shrinkage strain expected when a casting cools to ambient temperature. What are the implications of your findings, particularly for the design of the ceramic mould?

3.11   In recent years, attempts have been made to improve the thermal gradients typical of an investment casting furnace. Developments such as liquid metal cooling (LMC) and gas cooled casting (GCC) have been made. It has proven possible to reduce the primary dendrite arm spacing by about a factor of 2, from about 500 μm to 250 μm. How much shorter might then one expect the heat treatment time to be?

3.12 During directional solidification, heat is lost by (i) conduction to the chill through the solidified portion of the ingot, and (ii) radiation to the surroundings via the emissivity of the ceramic shell. Since these two contributions to the total heat loss, q_{total}, are additive, one has

$$q_{total} = (h_c + h_r)(T - T_0), \qquad \text{where} \qquad h_c = \frac{\kappa}{L}, \qquad h_r = \frac{\sigma(\epsilon T^4 - \alpha T^4)}{(T - T_0)}$$

where h_r and h_c are the radiative and conductive heat transfer coefficients, respectively, and L is the length of the solidified portion of the casting. The constants ϵ and α can be taken to be 0.5. The Stefan–Boltzmann constant $\sigma = 5.67 \times 10^{-8}$ W/(m^2 K^4).

Using reasonable values for κ, T and T_0, find the length of the solidified ingot at which the contributions to the heat loss from radiation and conduction are equal. Hence show that, for a casting of any significant size, the majority of heat is lost by radiation rather than by conduction.

3.13 After casting, components fabricated from single-crystal superalloys undergo a complicated heat treatment designed to remove the microsegregation inherited from the casting process. What would be the implications of not doing this?

3.14 For the single-crystal superalloy CMSX-4, the density of the liquid phase has been measured to be 7.754 g/cm^3 at the melting point; it varies at a rate -0.0009 g/cm^3 K with temperature. Using this information, compare the freckling susceptibility of the first-, second- and third-generation single-crystal superalloys Rene N4, CMSX-4 and CMSX-10. Do this by evaluating the differences in density between the liquid at the melting point and that at a (lower) temperature at which solidification is 10% completed. Use the Thermo-Calc software for making the equilibrium calculations. What assumptions have been invoked?

Hence show that there is a tendency for the later generations of single-crystal alloys to be prone to freckling.

3.15 Demonstrate that, for a binary phase diagram which possesses a liquidus slope of m, the undercooling, ΔT, experienced by an alloy of mean composition c_0 at a supersaturation Ω close to zero is given by

$$\Delta T \sim mc_0\Omega(1 - k)$$

where k is the solid/liquid partitioning coefficient. (Hint: Note that $\Delta T = (c_1^* - c_0)m$, where $c_1^* = c_0/(1 - \Omega(1 - k))$. The factor c_1^* is the solute concentration in the liquid at the solid/liquid interface.)

3.16 The alloy development programmes have contributed greatly to the improved creep strengths of modern single-crystal superalloy materials. What have been the physical principles which underpin this work?

3.17 Harada has carried out an elegant experiment in which the creep lifetimes of the single-crystal superalloys TMS-75 and TMS-82+ were measured as a function of the fraction of the γ' phase present. The alloys were cast such that the compositions of the γ and γ' phases were on a common tie-line, so that the phase compositions remain invariant. The results from this experiment are given in Figure 3.29.

What do the results from this experiment tell us about the creep deformation behaviour of superalloys? Why is the maximum creep resistance not imparted at a 50% fraction of γ'?

3.18   When designing a new single-crystal superalloy composition, what are the properties which are desirable? Give examples of trade-offs between the various properties that are required.

3.19   In Figure 3.33, the evolution of the lattice misfit, δ, in the single-crystal superalloy SC16 is plotted during temperature decrements of $100\,°C$ between $1200\,°C$ and $800\,°C$. It is found that, after the temperature is stabilised, some further time is required before δ reaches its equilibrium value. What are the physical effects occurring during the periods whilst equilibrium is being attained?

3.20   It has been reported that the rate of oxide formation in Al_2O_3-forming single-crystal superalloys is greatly increased with additions of Ti to the alloy chemistry. Explain why this effect occurs.

3.21   The Cannon-Muskegon company has made great strides in recent years in the reduction of sulphur levels in the casting stock used for single-crystal blading. Levels are now at the 5 ppm level or lower, whereas ten years ago they were at about 40 ppm. What advantages does this confer?

3.22   Hot corrosion can severely degrade a turbine blade aerofoil. Under what conditions can this occur? In practice, one can distinguish between type I and type II hot corrosion. Distinguish between these two modes of attack, noting the principle reactions which are responsible for them.

3.23   In designing the second-generation single-crystal superalloy PWA1484, Pratt & Whitney improved the oxidation resistance of the alloy by about a factor of 1.6 when compared against their first-generation alloy PWA1480. Compare the chemical compositions of these two alloys, and hence explain how this improvement was achieved.

3.24   Use Thermo-Calc to investigate the manner in which the elements in a typical nickel-based superalloy influence the size of the heat-treatment window. Develop an empirical equation such as

$$T_{window}(°C) = A + B\,wt\%Al + C\,wt\%Ta + D\,wt\%W + \cdots$$

where A, B, C, ... are parameters to be determined. Base the calculations on a typical second-generation superalloy. Once completed, correlate the sign and size of the parameters with the efficacy of the corresponding element to promote the γ' phase. Do any trends emerge?

3.25   Explain why it might not be sensible, even for single-crystal superalloys, to eliminate completely the grain-boundary strengtheners such as carbon and boron from the melt chemistry.

3.26   The gamma prime, γ', phase in nickel-based superalloys is often cuboidal in form. Why?

3.27 At temperatures of around 950 °C, creep deformation of single-crystal superalloys occurs by thermally activated slip of octahedral dislocations on the $\{111\}\langle110\rangle$ system. However, the Schmid factor for this system is not a maximum on the $\langle100\rangle$ pole; rather, the maximum is at angle of $\sim9°$ from $\langle100\rangle$ on the $\langle100\rangle/\langle111\rangle$ axis of the stereographic triangle. Quantify the percentage difference in applied stress level which can be tolerated in two single components: the first aligned exactly along $\langle100\rangle$ and the second at the point of maximum Schmid factor.

3.28 A single-crystal superalloy contains a volume fraction of 70% of the γ' phase. The misfit between the γ and γ' phases is 0.01. Estimate the inelastic (creep) strain necessary to relax the γ/γ' coherency stresses in this alloy.

3.29 Why is it fortuitous, particularly in terms of the mechanical properties of single-crystal blading, that the growth direction $\langle100\rangle$ is the least stiff from an elastic point of view?

3.30 A turbine blade cast from a polycrystalline nickel-based superalloy (density 8900 kg/m^3) is working at a temperature of 600 °C. The root and tip radii, r_r and r_t, of the blade are 0.2 and 0.25 m, respectively, measured from the centre of the rotation. The blade rotates at 10 000 rev/min. Show that the stress in an element a distance R from the centre of rotation of the blade is given by

$$\sigma = \int_R^{r_t} \rho\omega^2 r \, dr$$

Use this result to find the variation of the stress along the blade. From this derive an equation for the variation in strain rate along the blade, and hence the rate of change of the blade with time.

You may assume that the blade is made from a superalloy which creeps in steady-state according to

$$\dot{\epsilon} = A\sigma \exp\left\{-\frac{Q}{RT}\right\}$$

with $A = 4.3 \times 10^{-11}$ m^2/(N s) and $Q = 103$ kJ/mol.

3.31 MacKay and Meier have measured the orientation dependence of primary creep performance of MarM247 single-crystals for various locations in the standard triangle of the stereographic projection. The $\langle111\rangle$ orientation performs best, followed by $\langle001\rangle$; $\langle011\rangle$ displays the worst behaviour. Rationalise this observation by determining the Schmid factors for deformation on the $\langle11\bar{2}\rangle\{111\}$ and $\langle1\bar{1}0\rangle\{111\}$ slip systems.

As a further exercise, plot out the contours of the Schmid factor within the standard triangle for the two slip systems.

3.32 The creep performance of three generations of single-crystal superalloys can be compared by plotting data for the stress, temperature and rupture life on a Larson–Miller plot; see Figure 1.16.

The turbine blading of a modern jet engine experiences temperatures of 1000 °C and stresses of 175 MPa during take-off. Compare the rupture lives of turbine blades cast in SRR99, CMSX-4 and RR3000. Take $C = 20$. Thus convince yourself of the considerable advances in creep performance which have been made by the alloy designers.

At temperatures beyond 1050 °C the performance of RR3000 is only marginally superior to that of CMSX-4. What is the reason for this?

3.33   In this chapter, a model was presented for the evolution of the tertiary creep strain during deformation (see Equation (3.30)). Using the equations given, demonstrate that the creep strain is proportional to $\exp\{kt\}$, i.e. that it grows exponentially with time.

Unfortunately, one difficulty with the model is that the time to infinite strain is indeterminate, which means that it cannot be used to model data for time to rupture. Convince yourself that this is the case. Propose a modification to the analysis which circumvents this difficulty. In doing so, derive an expression for the time to rupture that is suitable for modelling purposes.

3.34   During the creep deformation of single-crystal superalloys at temperatures of around 750 °C, the $L1_2$ structure of γ' can be sheared by dislocations. Draw out a plan view of the {111} plane for the $L1_2$ structure, and indicate on it the locations of the Ni and Al atoms. Suggest likely Burgers vectors for these dislocations in γ' and determine their magnitudes. Thus explain why $a/2\langle 1\bar{1}0\rangle\{111\}$ dislocations gliding in γ are unable to enter the γ', unless the applied stress is very significant.

3.35   A typical single-crystal superalloy contains cuboidal γ' precipitates of edge length 0.5 μm. If the volume fraction of γ' particles is 70%, estimate the width of the γ channels which surround them. Hence convince yourself that any significant contribution to creep deformation made by $(111)\langle 1\bar{1}0\rangle$ dislocations in γ must involve significant cross-slip or climb, in addition to simple glide.

3.36   By modelling the microstructure of a single-crystal superalloy using springs of appropriate stiffnesses, compare the stress experienced by the horizontal and vertical channels of γ, under conditions of uniaxial loading. Use the dimensions determined in the previous question. Take E for γ to be 100 GPa and the value for γ' as 10% greater than this. Thus, rationalise the observation that creep deformation in these materials occurs first in the horizontal γ channels.

3.37   When single-crystal superalloys are tested at elevated temperatures under fatigue conditions, it is found that the fracture surface is either (a) quasi-brittle in form and characterised by crystallographic fracture along the {111} planes, or else (b) ductile in form with the fracture surface perpendicular to the applied stress.

Giving your reasoning, which mode of failure is expected at (i) higher rather than lower temperatures, (ii) LCF rather than HCF conditions and (iii) lower mean stresses?

When the crystallographic mode dominates, the fatigue life is rather insensitive to the alloy content, i.e. the first-, second- and third-generation alloys do not behave very differently. However, the converse is true for the ductile mode. Account for this observation.

3.38   The strain-controlled LCF data of Dalal and co-workers are given in Figure 3.71, for an experimental single-crystal superalloy. Considerable fatigue anisotropy is displayed. Demonstrate that if one plots the product $\Delta\epsilon_{total} E_{hkl}$ rather than merely $\Delta\epsilon_{total}$ against the N_f, where $\Delta\epsilon_{total}$ is the total strain range and N_f is the number of cycles to failure, then the fatigue data collapse onto a straight line. (Note that Dalal *et al.* report values of 97, 131, 165, 186 and 220 GPa for the Young's moduli, E_{hkl}, in the $\langle 001 \rangle$, $\langle 113 \rangle$, $\langle 011 \rangle$, $\langle 112 \rangle$ and $\langle 111 \rangle$ direction, respectively [71].)

3.39 The Goodman diagram accounts for the effect of mean stress, σ_{mean}, on the relationship between the fatigue life, N_f, and the alternating stress range, $\Delta\sigma$. An example is given in Figure 3.79 for the single-crystal superalloy PWA 1484 under HCF conditions at 1038 °C for a lifetime of 10^7 cycles.

Is a greater temperature dependence expected for the creep rupture or the fatigue mode of failure in this case? Hence draw a sketch illustrating how the curves on the Goodman diagram are expected to alter as the temperature is changed. Will the gradient be greater or less as the temperature is increased? How will the behaviour of (i) coated and (ii) uncoated specimens compare?

3.40 Consider what happens to the stress distribution in a hollow turbine blade during engine start-up. The blade starts cold and heats up from outside in; therefore creep is expected first near the outer surfaces. Examine what effect this will have on the load borne by the internal (cooler) webs of the blading as creep proceeds.

3.41 The number of turbine blades required is dictated by the pressure drop to be taken across any given stage; in fact, these two numbers are inversely proportional to each other. A suitable equation based upon idealised work is as follows:

$$\frac{\dot{M}\Delta V_w}{N_b} = C_L(P_{in} - P_{out})hb$$

where $\dot{M}$ is the total mass flow of air in kg/s, ΔV_w is the velocity change induced by the blading, N_b is the number of blades of height h and breadth b, C_L is a lift coefficient of about 0.8 and P_{in} and P_{out} are the pressures of gas entering and leaving the turbine stage, respectively.

Using this information, estimate the number of turbine blades, N_b, required in the first stage of the Trent 800 high-pressure (HP) turbine, using typical dimensions for the blading and making assumptions where necessary.

What is the weight of a typical HP blade, denoted M_b? Hence estimate the weight of a complete row of high-pressure turbine blades.

The following information is provided. The velocity of gas exiting from the nozzle guide vanes (NGVs) is about 0.3 km/s, and this can be assumed to be axial; the turbine blade accelerates it by a factor of $\sqrt{2}$. The pressure drop across the stage is about 800 kPa.

3.42 The HP turbine blades of Rolls-Royce's Trent 800 engine are shrouded, but in General Electric's GE90 they are shroudless. Estimate the weight penalty, and hence the extra loading on the bore of the turbine disc from the presence of the shrouds. Why might it not be sensible to remove them? List the advantages and disadvantages of shrouded and shroudless blades.

3.43 When designing the turbine stages of an aeroengine, weight must be minimised. However, to maintain efficiency, the non-dimensional loading of the turbine, given by $c_p\Delta T/U_m^2$, must not exceed 2; here ΔT is the temperature drop across the turbine stage and U_m is the velocity of the blade at mean height. The heat capacity, c_p, has a value of 1000 J/(kg K).

Use these considerations to find a suitable mean location for the HP and LP turbine blades, relative to the axis of the engine, given typical HP and LP shaft speeds and estimated values of ΔT. Establish whether there are benefits to designing a turbine consisting of multiple (two or more) stages, across which a corresponding fraction of ΔT is dropped.

3.44 In practice, a turbine blade might be considered to have 'failed' for a number of different reasons. What might these reasons be?

3.45 For the shroudless high-pressure blades commonly used in aeroengines manufactured in the USA, wear of the turbine tip and turbine seals can be considerable. What are the effects of this wear likely to be?

3.46 Convince yourself that, if a turbine blade is modelled as a uniform slab, the centrifugal stress at the blade root is given by $\rho\omega^2 h\bar{r}$, where $\bar{r}$ is the average of the blade root and tip radii, and that the peak bending stress can be estimated from

$$\frac{\sigma_{peak,bending}}{d/2} = \frac{\dot{M}\Delta V_w}{N_b}\left(\frac{2h}{3}\right)\Big/ bd^3/12$$

where b is the breadth, d the depth and h the height of the blade. The other symbols have the same meaning as in Question 3.41.

Suppose that after a preliminary turbine design is completed, the bending stresses are found to be too high. What simple measure could be taken to reduce them without significant alteration of the dimensions and weight of the blade?

3.47 The high-pressure (HP) stage of the CF6 turbine engine contains 82 blades. Each weighs about 200 g, although, in practice, each is not perfectly uniform – thus, each set of blades must be arranged on the turbine disc such that the turbomachinery is as balanced as possible. The mean radius, r_m, of the turbine blades is 0.5 m, measured from the axis of the turbine disc.

The degree of unbalance is expressed in units of g cm; thus 1 g of excess weight displaced 2 cm from ths axis is equivalent to 2 g cm of unbalance. Write a computer program which generates a distribution of masses for the 82 blades, consistent with a standard deviation in the mass equivalent to 0.1%. Then determine the order in which the blades should be placed on the disc, if the unbalance is to be minimised. What magnitude of unbalance is expected?

3.48 Many of the latest generations of industrial gas turbines (for example, GE Power Systems' 'H' class unit) employ combined-cycle operation such that closed-loop steam cooling is used to moderate the temperature of the high-pressure turbine blading. Explain why this situation provides an extra incentive for the use of single-crystal technology (for example, Rene N5) for the blading.

3.49 The Pratt & Whitney company uses the concept of a 'technical readiness level' (TRL) to define the status of a new material which is being considered for engine application. Thus, TR-1 corresponds to the 'eureka' moment, with TR-9 being the level just prior to service entry. During the development cycle of a new single-crystal turbine blade alloy, to what might the technical readiness levels TR-2 through to TR-8 correspond? Describe the work which needs to be done and the technical progress to be demonstrated during these periods.

References

[1]   D. C. Wright and D. J. Smith, Forging of blades for gas turbines, *Materials Science and Technology*, **2** (1986), 742–747.

[2]   G. A. Whittaker, Precision casting of aero gas turbine components, *Materials Science and Technology*, **2** (1986), 436–441.

[3]   M. McLean, *Directionally Solidified Materials for High Temperature Service* (London: The Metals Society, 1983).

[4]   P. R. Beeley and R. F. Smart, eds, *Investment Casting* (London: The Institute of Materials, 1995).

[5]   F. L. Versnyder and R. W. Guard, Directional grain structures for high temperature strength, *Transactions of the American Society for Metals*, **52** (1960), 485–493.

[6]   M. J. Goulette, P. D. Spilling and R. P. Arthey, Cost-effective single crystals, in R. H. Bricknell, W. B. Kent, M. Gell, C. S. Kortovich and J. F. Radavich, eds, *Superalloys 1984* (Warrendale, PA: The Metallurgical Society of AIME, 1984), pp. 167–176.

[7]   A. J. Elliott, S. Tin, W. T. King, S. C. Huang, M. F. X. Gigliotti and T. M. Pollock, Directional solidification of large superalloy castings with radiation and liquid metal cooling (LMC): a comparative assessment, *Metallurgical and Materials Transactions*, **35A** (2004), 3221–3231.

[8]   B. H. Kear and B. J. Piearcey, Tensile and creep properties of single crystals of the nickel-base superalloy Mar-M200, *Transactions of the Metallurgical Society of AIME*, **239** (1967), 1209–1218.

[9]   P. Carter, D. C. Cox, Ch.-A. Gandin and R. C. Reed, Process modelling of grain selection during solidification of superalloy single crystal castings, *Materials Science and Engineering*, **A280** (2000), 233–246.

[10] H. S. Carslaw and J. C. Jaegar, *Conduction of Heat in Solids*, 2nd edn (Oxford: Clarendon Press, 1959).

[11] J. D. Livingston, H. E. Cline, E. F. Koch and R. R. Russell, High speed solidification of several eutectic alloys, *Acta Metallurgica*, **18** (1970), 399–404.

[12] A. F. Giamei and J. G. Tschinkel, Liquid metal cooling: a new solidification technique, *Metallurgical Transactions*, **7A** (1976), 1427–1434.

[13] A. Lohmuller, W. Esser, J. Grossman, M. Hordler, J. Preuhs and R.F. Singer, Improved quality and economics of investment castings by liquid metal cooling, in T. M. Pollock, R. D. Kissinger, R. R. Bowman *et al.*, eds, *Superalloys 2000* (Warrendale, PA: The Minerals, Metals and Materials Society (TMS), 2000), pp. 181–188.

[14] M. Konter, E. Kats and N. Hofmann, A novel casting process for single crystal turbine components, in T. M. Pollock, R. D. Kissinger, R. R. Bowman *et al.*, eds, *Superalloys 2000* (Warrendale, PA: The Minerals, Metals and Materials Society (TMS), 2000), pp. 189–200.

[15]   D. C. Cox, B. Roebuck, C. M. F. Rae and R. C. Reed, Recrystallisation of single crystal superalloy CMSX-4, *Materials Science and Technology*, **44** (2003), 440–446.

[16]   P. Aubertin, S. L. Cockcroft and A. Mitchell, Liquid density inversions during solidification of superalloys and their relationship to freckle formation in castings, in R. D. Kissinger, D. J. Deye, D. L. Anton *et al.*, eds, *Superalloys 1996* (Warrendale, PA: The Minerals, Metals and Materials Society (TMS), 1996), pp. 443–450.

[17]   S. Tin and T. M. Pollock, Stabilization of thermosolutal convective instabilities in Ni-based single-crystal superalloys: carbide precipitation and Rayleigh numbers, *Metallurgical and Materials Transactions*, **34A** (2003), 1953–1967.

[18]   T. M. Pollock, W. H. Murphy, E. H. Goldman, D. L. Uram and J. S. Tu, Grain defect formation during directional solidification of nickel base single crystals, in S. D. Antolovich, R. W. Stusrud, R. A. MacKay, *et al.*, eds, *Superalloys 1992* (Warrendale, PA: The Minerals, Metals and Materials Society (TMS), 1992), pp. 125–134.

[19]   M. S. A. Karunaratne, C. M. F. Rae and R. C. Reed, On the microstructural instability of an experimental nickel-base superalloy, *Metallurgical and Materials Transactions*, **32A** (2001), 2409–2421.

[20]   C. Beckermann, J. P. Gu and W. J. Boettinger, Development of a freckle predictor via Rayleigh number method for single-crystal nickel-base superalloy castings, *Metallurgical and Materials Transactions*, **31** (2000), 2545–2557.

[21]   K. C. Mills, *Recommended Values of Thermophysical Properties for Selected Commercial Alloys* (Materials Park, OH: ASM International, 2001).

[22]   W. Kurz and D. J. Fisher, *Fundamentals of Solidification* (Aedermannsdorf, Switzerland: Trans Tech Publications, 1986).

[23]   H. S. Whitesell, L. Li and R. A. Overfelt, Influence of solidification variables on the dendrite arm spacings of Ni-based superalloys, *Metallurgical and Materials Transactions*, **31B** (2000), 546–551.

[24]   D. A. Ford and R. P. Arthey, Development of single crystal alloys for specific engine applications, in R. H. Bricknell, W. B. Kent, M. Gell, C. S. Kortovich and J. F. Radavich, eds, *Superalloys 1984* (Warrendale, PA: The Metallurgical Society of AIME, 1984), pp. 115–124.

[25]   K. Harris, G. L. Erickson, S. L. Sikkenga, W. D. Brentnall, J. M. Aurrecoechea and K. G. Kubarych, Development of the rhenium-containing superalloys CMSX-4 and CM 186 LC for single crystal blade and directionally solidified vane applications in advanced turbine engines, in S. D. Antolovich, R. W. Stusrud, R. A. MacKay *et al.*, eds, *Superalloys 1992* (Warrendale, PA: The Minerals, Metals and Materials Society (TMS), 1992), pp. 297–306.

[26]   T. M. Pollock and A. S. Argon, Creep resistance of CMSX-3 nickel base superalloy single crystals, *Acta Metallurgica et Materialia*, **40** (1992), 1–30.

[27]   T. Murakumo, T. Kobayashi, Y. Koizumi and H. Harada, Creep behaviour of Ni-base single-crystal superalloys with various γ' volume fractions, *Acta Materialia*, **52** (2004), 3737–3744.

[28] P. Caron, High γ' solvus new generation nickel-based superalloys for single crystal turbine blade applications, in T. M. Pollock, R. D. Kissinger, R. R. Bowman *et al.*, eds, *Superalloys 2000* (Warrendale, PA: The Minerals, Metals and Materials Society (TMS), 2000), pp. 737–746.

[29] H. Harada and H. Murakami, Design of Ni-base superalloys, in T. Saito, ed., *Computational Materials Design* (Berlin: Springer-Verlag, 1999), pp. 39–70.

[30] G. Bruno and H. C. Pinto, The kinetics of the γ' phase and its strain in the nickel-base superalloy SC16 studied by in-situ neutron and synchrotron radiation diffraction, in K. A. Green, T. M. Pollock, H. Harada *et al.*, eds, *Superalloys 2004* (Warrendale, PA: The Minerals, Metals and Materials Society (TMS), 2004), pp. 837–848.

[31] H. A. Kuhn, H. Biermann, T. Ungar and H. Mughrabi, X-ray study of creep deformation induced changes of the lattice misfit in the gamma prime hardened monocrystalline nickel-base superalloy SRR99, *Acta Metallurgica et Materialia*, **11** (1991), 2783–2794.

[32] Y. Murata, S. Miyazaki, M. Morinaga and R. Hashizume, Hot corrosion resistant and high strength nickel-based single crystal and directionally-solidified superalloys developed by the d-electrons concept, in R. D. Kissinger, D. J. Deye, D. L. Anton *et al.*, eds, *Superalloys 1996* (Warrendale, PA: The Minerals, Metals and Materials Society (TMS), 1996), pp. 61–70.

[33] A. C. Yeh, C. M. F. Rae and S. Tin, High temperature creep behaviours of Ru-bearing Ni-based single crystal superalloys, in K. A. Green, T. M. Pollock, H. Harada *et al.*, eds, *Superalloys 2004* (Warrendale, PA: The Minerals, Metals and Materials Society (TMS), 2004), pp. 677–686.

[34] S. Walston, A. Cetel, R. Mackay, K. O'Hara, D. Duhl and R. Dreshfield, Joint development of a fourth generation single crystal superalloy, in K. A. Green, T. M. Pollock, H. Harada *et al.*, eds, *Superalloys 2004* (Warrendale, PA: The Minerals, Metals and Materials Society (TMS), 2004), pp. 15–24.

[35] A. P. Ofori, C. J. Humphreys, S. Tin and C. N. Jones, A TEM study of the effect of platinum group metals in advanced single crystal nickel-base superalloys, in K. A. Green, T. M. Pollock, H. Harada *et al.*, eds, *Superalloys 2004* (Warrendale, PA: The Minerals, Metals and Materials Society (TMS), 2004), pp. 787–794.

[36] A. D. Cetel and D. N. Duhl, Second generation nickel-base single crystal superalloy, in S. Reichman, D. N. Duhl, G. Maurer, S. Antolovich and C. Lund, eds, *Superalloys 1988* (Warrendale, PA: The Metallurgical Society, 1988), pp. 235–244.

[37] C. Sarioglu, C. Stinner, J. R. Blachere *et al.*, The control of sulphur content in nickel-base single crystal superalloys and its effects on cyclic oxidation resistance, in R. D. Kissinger, D. J. Deye, D. L. Anton *et al.*, eds, *Superalloys 1996* (Warrendale, PA: The Minerals, Metals and Materials Society (TMS), 1996), pp. 71–80.

[38] K. Harris and J. B. Wahl, Improved single crystal superalloys, CMSX-4 (SLS) [La+Y] and CMSX-486, in K. A. Green, T. M. Pollock, H. Harada *et al.*, eds, *Superalloys 2004* (Warrendale, PA: The Minerals, Metals and Materials Society (TMS), 2004), pp. 45–52.

[39] A. Akhtar, M. S. Hook and R. C. Reed, On the oxidation of the third generation superalloy CMSX-10, *Metallurgical and Materials Transactions*, **36A** (2005), 3001–3017.

[40] F. S. Pettit and C. S. Giggins, Hot corrosion, in C. T. Sims, N. S. Stoloff and W. C. Hagel, eds, *Superalloys II* (New York: John Wiley, 1987), pp. 327–358.

[41] N. Birks, G. H. Meier and F. S. Pettit, High temperature corrosion, in J. K. Tien and T. Caulfield, eds, *Superalloys, Supercomposites and Superceramics* (San Diego: Academic Press Inc., 1989), pp. 439–489.

[42] T. I. Barry and A. T. Dinsdale, Thermodynamics of metal gas liquid reactions, *Materials Science and Technology*, **3** (1987), 501–511.

[43] S. Mrowec, T. Werber and M. Zastawnik, The mechanism of high temperature sulphur corrosion of nickel-chromium alloys, *Corrosion Science*, **6** (1966), 47–68.

[44] R. C. Reed, N. Matan, D. C. Cox, M. A. Rist and C. M. F. Rae, Creep of CMSX-4 superalloy single crystals: effects of rafting at high temperature, *Acta Materialia*, **47** (1999), 3367–3381.

[45] N. Matan, *Rationalisation of the Creep Performance of the CMSX-4 Single Crystal Superalloy*. Unpublished Ph.D. Thesis, University of Cambridge (1999).

[46] N. Matan, D. C. Cox, P. Carter, M. A. Rist, C. M. F. Rae and R. C. Reed, Creep of CMSX-4 superalloy single crystals: effects of misorientation and temperature, *Acta Materialia*, **47** (1999), 1549–1563.

[47] V. Sass, U. Glatzel and M. Feller-Kniepmeier, Creep anisotropy in the monocrystalline nickel-base superalloy CMSX-4, in R. D. Kissinger, D. J. Deye, D. L. Anton *et al.*, eds, *Superalloys 1996* (Warrendale, PA: The Minerals, Metals and Materials Society, 1996), pp. 283–290.

[48] F. R. N. Nabarro and H. L. de Villiers, *The Physics of Creep* (London: Taylor and Francis, 1995).

[49] T. M. Pollock and R. D. Field, Dislocations and high temperature plastic deformation of superalloy single crystals, in F. R. N. Nabarro and M.S. Duesbery, eds, *Dislocations in Solids*, vol. 11, (Amsterdam: Elsevier, 2002), pp. 593–595.

[50] R. N. Ghosh, R. V. Curtis and M. McLean, Creep deformation of single crystal superalloys – modelling the crystallographic anisotropy, *Acta Metallurgica et Materialia*, **38** (1990), 1977–1992.

[51] M. McLean, Nickel-base superalloys: current status and potential, *Philosophical Transactions of the Royal Society of London A*, **351** (1995), 419–433.

[52] T. M. Pollock and A. S. Argon, Directional coarsening in nickel base superalloy single crystals with high volume fractions of coherent precipitates, *Acta Metallurgica et Materialia*, **42** (1994), 1859–1874.

[53] D. M. Shah, S. Vega, S. Woodard and A. D. Cetel, Primary creep in nickel-base superalloys, in K. A. Green, T. M. Pollock, H. Harada *et al.*, eds, *Superalloys 2004* (Warrendale, PA: The Minerals, Metals and Materials Society (TMS), 2004), pp. 197–206.

[54] G. R. Leverant and B. H. Kear, The mechanism of creep in gamma prime precipitation hardened nickel-base alloys at intermediate temperatures, *Metallurgical Transactions*, **1** (1973), 491–498.

[55] G. L. Drew, R. C. Reed, K. Kakehi and C. M. F. Rae, Single crystal superalloys: the transition from primary to secondary creep, in K. A. Green, T. M. Pollock, H. Harada *et al.*, eds, *Superalloys 2004* (Warrendale, PA: The Minerals, Metals and Materials Society (TMS), 2004), pp. 127–136.

[56] R. A. MacKay and R. D. Maier, The influence of orientation on the stress rupture properties of nickel-base superalloy single crystals, *Metallurgical Transactions*, **13A** (1982), 1747–1754.

[57] C. M. F. Rae, N. Matan and R. C. Reed, On the role of stacking fault shear during the primary creep of CMSX-4 superalloy single crystals, *Materials Science and Engineering*, **300** (2001), 127–135.

[58] H. Murakami, T. Yamagata, H. Harada and M. Yamazaki, The influence of Co on creep deformation anisotropy in Ni-base single crystal superalloys at intermediate temperatures, *Materials Science and Engineering*, **A223** (1997), 54–58.

[59] T. P. Gabb, S. L. Draper, D. R. Hull, R. A. MacKay and M. V. Nathal, The role of interfacial dislocation networks in high temperature creep of superalloys, *Materials Science and Engineering*, **A118** (1989), 59–69.

[60] J. Y. Buffiere and M. Ignat, A dislocation based criterion for the raft formation in nickel-based superalloy single crystals, *Acta Metallurgica et Materialia*, **43** (1995), 1791–1797.

[61] M. Veron, Y. Brechet and F. Louchet, Strain induced directional coarsening in Ni-base superalloys, *Scripta Metallurgica et Materialia*, **34** (1996), 1883–1886.

[62] N. Matan, D. C. Cox, C. M. F. Rae and R. C. Reed, On the kinetics of rafting in CMSX-4 superalloy single crystals, *Acta Materialia*, **47** (1999), 2031–2045.

[63] R. D. Field, T. M. Pollock and W. H. Murphy, The development of γ/γ' interfacial dislocation networks during creep in Ni-base superalloys, in S. D. Antolovich, R. W. Stusrud, R. A. MacKay *et al.*, eds, *Superalloys 1992* (Warrendale, PA: The Minerals, Metals and Materials Society, 1992), pp. 557–566.

[64] Y. Koizumi, T. Kobayashi, T. Yokokawa *et al.*, Development of next-generation Ni-base single crystal superalloys, in K. A. Green, T. M. Pollock, H. Harada *et al.*, eds, *Superalloys 2004* (Warrendale, PA: The Minerals, Metals and Materials Society (TMS), 2004), pp. 35–43.

[65] J. X. Zhang, T. Murakumo, H. Harada, Y. Koizumi and T. Kobayashi, Creep deformation mechanisms in some modern single-crystal superalloys, in K. A. Green, T. M. Pollock, H. Harada *et al.*, eds, *Superalloys 2004* (Warrendale, PA: The Minerals, Metals and Materials Society (TMS), 2004), pp. 189–195.

[66] R. C. Reed, N. Matan, D. C. Cox, M. A. Rist and C. M. F. Rae, Creep of CMSX-4 superalloy single crystals: effects of rafting at high temperature, *Acta Materialia*, **47** (1999), 3367–3381.

[67] J. Lecomte-Beckers, Relation between chemistry, solidification behaviour, microstructure and microporosity in nickel-base superalloys, in S. Reichman, D. N. Duhl, G. Maurer, S. Antolovich and C. Lund, eds, *Superalloys 1988* (Warrendale, PA: The Metallurgical Society, 1988), pp. 713–721.

[68] S. Suresh, *Fatigue of Materials*, 2nd edn (Cambridge: Cambridge University Press, 1998).

[69] P. Lukas and L. Kunz, Cyclic slip localisation and fatigue crack initiation in FCC single crystals, *Materials Science and Engineering*, **A314** (2001), 75–80.

[70] V. Brien and B. Decamps, Low cycle fatigue of a nickel-based superalloy at high temperature: deformation microstructures, *Materials Science and Engineering*, **A316** (2001), 18–31.

[71] R. P. Dalal, C. R. Thomas and L. E. Dardi, The effect of crystallographic orientation on the physical and mechanical properties of an investment cast single crystal nickel-base superalloy, in R. H. Bricknell, W. B. Kent, M. Gell, C. S. Kortovich and J. F. Radavich, eds, *Superalloys 1984* (Warrendale, PA: The Metallurgical Society of AIME, 1984), pp. 185–197.

[72] T. P. Gabb and G. Welsch, The high temperature deformation in cyclic loading of a single crystal nickel-base superalloy, *Acta Metallurgica*, **37** (1989), 2507–2516.

[73] T. P. Gabb, J. Gayda and R. V. Miner, Orientation and temperature dependence of some mechanical properties of the single crystal nickel-base superalloy Rene N4. Part II: Low cycle fatigue behaviour, *Metallurgical Transactions*, **17A** (1986), 497–505.

[74] H. Mughrabi, S. Kraft and M. Ott, Specific aspects of isothermal and anisothermal fatigue of the monocrystalline nickel-base superalloy CMSX-6, in R. D. Kissinger, D. J. Deye, D. L. Anton *et al.*, eds, *Superalloys 1996* (Warrendale, PA: The Minerals, Metals and Materials Society, 1996), pp. 335–344.

[75] D. W. Maclachlan and D. M. Knowles, Fatigue behaviour and lifing of two single crystal superalloys, *Fatigue and Fracture of Engineering Materials and Structures*, **24** (2001), 503–521.

[76] P. K. Wright, M. Jain and D. Cameron, High cycle fatigue in a single crystal superalloy: time dependence at elevated temperature, in K. A. Green, T. M. Pollock, H. Harada *et al.*, eds, *Superalloys 2004* (Warrendale, PA: The Minerals, Metals and Materials Society (TMS), 2004), pp. 657–666.

[77] N. A. Cumpsty, *Jet Propulsion: A Simple Guide to the Aerodynamic and Thermodynamic Design and Performance of Jet Engines* (Cambridge: Cambridge University Press, 1997).

[78] A. G. Dodd, Mechanical design of gas turbine blading in cast superalloys, *Materials Science and Technology*, **2** (1986), 476–485.

[79] Rolls-Royce, *The Jet Engine*, 4th edn (Derby, UK: The Technical Publications Department, Rolls-Royce plc, 1992).

4 Superalloys for turbine disc applications

The turbine discs are amongst the most critical of components in the aeroengine. To put this into perspective, consider a typical modern civil turbofan such as the Trent 800 – the turbine discs represent about 20% of its total weight and their cost accounts for about 10% of the engine's value upon entry into service. For a military engine, such as the EJ200, the figures are nearer 5% and 25%. The primary function of the turbine discs is to provide fixturing for the turbine blades located in the gas stream, from which mechanical energy is extracted; the complete assembly of discs and blades is then capable of transmitting power to the fan and compressor sections, via the shafts which run along almost the complete length of the engine. During the design stages, their geometry must be optimised by balancing the competing demands of minimisation of weight, dimensional stability and mechanical integrity. Thus risk mitigation by an appropriate disc lifing strategy is vital, since a disc failure represents a potentially fatal hazard to the aircraft and its occupants. Metallurgical damage due to fatigue during service and its acceleration due to oxidation and/or corrosion must therefore be quantified and predicted, so that each disc is withdrawn from service after a prescribed number of take-off/landing cycles, known as the *safe working life*.

To convince oneself of the properties required of a turbine disc alloy, it is instructive to consider the operating conditions experienced by a typical high-pressure (HP) turbine disc. The HP turbine blades project directly into the gas stream in excess of 1550 °C, causing the temperatures in the rim of the turbine disc to approach 650 °C or beyond during service. The rotational speeds are very significant, the exact value tending to increase as the disc diameter decreases; for the Trent 800 engine 10 500 revolutions per minute is typical, whilst for helicopter engines the figure may be in excess of four times this. Consequently, the mechanical stresses generated in the bore region may reach 1000 MPa during take-off, i.e. they may exceed the uniaxial tensile yield strength of the material. Higher stresses will be present in the overspeed condition which arises if the shaft fails – a theoretical possibility which must be acknowledged in the design process. The energy stored in a HP turbine disc under such conditions is very significant: equivalent, in fact, to 0.75 kg of high explosive, or a saloon car travelling at over 100 mph.

4.1 Processing of the turbine disc alloys

The discs for the gas turbine engine are fabricated by the machining of superalloy forgings, but two distinct approaches are available for the preparation of the billets to be operated on

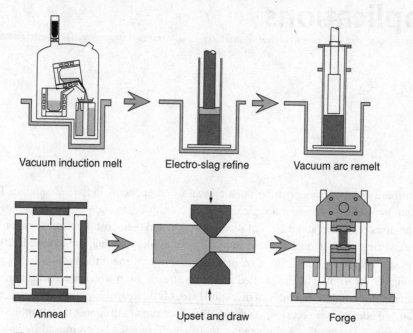

Fig. 4.1. The sequence of processes used for the production of turbine disc alloys by ingot metallurgy. Adapted from ref. [3].

by the forgemaster. The first is conventional ingot metallurgy, which involves the thermal-mechanical working of material produced by vacuum induction melting, electro-slag remelting and vacuum arc remelting [1]. Billets processed in this way are referred to as *cast-and-wrought* product. A second route involves powder metallurgy. The choice of route depends upon a number of factors, but is largely dictated by the chemistry of the chosen superalloy [2]. Alloys such as Waspaloy and IN718 are generally prepared by ingot metallurgy, since the levels of strengthening elements Al, Ti and Nb are relatively low, so that the additional cost associated with powder processing cannot be justified. Figure 4.1 illustrates the various steps taken in this case [3]. Vacuum induction melting is followed by electro-slag refining and vacuum arc remelting; after annealing to improve the compositional homogeneity, the billet is thermal-mechanically worked before undergoing a series of forging operations. Unfortunately, ingot metallurgy cannot be applied to the heavily alloyed grades such as Rene 95 and RR1000, since the levels of segregation arising during melt processing and the significant flow stress at temperature cause cracking during thermal-mechanical working; instead, powder-processing is preferred, see Figure 4.2 [3]. Here, vacuum induction melting (VIM) is used as before, followed by remelting and inert gas atomisation to produce powder – this is then sieved to remove any large non-metallic inclusions inherited from the processing. This step is important since it improves, in principle, the cleanliness of the product. To prepare a billet ready for forging, the powder is consolidated by sealing it into a can, which is then degassed and sealed, and hot isostatic pressing and/or extrusion follows. In principle therefore the concentration of inclusions will be lower in the powder metallurgy

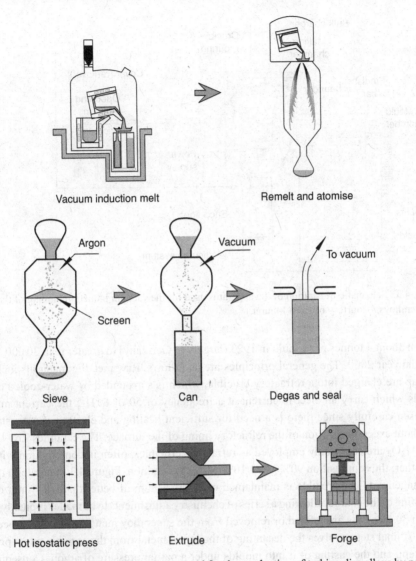

Vacuum induction melt   Remelt and atomise

Argon   Vacuum   To vacuum

Screen

Sieve   Can   Degas and seal

or

Hot isostatic press   Extrude   Forge

Fig. 4.2. The sequence of processes used for the production of turbine disc alloys by powder metallurgy. Adapted from ref. [3].

product than in that produced by ingot metallurgy, although this is achieved at greater cost and by a processing route which is significantly more complex.

Regardless of the route to be employed for the processing of the turbine disc material, VIM is used as the standard melting practice for the preparation of the superalloy stock. Vacuum metallurgy dates from the 1840s, but considerable advances were made in Germany during World War I when Ni–Cr alloys were produced as alternatives to platinum-based thermocouple alloys. The first commercial application of the superalloys came in the 1950s with the production of 5 kg heats of Waspaloy; each heat was forged to a single turbine blade. The success of the industry can be judged by noting that worldwide production has expanded

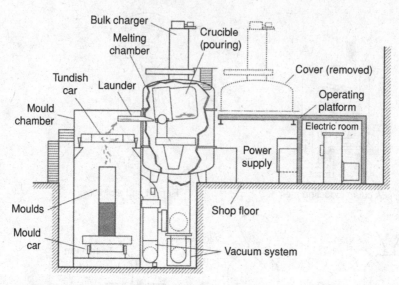

Fig. 4.3. Schematic illustration of the apparatus used for the vacuum induction melting of the superalloys. (Courtesy of Gern Maurer.)

from about 4 tonnes per annum in 1923 (largely in Germany) to greater than 30 000 tonnes in the year 2000. The general principles are as follows. Raw metallic materials including scrap are charged into a refractory crucible, which is surrounded by water-cooled copper coils which carry an electric current at a frequency of 50 or 60 Hz; the current must be chosen carefully since there is a need for sufficient heating and electromagnetic stirring, without excessive erosion of the refractory lining of the furnace. The ceramics ZrO_2, MgO or Al_2O_3 are commonly employed as refractory crucibles, often in combination with one another; the composition 90% MgO/10% Al_2O_3 is common. Figure 4.3 illustrates a typical arrangement. The crucible is maintained under a vacuum of better than 10^{-4} atm during melting of the charge, allowing a series of chemistry adjustments to be made. Typically, more than 30 elements are refined or removed from the superalloy melt during VIM processing [3]. A final step involves the decanting of the liquid metal from the crucible into a pouring system, and the casting of it into moulds under a partial pressure of argon. Consequently, VIM is a batch, rather than a continuous, process.

A number of important phenomena occur during VIM processing. First, substantial reductions in the concentrations of nitrogen and oxygen are achieved, since these degas and are thus removed by the vacuum system. This helps to prevent the oxidation of reactive elements such as Al, Ti, Zr and Hf; this is probably the primary justification for vacuum processing. Second, low-melting-point contaminants such as Pb, Bi are removed from the melt by volatilisation [4] on account of their substantial vapour pressures; see Figure 4.4. These elements commonly enter the process via recycled material, for example, because they are used to mount superalloys during machining processes. The rate-controlling step here appears to be mass transport in the liquid phase – consequently, electromagnetic stirring accelerates the removal of these 'tramp' elements. This is fortuitous, since the high-temperature properties of the superalloys are substantially impaired if these elements

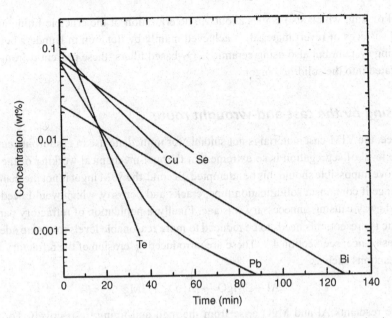

Fig. 4.4. Evaporation of elements from Ni–20 wt%Cr alloy in a vacuum induction melting furnace [4].

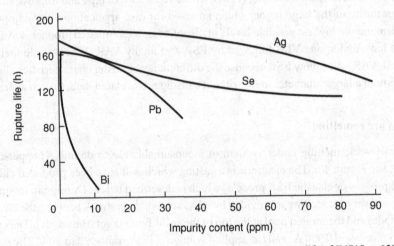

Fig. 4.5. Effect of various trace elements on the stress-rupture life of IN718 at 650 °C and 690 MPa. Data taken from ref. [5].

are present (see Figure 4.5); for IN718 tested in creep at 650 °C and 690 MPa, the rupture life is decreased ten-fold and two-fold by the addition of only 10 ppm of Bi and 40 ppm of Pb, respectively [5]. This explains why strict limits are placed on the levels of trace elements which are permissible in premium-grade superalloys, and justifies the emphasis placed on compositional analyses during processing for quality assurance purposes. Finally, desulphurisation of the melt is achieved typically by adding lime (CaO) and fluxing agents.

Removal of slag, which also comes about via deterioration of the crucible lining and the oxidised surfaces of revert material, is achieved mainly by flotation in launders built into the pouring system, but also using ceramic ZrO_2-based filters; these prevent it from being incorporated into the solidified ingot.

4.1.1   Processing by the cast-and-wrought route

In practice, the VIM-cast material is not suitable for immediate use in gas turbine engines. First, the level of segregation is so extreme that thermal-mechanical working of the ingot would prove impossible should this be attempted. Second, the VIM ingot is not mechanically sound, since it contains a solidification pipe, cracks and porosity, which would need to be machined away causing unnecessary wastage. Finally, a population of refractory particles of ceramic is present; this needs to be reduced to more reasonable levels to ensure adequate fatigue resistance; see Section 4.3. These are introduced by erosion of the refractory lining and by reactions such as

$$2Al + 3MgO \rightarrow Al_2O_3 + 3Mg \qquad (4.1)$$

where the reactants Al and MgO arise from the melt and lining, respectively. For these reasons, superalloy stock processed by conventional cast-and-wrought processing is subjected to further *secondary* melting processes such as vacuum arc remelting (VAR) and possibly also electro-slag remelting (ESR). These reduce ingot pipe and improve the yield and forgeability of the large ingots which are used for disc applications. For components which demand the highest possible levels of cleanliness, *triple*-melted product is available: this will have undergone VIM followed by ESR and finally VAR. Where triple melting is employed, VAR will follow ESR because the different heat transfer characteristics allow the production of a larger diameter ingot without causing melt-related defects such as freckles.

A   Vacuum arc remelting

This involves the melting under vacuum of a consumable electrode into a copper-cooled crucible; see Figure 4.6. The electrode is a casting which will have been produced either by the VIM process or else the ESR process, which is described below. During processing, the energy necessary for melting is provided by a DC arc, which is struck between the electrode (the cathode) and the molten pool at the top of the solidifying ingot (the anode). The current used is typically ~10 000 A, and the applied voltage is in the range 20 to 50 V. The scale of the process is impressive; finished VAR ingots are several metres in length and weigh up to 20 tonnes.

The heat and mass transfer processes occurring during melting have a profound influence on the quality of the resulting VAR ingot. During processing, a thin molten film is formed on the electrode, which flows under gravity to form hanging protuberances. Metal transfer occurs as droplets form and momentarily short circuit the arc as they fall from the molten film. To maintain a constant arc gap, a ram arrangement is used to drive the electrode into the VAR crucible as melting proceeds. Considerable emphasis is placed on control strategies and process monitoring. For example, dynamic load cell weighing systems have

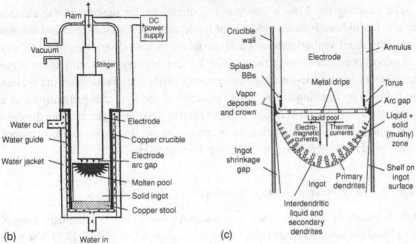

Fig. 4.6. (a) Industrial-scale vacuum arc remelting (VAR) furnace (courtesy of Jai Narayan, Consarc Corporation), (b) Schematic illustration of the VAR furnace. (c) Close-up of the details in the vicinity of the molten pool.

Table 4.1. *Chemical analysis of the bulk and the defect area in wt%*

Element	Alloy	White spot
Al	0.49	0.41
Si	0.20	0.19
Ti	0.81	0.62
Cr	17.50	17.71
Fe	17.75	19.21
Ni	54.97	55.70
Nb	5.10	2.96
Mo	3.18	3.20
$T_{solidus}$	1242 °C	1309 °C
$T_{liquidus}$	1348 °C	1373 °C

Courtesy of Laurence Jackman

been developed to enable computerised control of the melt rate, which is typically in the range 2–6 kg/min. As the electrode is melted, a liquid pool is produced which resides on the already solidified material; the size and shape of the molten zone remains approximately constant during processing, apart from brief transients associated with start-up and finishing, i.e. a steady-state is achieved. For a given electrode diameter, the depth of the pool increases with the power input, the product of the current and the applied voltage; modelling work [6] indicates that the efficiency of this process is in the range 50–75%, and that the current enters the ingot at the meniscus and at the stool at the bottom of the copper crucible on which the solidified material sits. The molten pool is subjected to complex flow patterns, which are due to a combination of the buoyancy forces, electromagnetic stirring and the Marangoni effect arising from the temperature dependence of the surface tension. The velocity of the molten material on the surface can reach 0.01 m/s; the flow pattern is radial, consistent with the symmetry of the situation, and outward from the centre of the pool; see Figure 4.7.

The resulting ingot has a structure and chemistry far superior to the electrode being melted. It will have reduced amounts of tramp elements, such as lead and bismuth, which have substantial vapour pressures and are therefore volatilised. Degassing of oxygen and nitrogen also occurs through the vacuum system, although it is difficult to remove them completely when nitride-forming elements such as titanium and niobium are present. Oxides are removed by flotation and thermal dissociation; the former is generally more effective, with surface tension and Marangoni forces playing a part in agglomerating the oxide debris. On account of the enhanced cleanliness produced, VAR remains the most important and widely used of the remelting practices for the superalloys.

Case study. Melt-related 'white spot' defects

Defects known as *white spots* are solute-lean regions which lack the hardening phases γ' and γ'' and the associated hardening elements such as Al, Ti and Nb [7,8]. The terminology comes from the metallographical etching techniques which provide preferential attack, leaving the solute-lean regions bright or white; see Figure 4.8 and Table 4.1. Experiments have shown that mechanical properties, such as the yield stress and ultimate tensile strength

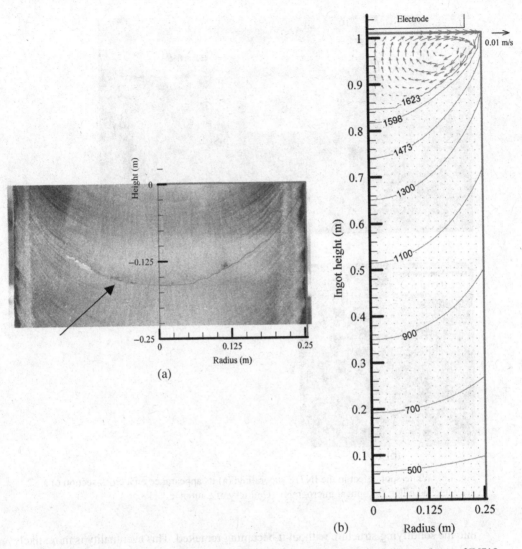

Fig. 4.7. Results from the modelling of the VAR processing of a 0.5 m diameter ingot of IN718 superalloy: (a) comparison between predicted and measured pool profile; (b) computed temperature distribution in the ingot and the flow in the molten pool [6].

(UTS), are reduced when white spots are present, and this is obviously a cause for concern. The VAR process is very much more prone to white spot than the ESR process, described below. It should be noted that there is no known method for assuring the complete absence of white spot from superalloy ingots for which VAR is the final remelting process. However, careful attention to process control for both VIM electrode production and VAR processing reduces the probability of its occurrence.

Various mechanisms have been proposed for the formation of white spot, and it seems probable that each occurs to some degree. However, a common feature is the falling of material into the molten pool, such that its differing chemistry allows it to be incorporated

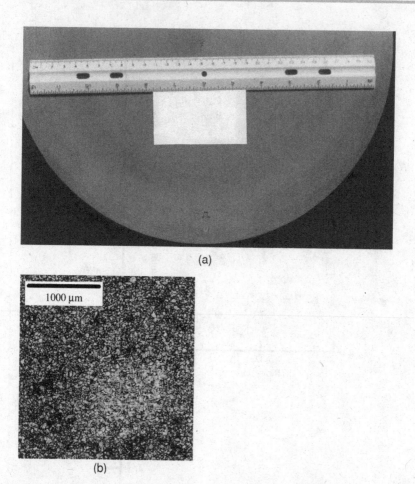

(a)

(b)

Fig. 4.8. White-spot defect in the IN718 superalloy: (a) its appearance on a cross-section of a macroetched billet; (b) optical micrograph. (Courtesy of Laurence Jackman.)

into the solidifying structure without it becoming remelted. This eventuality is more likely for (i) solute-lean material with a melting temperature that is significantly larger than for the bulk composition, (ii) larger particles which do not dissolve as readily as smaller ones and (iii) larger local solidification times. Hence the frequency of white-spot occurrence is dependent upon the resistance of the process to instabilities in the process.

Where does this material originate from? Studies have shown that one source is the crown and shelf region, which occurs in the VAR process at the pool/crucible interface. During processing, small changes in current distribution cause the centre of the pool to shift, such that the remelting of some fraction of the crown is promoted; this is then under-cut and material is free to drop into the molten pool. A second possibility is fragmented dendritic material from the electrode itself: obviously the interdendritic regions will melt preferentially because of their lower melting point; this emphasises the importance of the mechanical integrity of the electrode. Modelling studies [9] indicate that particles falling from the electrode are likely to melt completely before entrapment in the mushy zone,

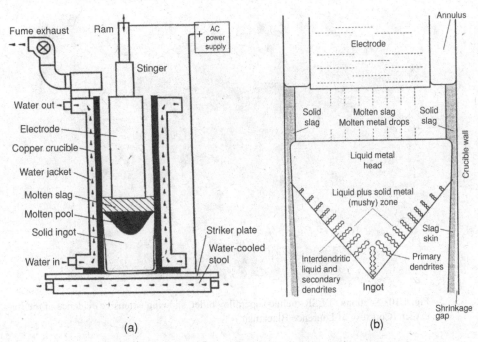

Fig. 4.9. (a) Schematic illustration of the apparatus used for the electro-slag remelting of superalloys. (b) Close-up of the details in the vicinity of the molten pool.

provided that their diameter is greater than 6 mm; crown fall-in was considered to be a more likely source of white spot, since the maximum safe diameter of material falling into the pool was found to be in the range 2 to 3 mm.

B Electro-slag remelting

Electro-slag remelting (ESR) is used widely as a secondary melting process for the production of the material for turbine discs; see Figure 4.9. It has similarities to the VAR process in that a consumable electrode is remelted, but here a molten slag pool sits between the electrode and the solidifying ingot – hence processing can be carried out in air. A major significant advantage over VAR processing then arises: molten droplets from the electrode pass through the slag, allowing them to react with it such that impurities such as oxides and sulphur are removed. However, unlike the VAR process, an alternating current (AC) is applied; the electrode immersion plus the slag cap must be regarded as the major resistance in the circuit; consequently, the majority of the heat generation occurs in this region. Changing the weight of slag in the system alters the resistance of the circuit and thus the current and voltage required to maintain a given power input/melt rate.

A number of processing controls are used which aid in the production of ingots of high quality and consistency. First, a constant immersion of the electrode in the slag is maintained, via control of the electrode voltage. Second, one monitors the rate at which the electrode melts by measuring the loss of electrode weight during processing, and adjusts the current accordingly. During solidification, slag may be added or removed as processing proceeds

Fig. 4.10. Sections of ESR-melted superalloy billet, showing extensive evidence of the freckle defect. (Courtesy of Laurence Blackman.)

to maintain a uniform head over the pool. The choice of slag is critical, since it determines the reactivity with the molten droplets via different melting temperatures and viscosities. The slags are normally based upon fluorite CaF_2 with additions of CaO, MgO and Al_2O_3. Impurities of SiO_2 can cause problems, since Ti reduces it to metallic Si which will be incorporated into the ingot. This is managed by recycling the slag, which then approaches an equilibrium with Ti and Si pickup, thus minimising elemental exchange. In practice, minor changes in alloy chemistry are caused in this way; ESR is much more prone to this effect than VAR.

One disadvantage of the molten slag is that it acts as a heat reservoir at the top of the ingot/pool boundary and this makes the fusion profile obtained from the ESR process different from that obtained from VAR; generally with steeper sides, larger mushy zones and decreased cooling rates for comparable power input. These factors promote levels of chemical heterogeneity that are greater in ESR than in VAR processing; hence where triple melting is employed, ESR will always precede VAR. Additionally, the maximum ingot diameter which can be melted by ESR is dependent upon alloy composition. This is because large diameters promote a greater mushy zone and enhanced local solidification times, due to the increased distances. For this reason the processing of higher-strength alloys such as U720Li is limited to diameters of about 500 mm or 20 inches. Attempts have been made [10] to use ESR processing for the production of very large 890 mm (35 inch) IN718 ingots of weight approaching 14 tonnes, each of which is sufficient for just two turbine discs for a large land-based turbine. It was reported that no conditions could be found which yielded ingots that were free of extensive freckling; the heavy freckle channels were found to be rich in elements such as Nb, Mo and Ti, which are known to partition to the liquid phase. Subsequent VAR processing was used to produce sound triple-melted (VAR + ESR +

VAR) 760 mm (30 inch) product for this application. Figure 4.10 illustrates the formation of freckle defects in ESR-processed ingots of IN718; their occurrence seems to be most likely at the mid-radius position, where the permeability of the mushy zone is greater than at the centre of the ingot and is deep enough to support the interdendritic fluid flow required for their formation [11].

C Ingot-to-billet conversion: the cogging process

The ingots produced by the remelting processes are unsuitable for mechanical applications: they must undergo thermal-mechanical working in order to break down the as-cast structure and to reduce the grain size to acceptable levels, i.e. from a few tens of millimetres to a few tens of microns. This is known as ingot conversion and it is usually achieved by a process known as *cogging*, during which the diameter of the cylindrical ingot is reduced in size by a factor of approximately 2, so that its length increases four-fold, see Figure 4.11. The conversion is achieved in a series of stages, or heats. The ingot is placed in a furnace and allowed to reach an appropriate temperature for forging. Upon removal, it is manipulated between two horizontal dies which are driven hydraulically. These squeeze (or bite) the material, causing it to deform both in and out of the plane of the dies. Typically the ingot is deformed 20 to 30 times at various points along its length, with the pattern repeated with the ingot rotated through 90°, 45° and one final turn of 90°. At the end of this procedure, the ingot is returned to the furnace now with a characteristic octagonal cross-section. The deformation applied to the ingot causes substantial recrystallisation to a finer grain structure. After reheating the whole process is repeated.

To reduce a 60 cm diameter ingot down to the required 30 cm or so as many as seven heats may be required. In the later heats a temperature below the relevant solvus (for example, 1040 °C for the γ' in Waspaloy, 1010 °C for the δ in IN718) is employed to ensure that grain growth following deformation is inhibited. Usually, a final rounding operation is performed prior to slicing the billet for the subsequent forging operations. Typically the starting grain size is around ASTM000 (1 mm+), but, by the start of the third heat, the grain size has been reduced to around ASTM0 (0.35 mm). At the end of the process the grain size is ASTM5 or 6 (40–60 µm). Other ingot-to-billet conversion processes are possible, for example, gyratory forging machine (GFM) or radial forging, the principal difference being the strain profile imparted during processing.

D Open- and closed-die forging

Having achieved a fine-grained uniform structure through the conversion process, the objective of the forgemaster is to turn the cylinder of metal into a shape close to that of the turbine disc. The principal concern at this stage is not to promote excessive grain growth. Again the production of the disc is usually a three-stage process, typically comprising (i) upsetting by open-die forging, to produce an axisymmetric ingot of reduced thickness and greater cross-sectional area, (ii) blocking by closed-die forging, to place thicker and thinner sections in the appropriate place, and (iii) finishing, again by closed-die forging, to produce the desired disc shape. In these three operations the temperature is typically kept

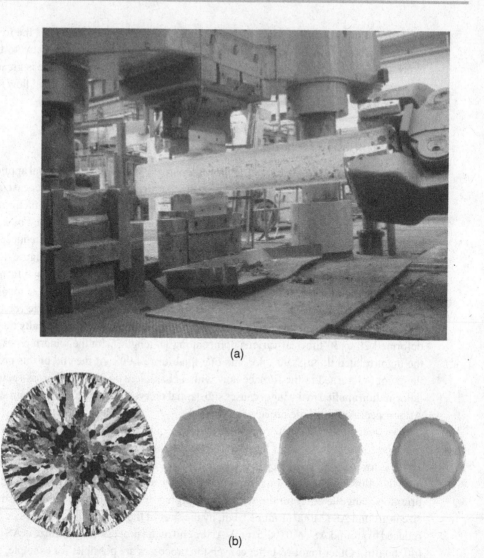

Fig. 4.11. (a) Cogging operation for the thermal-mechanical working of a superalloy billet.
(b) Cross-sections of the worked billet after various stages of working – with the unworked billet
at left. (Courtesy of Mark Roberts.)

very close to the appropriate solvus of the alloy being forged, thereby ensuring that any
recrystallisation that does occur results in a finer grain structure than that achieved in the
conversion stage. At the end of processing the forged discs are either quenched into a bath
of oil or water, or else left to cool on the shop floor.

A heat treatment is then applied to encourage the appropriate strengthening phases (for
example, γ') to precipitate in sufficient quantities to give the necessary high-temperature
properties. A quenching operation, typically into oil, is employed following the heat treatment to 'freeze in' the precipitation structure. Next, the discs are machined to a pre-specified
geometry which is known as the 'condition of supply'; this is usually rectilinear to allow

non-destructive evaluation (NDE) of the forging via ultrasonic testing. Finally, the rectilinear forging is machined to the final shape required for the turbine disc. These steps are obviously such that the mass of material in the final machined disc is only a small fraction of that in the turbine disc forging from which it is made. Consider, for example, an HP turbine disc for Rolls-Royce Corporation's AE3007 engine which powers the Cessna Citation X business jet. The final disc weighs about 8 kg, but it is machined from a disc of mass 45 kg.

Case study. Process modelling of forging operations

Although superalloys have been used for the turbine discs in jet engines for about 50 years, the processing is not without some challenges. The manufacturing stream is quite lengthy, involving a number of material suppliers, melters and forgemasters; at each stage there is a risk that defects will be introduced. Whilst these are usually detected by quality control procedures such as ultrasonic inspection, they nonetheless contribute to material wastage and hence to extra cost. Possible defects include (i) inadequate control of grain size and (ii) forging defects such as laps or inadequate filling of the dies.

Process modelling is one way of responding to this situation. Here, one builds mathematical models of the phenomena occurring during the manufacturing route. This should include continuum phenomena such as heat flow as well as the stresses and strains which are developed during deformation. However, it is becoming increasingly possible to simulate microstructure evolution such as recrystallisation, grain growth and texture. The emphasis is always on including as much as possible of the underlying theory, without making the simulations unwieldy, intractable or lengthy. It is important to have accurate thermophysical properties of the material, and a good estimate of the stress–strain behaviour at elevated temperatures.

There are many advantages of building a process model for the forging processes, as follows. A model, if sufficiently accurate, will have considerable predictive capability; see Figure 4.12. Models can be used to study the effect of changing process variables (such as die geometry, bite size, quenching rate) without resorting to experimental trials. They can therefore be used to optimise the processing in order to eliminate or reduce manufacturing waste. Furthermore, they generally help to reduce the number of design/make iterations required during component prototyping – this leads to savings in both time and money.

4.1.2 *Processing by the powder route*

The need for powder metallurgy (P/M) [12] for the production of superalloy turbine discs became apparent in the 1970s due to the development of alloys such as Rene 95, Astroloy and MERL 76 (with strength levels higher than earlier alloys such IN718 and Waspaloy), which were found to be unamenable to processing into the cast-and-wrought form. Initially, the use of P/M product was limited to gas turbines for military purposes, for example, the Pratt & Whitney F100 engine for the F-15 Eagle fighter, which entered service in 1974 [13], but commercial applications such as General Electric's CF6-80 engine for the Boeing 747-400 and the GE90 for the Boeing 777 have emerged as experience has been accumulated.

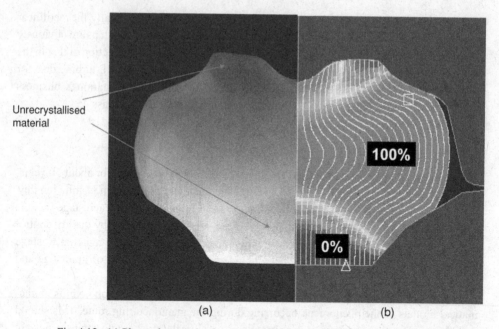

Unrecrystallised
material

(a)   (b)

Fig. 4.12. (a) Photomicrograph of a closed-die forging in the IN718 superalloy; (b) computer
simulation of the recrystallised regions in the forging, with the dead zones at top and bottom.
(Courtesy of Rob Guest.)

Advantages of powder-processing include improved property uniformity due to the elimination of macrosegregation and the development of a fine grain size. The disadvantages
are the extra cost associated with the production of the powder (this is offset somewhat
by the number of processing steps being fewer) and, arguably, the extra complexity of the
probabilistic lifing method required; see Section 4.3.

The first step in the production of P/M product is inert gas atomisation from master
melts produced by VIM processing; clearly, the cleanliness of these is important since
any contamination will be inherited by the powder. A typical arrangement is shown in
Figure 4.13. Molten metal is poured into a tundish, which contains a carefully designed
ceramic nozzle surrounded by one or more inert gas jets; usually argon is preferred. A
continuous stream of gas at high pressure is delivered to the stream of molten metal arriving
at the nozzle, the action of which causes its disintegration into spherical particles of diameter
30 to 300 µm. The atomisation chamber is designed to be long enough to allow solidification
of the particles before they reach the outlet of the chamber. Cooling rates are typically in
excess of 100 °C/s, being greater for finer particles. Figure 4.14 shows a SEM micrograph
of powder produced in this way. A number of variants of the process have been developed
with varying degrees of success, for example, soluble-gas atomisation in which atomisation
is induced by inserting a ceramic tube into the melt, the open end of which is connected
to a vacuum chamber; the dissolution of a soluble gas (usually hydrogen) over the molten
metal aids the atomisation process.

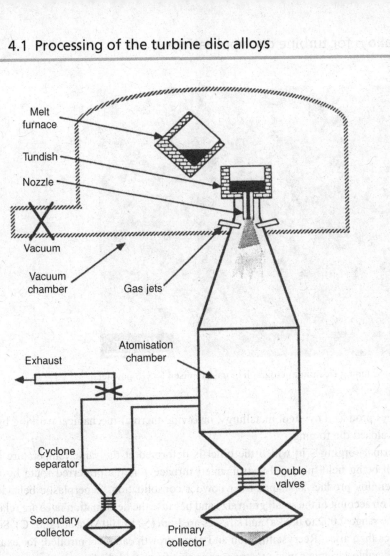

Fig. 4.13. Typical arrangement for the Ar atomisation used for the production of superalloys by powder metallurgy. (Courtesy of Tim Howson.)

Consolidation of the powder involves processing by extrusion. The fundamental challenge is to obtain good metallurgical bonding across the prior particle boundaries; to ensure bonding across oxide films which are inevitably present, the particles must be brought into contact at temperature and subjected to pressure and mechanical deformation. The powder is first packed into a carefully cleaned steel container. This is evacuated to encourage outgassing, sealed and then compacted either by hot isostatic pressing (HIP) or occasionally using closed-die forging. A very high-tonnage press with specialised tooling is then used to extrude the material to the high reduction ratios necessary for fine-grained billet; this step has been shown to disperse any non-metallic inclusions still present, thus improving the defect tolerance of the material. Some attempts have been made to eliminate the extrusion step, replacing it, for example, with a sub-solidus, hot isostatic pressing (SSHIP) process [14]; this is followed by conventional ingot-to-billet conversion of the type employed for

Fig. 4.14. Scanning electron micrograph of gas atomised IN718 powder.

superalloys produced by ingot metallurgy, involving thermal-mechanical working by cogging and closed-die forging.

Isothermal forging – in which the billet is deformed at the same temperature as the dies, both being held in a well-instrumented furnace – is the preferred route by which P/M superalloy product is shaped after powder consolidation. Superplastic behaviour is observed on account of the small grain size and because the deformation rates are relatively low (in the range 0.002 to 0.03/s) and stresses are high (50 to 100 MPa at 1200 °C). Several advantages then arise. Recrystallisation and grain growth can be controlled; for example, the considerable variations in microstructure arising from cold-die forging, due to excessive grain growth and die-chill, are avoided. Second, significant savings in material can be made since forgings can be produced of a geometry very close to the 'sonic shape' required by ultrasonic testing. Figure 4.15 illustrates possible processing sequences for the compressor discs of the GE F-101 engine, which was one of the first applications of the Rene 95 P/M superalloy [13].

When P/M processing was first applied to the production of turbine discs, the material was susceptible to the presence of defects – this was problematical in view of the need for fatigue resistance. Although the situation has improved very markedly, it is imperative to guard against the possibility of their presence. Consequently, the highest standards of handling are required, processing steps need to be strictly adhered to and quality assurance procedures must be enforced. For example, residual voids, pores and prior particle boundaries (PPBs) will be present if the steps taken to consolidate the powder are inadequate. Contamination by metallic inclusions remains a possibility, for example, by cross-contamination with a foreign alloy powder; this eventuality can be minimised by cleaning the chamber using washing runs. Of greatest concern is the possible presence of ceramic inclusions in the material,

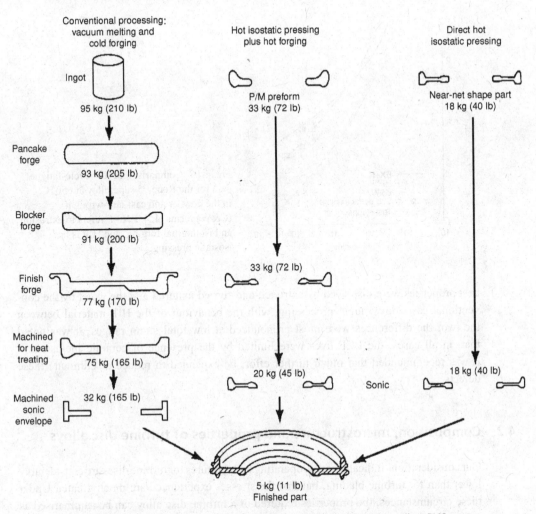

Fig. 4.15. Processing sequences for the production of F-101 compressor discs [13].

due to erosion of the melt crucible, tundish or nozzle. This is the major reason for the use of screening, which involves the passing of powder through filters of different mesh sizes; this removes the oversized particles, into which category the most damaging non-metallic inclusions fall. For most applications, the allowable size of powder is less than 230 mesh (63 μm), but, for the most stringent applications, it can be as low as 325 mesh (45 μm); obviously at such fine sizes the efficiency of the process is necessarily compromised. Blending of different powder sizes is commonplace.

The benefits of powder-processing over ingot metallurgy which produces cast-and-wrought product are now well established. For example, the low-cycle-fatigue (LCF) properties of Rene 95 at 650 °C have been compared [15] in the HIP, HIP + isothermally forged, extruded + isothermally forged and cast-and-wrought conditions; see Figure 4.16. The

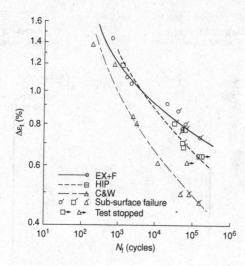

Fig. 4.16. Comparison of low-cycle-fatigue data for the Rene 95 superalloy at 650 °C, in the powder and cast-and-wrought (C&W) forms [15]; EX+F refers to extrusion and isothermal forging, and HIP is hot isostatic pressing.

best properties were displayed by extruded-and-forged material and the worst by the conventional cast-and-wrought processing, with the behaviour of the HIP material between the two; the differences were most pronounced at low total strain ranges. It was noted that, in all cases, the LCF lives were limited by the presence of ceramic inclusions – it was recommended that much further effort be expended to reduce or eliminate these defects.

4.2   Composition, microstructure and properties of turbine disc alloys

Our considerations indicate that the operating temperatures for turbine discs are considerably lower than for turbine blading, but that the stresses experienced are much greater. Under these circumstances, the properties required of a turbine disc alloy can be summarised as follows: (i) high yield stress and tensile strength to prevent yield and fracture, (ii) ductility and fracture toughness, to impart tolerance to defects, (iii) resistance to the initiation of fatigue cracking and (iv) fatigue crack propagation rates that are as low as possible. Creep resistance is also important, but traditionally it has been given less emphasis due to the lower operating temperatures and because a stress relaxation capability around notches and features of stress concentration is desirable.

What factors must be accounted for when designing a turbine disc alloy, or else when choosing one from the list of those available? Table 4.2 provides the compositions of some turbine disc alloys which are used in gas turbine engines [16]. Unsurprisingly, the chemical composition is found to be of crucial importance; however, just as importantly, if the best properties are to be attained then the microstructure must be conditioned in an optimum way. An added complexity, which does not apply to the case of the single-crystal superalloys considered in the previous chapter, is the presence of grain boundaries – these must be carefully engineered. Thus a strong appreciation of the relationship between alloy chemistry

Table 4.2. *The chemical composition of some common turbine disc alloys, in wt% [16]*

Alloy	Cr	Co	Mo	W	Nb	Al	Ti	Ta	Fe	Hf	C	B	Zr	Ni
Alloy 10	11.5	15	2.3	5.9	1.7	3.8	3.9	0.75	—	—	0.030	0.020	0.05	Bal
Astroloy	15.0	17.0	5.3	—	—	4.0	3.5	—	—	—	0.06	0.030	—	Bal
Inconel 706	16.0	—	—	—	2.9	0.2	1.8	—	40.0	—	0.03	—	—	Bal
Inconel 718	19.0	—	3.0	—	5.1	0.5	0.9	—	18.5	—	0.04	—	—	Bal
ME3	13.1	18.2	3.8	1.9	1.4	3.5	3.5	2.7	—	0.35	0.030	0.030	0.050	Bal
MERL–76	12.4	18.6	3.3	—	1.4	0.2	4.3	—	—	0.45	0.050	0.03	0.06	Bal
N18	11.5	15.7	6.5	0.6	—	4.35	4.35	—	—	—	0.015	0.015	0.03	Bal
Rene 88DT	16.0	13.0	4.0	4.0	0.7	2.1	3.7	—	—	—	0.03	0.015	0.03	Bal
Rene 95	14.0	8.0	3.5	3.5	3.5	3.5	2.5	—	—	—	0.15	0.010	0.05	Bal
Rene 104	13.1	18.2	3.8	1.9	1.4	3.5	3.5	2.7	—	—	0.030	0.030	0.050	Bal
RR1000	15.0	18.5	5.0	—	1.1	3.0	3.6	2.0	—	0.5	0.027	0.015	0.06	Bal
Udimet 500	18.0	18.5	4.0	—	—	2.9	2.9	—	—	—	0.08	0.006	0.05	Bal
Udimet 520	19.0	12.0	6.0	1.0	—	2.0	3.0	—	—	—	0.05	0.005	—	Bal
Udimet 700	15.0	17.0	5.0	—	—	4.0	3.5	—	—	—	0.06	0.030	—	Bal
Udimet 710	18.0	15.0	3.0	1.5	—	2.5	5.0	—	—	—	0.07	0.020	0.020	Bal
Udimet 720	17.9	14.7	3.0	1.25	—	2.5	5.0	—	—	—	0.035	0.033	0.03	Bal
Udimet 720LI	16.0	15.0	3.0	1.25	—	2.5	5.0	—	—	—	0.025	0.018	0.05	Bal
Waspaloy	19.5	13.5	4.3	—	—	1.3	3.0	—	—	—	0.08	0.006	—	Bal

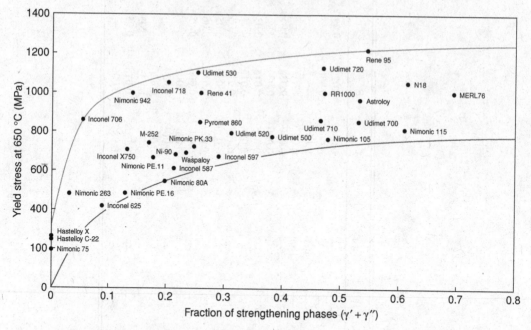

Fig. 4.17. Variation of the yield stress at 650 °C with the total fraction of the γ' and γ'' strengthening phases, for a number of common turbine disc alloys.

and microstructure is required if an optimum combination of properties is to be achieved. The experience gained so far indicates that the following three guidelines should be adhered to.

4.2.1   Guideline 1

> To impart strength and fatigue resistance, the fraction of the γ' phase should be optimised by appropriate choice of the γ'-forming elements (Al, Ti and Ta) – placing it in the range 40% to 55% – and heat treatments chosen to promote a uniform distribution of γ' particles.

The yield stresses of the turbine disc alloys correlate very strongly with the proportions of the strengthening phases γ' and γ'', and thus a first overriding concern when designing a turbine disc alloy is the choice of the concentration of elements which promote these phases, such that the desired strength levels are achieved. One can see this by using thermodynamic software such as Thermo-Calc to estimate the fractions of γ' and γ'' present in the alloys given in Table 4.1, and comparing the predictions with mechanical property data. The results, see Figure 4.17, confirm that the yield stress at 650 °C correlates very strongly with the sum of the fractions of the strengthening phases; the yield stress at ambient temperature behaves in the same way. The creep performance, for example the stress for 1000 h life at

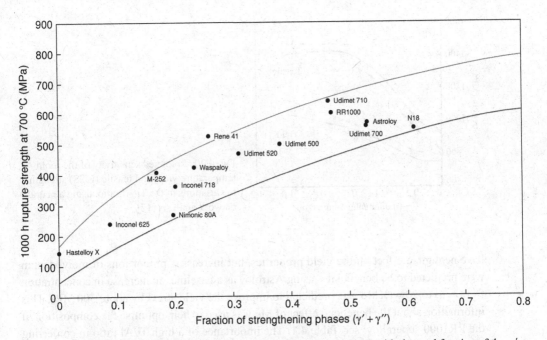

Fig. 4.18. Variation of the 1000 h creep rupture strength at 700 °C with the total fraction of the γ' and γ'' strengthening phases, for a number of common turbine disc alloys.

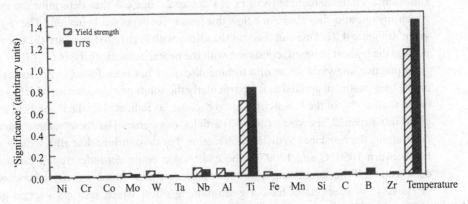

Fig. 4.19. The relative 'significance' of the elemental concentrations and temperature for a neural network analysis of the yield and ultimate tensile strength (UTS) of turbine disc superalloys [17].

700 °C, shows a similar correlation; see Figure 4.18. These findings have been confirmed by a neural network analysis [17] on a dataset containing information on the compositions and strength levels of over 200 polycrystalline superalloys; this has confirmed that the concentrations of Al, Ti and Nb are the most influential in improving the yield properties; see Figure 4.19. Interestingly, when the neural network was interrogated after being trained, it predicted that higher Ti/Al ratios are important in conferring strength (see Figure 4.20); this is consistent with the hardening theories of Chapter 2, which emphasise the importance of the APB energy, which increases as the Ti/Al ratio is raised [18]. Cobalt was found to

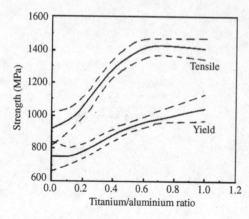

Fig. 4.20. Predicted sensitivity of the room temperature yield and tensile (UTS) strength to variations in the Ti/Al ratio, using a neural network analysis [17].

have negligible effect on the yield properties, but increasing proportions of molybdenum were predicted to be beneficial – using Astroloy as a baseline, an increase in concentration from zero to 20 wt%Mo was predicted to improve the yield stress by about 500 MPa. This information about the beneficial effect of Mo was used to help optimise the composition of the RR1000 superalloy; see Table 4.1. The importance of a high Ti/Al ratio in conferring strength can be further demonstrated by considering the data; if one plots the yield stress against the root of the sum of the strengthening phases, a reasonably straight line is found, consistent with the hardening theories of Chapter 2 – one can then determine the extent to which any one alloy lies above or below this 'mean', or expected, behaviour. The results are given in Figure 4.21. One can see that the alloys with high Ti/Al ratio are the ones which possess the highest strength, consistent with the neural network analysis of [17].

In practice, the yield stress of a turbine disc alloy has been found to depend critically on the heat treatment applied to it – particularly the solutioning treatment, which is carried out in the vicinity of the γ' solvus. This is because, as indicated by the theoretical analysis given in Chapter 2, the size of the γ' particles is as crucial as their volume fraction in determining the resistance to dislocation motion. For most turbine disc alloys, the γ' solvus lies between 1050 °C and 1200 °C, the exact value being dependent upon the chemical composition. If the solutioning heat-treatment temperature is sub-solvus, the undissolved, or *primary*, γ' precipitates have a grain-pinning effect which restricts γ-grain growth; see Figure 4.22. Super-solvus heat treatment eliminates the primary γ' and the γ grains are correspondingly coarser. It follows that the γ' phase distribution is, in practice, rather complex; see Figure 4.23, which is a schematic illustration of the microstructure of a turbine disc alloy [19]. In addition to the primary γ', intragranular secondary and tertiary γ' particles are present within the γ grains, typically at particle diameters of $\sim$100 nm and $\sim$50 nm, respectively. This bimodal distribution occurs because of the interplay between the kinetics of γ' particle nucleation, growth and coarsening, which occurs during isothermal solutioning and ageing heat treatments; the cooling rate is also important since the equilibrium fraction of γ' is rarely attained during these, so that further γ' nucleation and growth occurs during cooling. During the ageing heat treatment, which is typically carried out at around 700 °C for several hours, an age-hardening response is observed, as in many precipitation-hardened

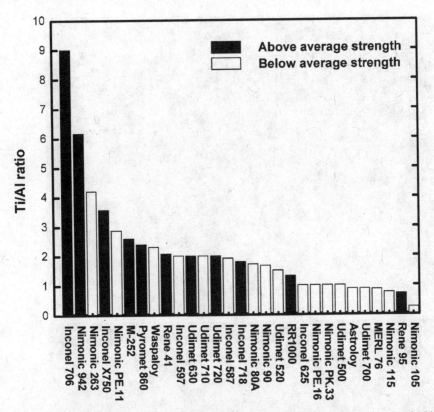

Fig. 4.21. Variation of the Ti/Al ratio in a number of polycrystalline superalloys, ranked in order of strength consistent with a deviation away from a perfect correlation of yield stress and the square root of the fraction of strengthening phases.

systems. The data in Figure 4.24 for the yield stress and UTS of the Udimet 720Li superalloy illustrate this; in this case, the length of the heat treatment is optimal at about 30 h [19]. When the γ' particle distributions giving rise to this behaviour are examined, one finds that the size of the secondary γ' particles is unaltered by the heat treatment; see Figure 4.25. On the other hand, the tertiary γ' coarsens considerably, and this proves unambiguously that it controls the strength level achieved. Clearly, the heat treatment must be chosen carefully if optimum properties are to be attained. On the basis of these observations, one can argue that, for absolute static strength, higher solutioning temperatures are to be preferred, since the proportion of primary γ' is then reduced, so that a greater proportion of the available γ' is then placed in the range 50 nm to 100 nm where it has the greatest effect.

The situation, however, is complicated because the fatigue properties must also be considered. The distribution of the γ' precipitates has been found to have a significant influence on the dwell-time fatigue crack propagation (FCP) rate; for example, an improvement in FCP rate in the Paris regime of about an order of magnitude has been reported in ref. [20] when heat treatment was super-solvus rather than sub-solvus, this microstructural effect being more potent than minor changes in chemistry involving Nb/Ta ratio and Co content

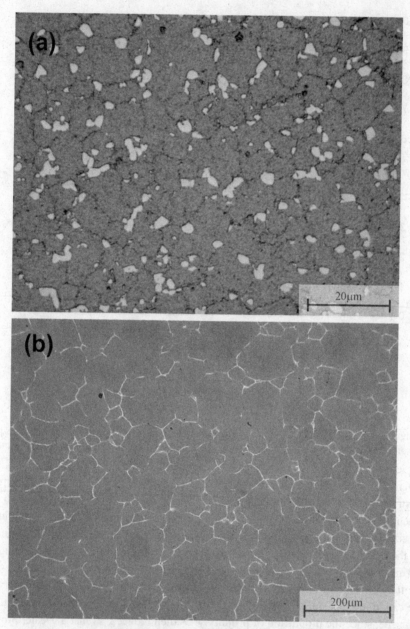

Fig. 4.22. Optical micrographs of RR1000 superalloy: (a) sub-solvus heat-treated for 4 h at 1130 °C; (b) super-solvus heat-treated for 4 h at 1170 °C. The solvus temperature in this alloy is about 1150 °C. Note the grain-pinning effect, which occurs when the heat-treatment temperature is sub-solvus. (Courtesy of Rob Mitchell.)

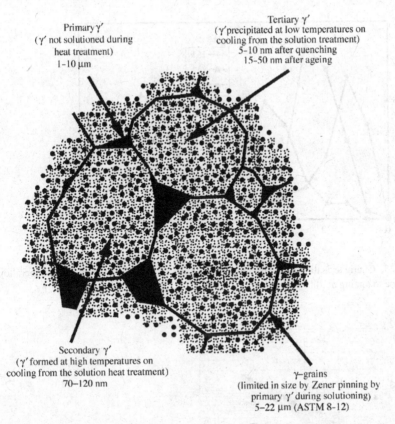

Primary γ′
(γ′ not solutioned during
heat treatment)
1–10 μm

Tertiary γ′
(γ′ precipitated at low temperatures on
cooling from the solution treatment)
5–10 nm after quenching
15–50 nm after ageing

Secondary γ′
(γ′ formed at high temperatures on
cooling from the solution heat treatment)
70–120 nm

γ–grains
(limited in size by Zener pinning by
primary γ′ during solutioning)
5–22 μm (ASTM 8-12)

Fig. 4.23. Schematic illustration of the distribution of the γ′ phase in a turbine disc alloy [19].

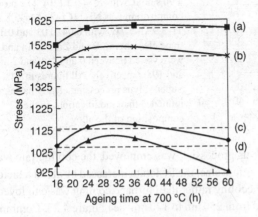

Fig. 4.24. Hardening response of the Udimet 720Li superalloy due to heat treatment at 700 °C [19].
(a) Room temperature, UTS; (b) room temperature, 0.2% proof stress; (c) UTS at 600 °C; (d) 0.2%
proof stress at 600 °C.

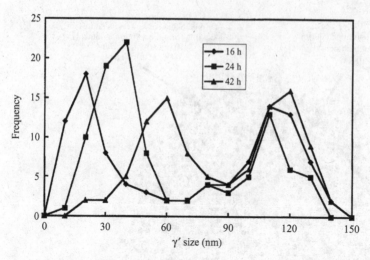

Fig. 4.25. Characterisation of the bimodal distribution of γ' in the Udimet 720Li superalloy and its response to ageing at 700 °C [19].

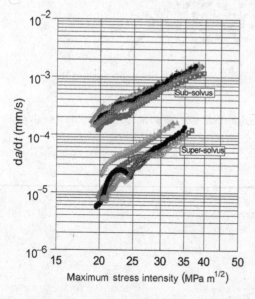

Fig. 4.26. Dwell fatigue crack propagation data for seven P/M superalloys at 704 °C and a 90 s hold, with $R = 0.1$ [20]. The base composition was Ni–11Cr–2.6Mo–3.8 Ti–3.9Al–5.7W, with 0.04, 0.03 and 0.1 of the trace elements C, B and Zr; Nb, Ta and Co were varied between 0 and 1, 1 and 2 and 15 and 19, respectively. All figures in wt%. Super-solvus processing clearly has a greater influence than modifications in the composition of the alloy.

(see Figure 4.26). When super-solvus processing was employed, the cooling rate was also found to be important at constant γ-grain size; the FCP rate increased by about a factor of 4 at a K_{max} of 25 MPa m$^{1/2}$ when air cooling rather than furnace cooling was employed – because the γ' particle size decreased from 250 nm to 170 nm; see Figure 4.27. Combinations of the original super-solvus heat treatment followed by subsequent sub-solvus heat treatment were shown to be very detrimental, with a further increase in FCP rate of at least 3 – this can be attributed to the dissolution of the larger γ' particles and the reprecipitation of smaller ones upon cooling. From these observations it seems that the γ' particle size

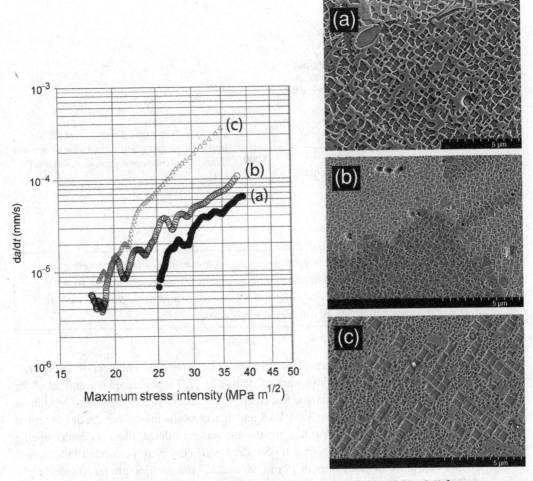

Fig. 4.27. Dwell fatigue crack propagation data at 704 °C (90 s hold and $R = 0.1$) for an experimental P/M superalloy, heat-treated to give different γ' sizes [20]: super-solvus solutioning, followed by (a) furnace cooling, no resolutioning (b) air cooling, no resolutioning and (c) air cooling, sub-solvus resolutioning at 1171 °C.

distribution for optimum FCP resistance is different from that required for optimum static strength. From a practical standpoint, however, it is not usually possible to manipulate the γ' particle distribution independently of the size of the γ grains, and this represents a dilemma when designing the heat treatment for a turbine disc superalloy. Tensile strength and low-cycle-fatigue (LCF) life (controlled largely by fatigue initiation) are best when the grain size is small, but the creep properties and fatigue crack propagation rates are not optimum; these are better at large grain sizes. This is illustrated in Figure 4.28, which is adapted from ref. [21]. For a balanced portfolio of properties, a compromise needs to be struck between the need for absolute static strength and resistance to cyclic fatigue crack propagation. This challenge is discussed further in the following section.

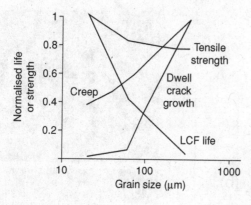

Fig. 4.28. Schematic illustration of the important properties of turbine disc alloys and their dependence upon grain size. After ref. [21].

4.2.2   Guideline 2

> *The grain size should be chosen for the desired combination of yield strength, resistance to fatigue crack initiation (both of which scale inversely with grain size), creep strength and resistance to fatigue crack growth (which scale directly with it). A size in the range* 30 μm *to* 50 μm *is commonly chosen.*

The necessity to control the grain size has been demonstrated by a number of researchers, but a consideration of the careful work of Bain *et al.* [22] is warranted because all of the important properties – yield, fatigue and creep – were measured and reported. The Udimet 720 superalloy, processed using VIM/VAR and then combinations of hot-die or isothermal forging and sub- or super-solvus heat treatments, was considered. Identical double ageing treatments of 760 °C/8 h/air cool and 650 °C/24 h/air cool were performed in all cases. In this way, uniform grain sizes of 19 μm, 90 μm, 127 μm and 360 μm (corresponding to ASTM grain sizes of 8.5, 4, 3 and 0, respectively) were produced; unfortunately, one cannot rule out possible differences in the distributions of secondary and tertiary γ', although the micrographs presented [22] indicate that no primary γ' was present other than in the 19 μm material. The testing of tensile specimens indicated that the finest-grained material exhibited a room temperature yield stress (YS) of about 1200 MPa, approximately 200 MPa greater than the coarsest-grained material – a figure which is approximately equal to the difference in the YS reported for Udimet 720Li in the peak-aged and over-aged conditions; see Figure 4.29 (note that the Udimet 720 and 720Li superalloys have marginally differing chemistries (see Table 4.1)). The improvement reported in the ultimate tensile strength (UTS) was of a similar magnitude. Strain-controlled LCF tests were also carried out on smooth, polished testpieces with $R = 0$ and a frequency of 0.33 Hz at a total strain range of 0.80%. The LCF results reflect the improvements in YS won by grain refinement; the LCF life was approximately one order of magnitude worse at the largest grain size of 360 μm than at 19 μm. These considerations indicate that for polycrystalline superalloys, for the best static yield and LCF properties, fine grain sizes are to be preferred. This grain size dependence of the LCF performance must be added to the intrinsic LCF strength due the

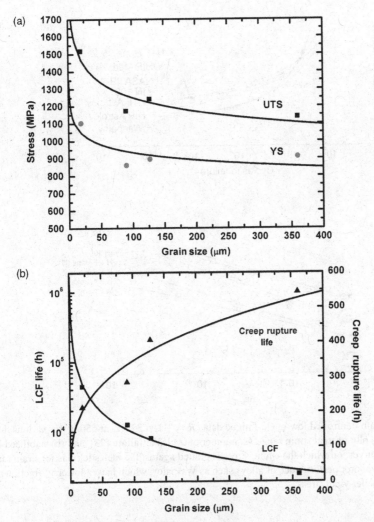

Fig. 4.29. Data for the mechanical properties of the Udimet 720 superalloy, as a function of the grain size developed by varying combinations of hot-die or isothermal forging and sub- or super-solvus heat treatments: (a) yield stress (YS) and ultimate tensile strength (UTS), (b) creep rupture life at 700 °C/690 MPa and low-cycle-fatigue (LCF) life at 650 °C at a total strain range of 0.8% with $R = 0$. Data from ref. [22].

alloy composition [23,24], which correlates strongly with the γ' fraction and thus also the yield stress; see Figure 4.30.

The influence of the grain size is very different when one considers the rates of fatigue crack growth and the creep rupture performance. Bain *et al.* [22] performed constant load creep tests at 700 °C and 690 MPa; for the largest grain size of 360 μm, the rupture life approached 600 h under these conditions, but at 19 μm it decreased to about 200 h. Similar observations have been made [25] in the NR3 superalloy heat treated at super- and sub-solvus temperatures to give coarse- and fine-grain structures, respectively (see Figure 4.31);

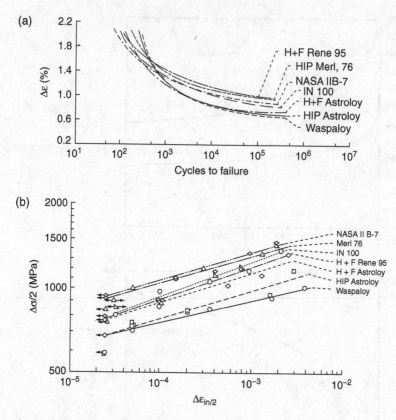

Fig. 4.30. Strain-controlled low-cycle-fatigue data ($R = -1$, 0.33 Hz) at 650 °C for various turbine disc alloys: (a) plot of total strain range vs. number of cycles to failure [23], and (b) stabilised cyclic stress–strain curves, on which the cyclic stress is plotted against the inelastic (plastic) strain range [24]. Note the poorer performance of alloys such as Waspaloy, which have a lower γ' fraction and therefore a smaller yield stress.

the minimum creep rate differed by about an order of magnitude with grain size when testing was carried out at 750 °C and 300 MPa, being better for the coarse-grain size, although the influence was less pronounced at lower temperatures and higher stresses. Thus larger grain sizes are beneficial for creep performance. The same is found to be true of the fatigue crack propagation rate; this is important if the damage-tolerant method for life assessment is being used (see Section 4.3) since its magnitude controls the fatigue life which can be declared. Fatigue cyclic crack growth rates were measured by Bain *et al.* [22] in air at 650 °C at an R ratio of 0.05, at a frequency of 3.3 Hz and with a 5 min dwell at the maximum load. A second set of tests was carried out at 425 °C without the dwell, but under otherwise identical conditions. Particularly at the higher temperature, the dependence of crack growth rate on grain size was considerable (see Figure 4.32), being about two orders of magnitude smaller for the larger grain size; consistent with the promotion of a tortuous crack path, this was attributed to rougher fracture surfaces which were largely intergranular in nature. At 425 °C, the dependence on grain size was less pronounced, but the

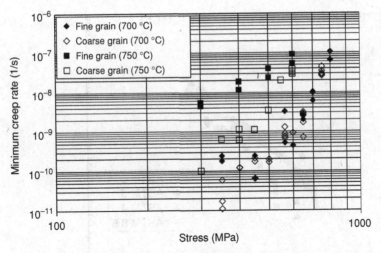

Fig. 4.31. Constant load creep data for the NR3 P/M superalloy, at temperatures of 700 °C and 750 °C in both the fine- and coarse-grained conditions [25].

smaller grain size was again found to perform better, particularly in the threshold regime at low ΔK; the fracture surfaces were transgranular and faceted. However, at both temperatures, the Paris exponents in stage II growth were not heavily influenced by grain size. These observations are consistent with the mechanisms of fatigue crack growth which are agreed upon [23,26]: at low temperatures, high strain rates and low ΔK, there is a strong tendency for crystallographic cracking on {111} planes due to the 'planar slip', which is then prevalent due to the limited number of slip systems that are activated. At higher temperatures and larger ΔK, which increases the size of the reversible plastic zone, slip is more homogeneous and 'wavy'. Li, Ashbaugh and Rosenberger [27] have characterised these effects by examining the fracture surfaces of IN100 under LCF conditions using high-resolution SEM microscopy; see Figure 4.33. At low temperatures, fatigue crack growth is found to occur parallel to slip band traces and thus along {111} planes (see Figure 4.33(c)); but, as the temperature increases, {001} facets become increasingly common, as evidenced by the cuboidal morphology of the γ' precipitates on the fracture surface. Related work on IN100 [28] has demonstrated a considerable influence of environment on the fatigue crack growth rate – at 620 °C with $R = 0.15$ and a 30 s dwell at the peak load, a decrease in the vacuum from 1 to 10^{-8} atm produced an order of magnitude decrease in the crack growth rate. When incipient fatigue cracks – after the initiation period – have dimensions on the scale of the grain size, their crack growth rates are several orders of magnitude larger than predicted from the linear elastic fracture mechanics, Paris Law regime [29–31]; see Figure 4.34. Such 'short-crack' growth, which can occur below the fatigue crack threshold for long cracks, is of importance for turbine disc applications; this is because of the necessity for an adequate lifing policy – for example, the life-to-first-crack method described in Section 4.3 – which considers a turbine disc to have 'failed' before the crack reaches a length consistent with the K_{1c} criterion. These effects are not accounted for in the traditional damage-tolerant approach.

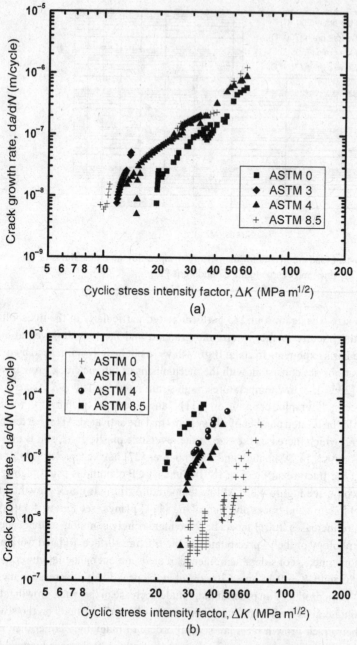

Fig. 4.32. Dwell fatigue crack propagation data for the Udimet 720 superalloy, as a function of the grain size developed by varying combinations of hot-die or isothermal forging and sub- or super-solvus heat treatments [22]: (a) 650 °C, with a 5 min dwell at peak load and (b) 425 °C. In both cases, $R = 0.05$.

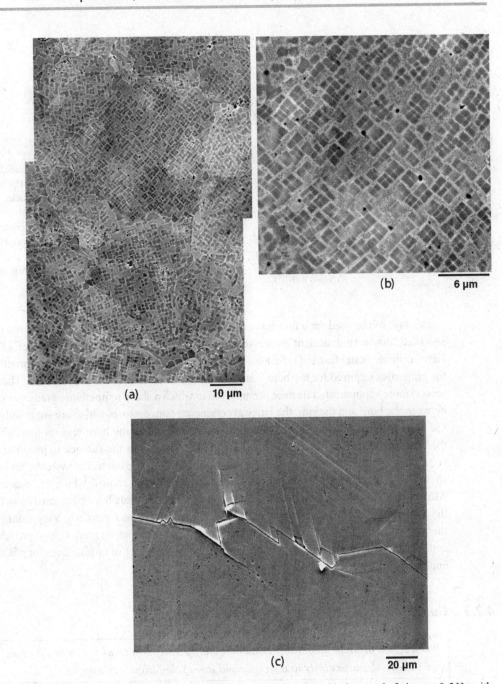

Fig. 4.33. Fracture surfaces for the IN100 P/M superalloy tested in low-cycle fatigue at 0.5 Hz with $R = 0.05$ [27]. (a), (b) SEM images of the {001} fracture surfaces typical of failure at 538 °C, showing the presence of cuboidal facets belonging to the γ' phase. (c) {111} fracture surface typical of fatigue crack growth at room temperature – note that the fracture path is parallel to the traces of the slip lines in the top right of the image.

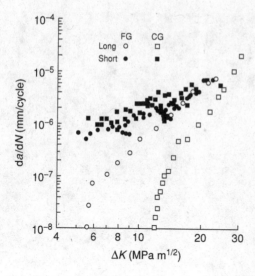

Fig. 4.34. Room-temperature fatigue crack propagation data for Astroloy at $R = 0.1$ and 40 Hz [29], for both long and short cracks; FG and CG refer to fine-grained and coarse-grained material, respectively. Note that the fatigue crack growth occurs more rapidly for short than for long cracks; also, when cracks are short, the effect of grain size is much reduced.

Because of the need for a fine grain size in the bore regions and a coarser one at the rim, so-called 'dual heat treatment' processes have been developed (for example, see ref. [32]). Here, a conventional furnace is first used to heat-treat the disc to a grain size appropriate for the properties required for the bore – at a temperature which is usually sub-solvus. Then, a second more sophisticated furnace is employed in which a sharp temperature gradient exists between the bore and the rim, the latter experiencing temperatures which are super-solvus; see Figure 4.35. For this purpose, cooling gas is directed at the bore regions throughout the heat treatment step, and insulating materials are placed in the furnace to promote the required temperature gradients. The use of thermocoupling and controller systems has been shown to be necessary to achieve the levels of temperature control which are required. Although this technology has yet to be used in production, trials have been carried out on the Rene 104 powder superalloy [32]; the ASTM grain size was shown to vary from 6 at the rim to about 11 at the bore, over a transition zone of about 50 mm. This approach is considered by the inventors to be advantageous for the processing of turbine discs for military applications.

4.2.3   Guideline 3

> *When added in small quantities, grain-boundary elements such as boron and carbon are beneficial, particularly to the creep and low-cycle-fatigue resistances.*

Grain-boundary elements such as boron and carbon are beneficial when added in small quantities, since they segregate to the γ/γ interfaces and increase the work of cohesion. Improvements in the creep rupture strength, creep ductility and low-cycle-fatigue behaviour then result. However, when added in excessive quantities, the precipitation of borides and carbides is promoted; it seems likely that at these levels any further additions of the

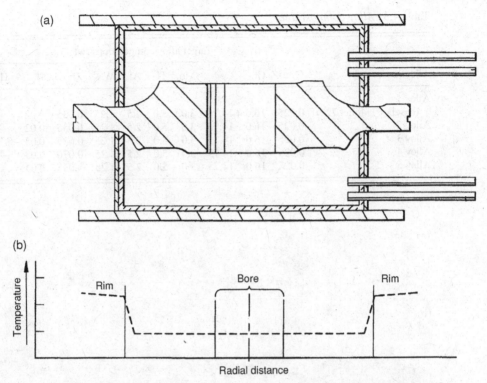

Fig. 4.35. Illustration of the arrangement for the dual heat-treatment process [32], in which a turbine disc forging is subjected to a temperature field which is deliberately engineered such that it varies from the bore to rim regions.

grain-boundary elements are ineffective. The optimum concentrations are approximately 0.03 wt%B and 0.025 wt%C; see Table 4.1. At these levels, it can be demonstrated that the γ–grain boundaries are covered with just a few monolayers of these grain-boundary elements. Other elements are often quoted as being effective at strengthening the grain boundaries. For instance, there is some evidence that zirconium, as a grain-boundary segregant, plays a role as a 'getter' of deleterious 'tramp' elements such as sulphur and phosphorus. However, it should be emphasised at this stage that the precise role played by the grain-boundary elements is controversial; more research is required.

To support these statements, consider the results reported by Garosshen et al. [33], who studied the mechanical properties of an experimental P/M alloy of base composition Ni–19Co–12.5Cr–5Al–4.4Ti–3.3Mo (wt%). When doped with 0.02 wt%B, 0.05 wt%Zr and 0.003 wt%C, the creep rupture life at 732 °C and 655 MPa improved remarkably from less than 0.1 h to 69 h; however, further additions of carbon to 0.06 wt% were ineffective, and additions of boron beyond 0.02 wt% reduced the rupture life, albeit only by a few hours. Boron additions over the solubility limit were shown to result in the intergranular precipitation of the M_3B_2 compound and caused no further improvement in mechanical properties, so it can be concluded that the strengthening effect of boron is not due to the formation of borides; furthermore, the additions of carbon, boron and zirconium had no effect on

Table 4.3. *Alloy compositions*

Alloy	Target alloy compositions (wt%)										
	C	Cr	Co	Mo	Tl	Al	W	Zr	B	Hf	NI
Alloy 1											
(baseline Udimet 720)	0.025	16.0	14.75	3.0	5.0	2.5	1.25	0.035	0.02	—	Bal
Alloy 2	0.025	16.0	14.75	3.0	5.0	2.5	1.25	0.035	0.02	—	Bal
Alloy 3	0.025	16.0	14.75	3.0	5.0	2.5	1.25	0.035	0.03	0.75	Bal
Alloy 4	0.025	16.0	14.75	3.0	5.0	2.5	1.25	0.070	0.03	—	Bal
Alloy 5	0.025	16.0	14.75	3.0	5.0	2.5	1.25	0.035	0.04	—	Bal

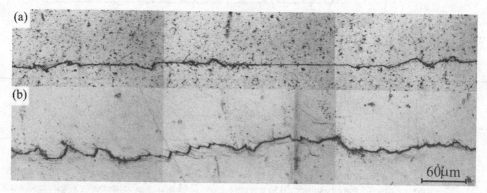

Fig. 4.36. Fatigue crack propagation paths in the IN718 superalloy doped with (a) 12 ppm B and (b) 29 ppm B [34]. Testing at room temperature in air, with $R = 0.05$.

the morphology of the γ' phase. Moreover, the fatigue crack progagation rates at 649 °C were improved marginally, but by a factor of no greater than 2; this finding is consistent with recent work on boron-doped IN718, which has demonstrated that the tortuosity of the fatigue crack path was greatly increased as the level of boron doping was increased from 12 ppm to 29 ppm; see Figure 4.36 [34]. The effect on fatigue initiation as manifested in the low-cycle-fatigue (LCF) life is very much more significant. Jain *et al.* [35] have doped Udimet 720Li in P/M form with varying concentrations of boron to 0.04 wt%, at a constant concentration of 0.025 wt%C; the effects of increasing the Zr concentration from 0.035 to 0.070 wt% and Hf from zero to 0.75 wt% were also studied. Testing was carried out at both 425 °C and 650 °C, at two grain sizes corresponding to ASTM 11 and ASTM 9, which were generated by sub- and super-solvus heat treatment, respectively. Un-doped Udimet 720Li in the cast-and-wrought condition was used as a control. The results are given in Figure 4.37 and Table 4.3. The results demonstrate that additions of the grain-boundary elements are beneficial to LCF life; improvements of at least 50% were noted for most conditions considered. Interestingly, the fatigue life for the fine-grained material (ASTM 11) was improved more substantially at 425 °C, whereas that for the coarse-grained material improved at 650 °C; no explanation of this finding was offered. In this study, no substantial

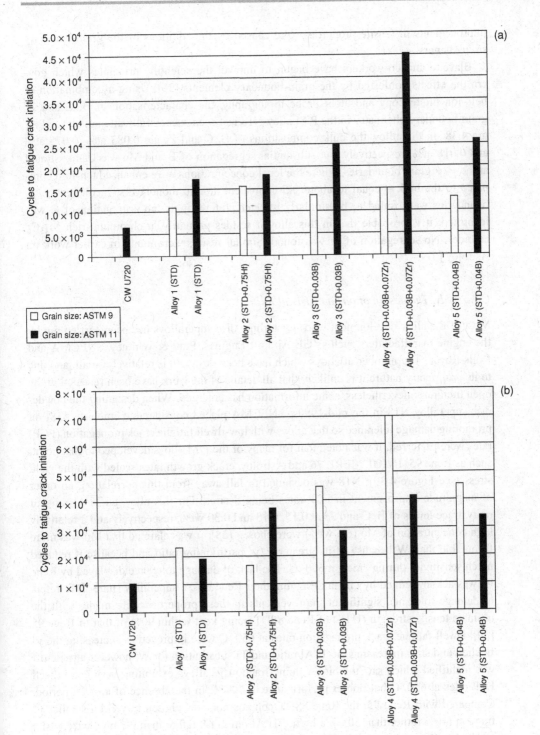

Fig. 4.37. Data showing the influence of small amounts of B, Zr and Hf on the LCF performance of the P/M U720Li turbine disc alloy [35]. LCF data at (a) 425 °C, $\Delta c_r = 0.9\%$, $R = 0.0$, air; (b) 650 °C, $\Delta c_r = 0.8\%$, $R = 0.0$, air. Note that CW U720 refers to cast-and-wrought material. For alloy compositions see Table 4.3.

improvements in tensile properties were caused by the additions of the grain-boundary strengtheners.

Blavette and co-workers have begun to unravel the scientific principles which govern the effects conferred by the grain-boundary elements [36]. Using a combination of field-ion microscopy and atom probe tomography, the characterisation of B, C and Zr grain-boundary chemistry in the P/M alloy N18 has been successfully attempted; see Figure 4.38. In this alloy, the bulk compositions of B, C and Zr are 0.083 at%, 0.075 at% and 0.018 at%, respectively, but substantial segregation of B and Mo was demonstrated at the γ/γ grain boundaries, where the local concentrations were enhanced ten-fold compared to the bulk – equal to about 0.5 equivalent atomic monolayers. Incoherent γ/γ' boundaries were found to be similarly affected. Interestingly, no segregation of C was identified; it is probable that in this alloy it resides primarily in chromium-rich $M_{23}C_6$ carbides. No segregation of Zr was found. Similar results were found in earlier work on Astroloy [37].

Case study. The design of turbine disc alloys

Most of the alloy development work for turbine disc superalloys has been carried out by the engine manufacturers such as GE Aircraft Engines, Pratt & Whitney, SNECMA and Rolls-Royce. The number of alloys which have been designed is relatively small, and due to its proprietary nature it is unlikely that all details of the work have been revealed in the open literature; nevertheless, some information has emerged. When designing the powder disc superalloy N18 in the mid-1980s, SNECMA placed considerable emphasis [38] on promoting damage tolerance so that alloys with low-dwell fatigue crack propagation (FCP) rates were preferred. It was noted that for many of the P/M alloys developed to that point, such as Rene 95, IN100, MERL 76 and Astroloy, crack growth rates scaled with the yield stress (see Figure 4.39); N18 was designed to fall away from this correlation, an order of magnitude improvement in FCP rates being desired. For this alloy, an Ti/Al ratio of unity, trace levels of B, C and Zr at 0.15, 0.15 and 0.30 wt%, respectively and a relatively high concentration of Mo (6.5 wt%) were chosen [38]; it was claimed that Mo was more favourable than W because of its greater γ/γ' partitioning ratio and because it reduced notch sensitivity during creep at 650 °C. The lack of damage tolerance displayed by Rene 95 was acknowledged by GE in designing their Rene 88DT superalloy (note 'DT' refers to damage tolerant); significant improvements in these properties were made, with the ultimate tensile strength (UTS) of Rene 88DT being kept within 90% of that of Rene 95 [39]. Dwell fatigue crack propagation rates at 650 °C were improved by increasing the γ' fraction and size, increasing the Ti/Al ratio and Ta substitution for W; however a trade-off was identified which saw the high-volume-fraction γ' alloys exhibiting very good dwell FCP rates at 650 °C but poorer performance at 400 °C in the absence of a dwell period. Compared with Rene 95, the Rene 88DT composition was chosen to yield a smaller γ' fraction (40% rather than 50%), a higher Ti/Al ratio (1.8 rather than 0.7 on a wt% basis) and higher Cr and Co contents (16 wt% rather than 13 wt%, and 13 wt% rather than 8 wt%, respectively) – this last change being made despite substitutions of Co and Cr providing no effect on dwell-time fatigue crack growth rate. The significant effect of Ta in reducing

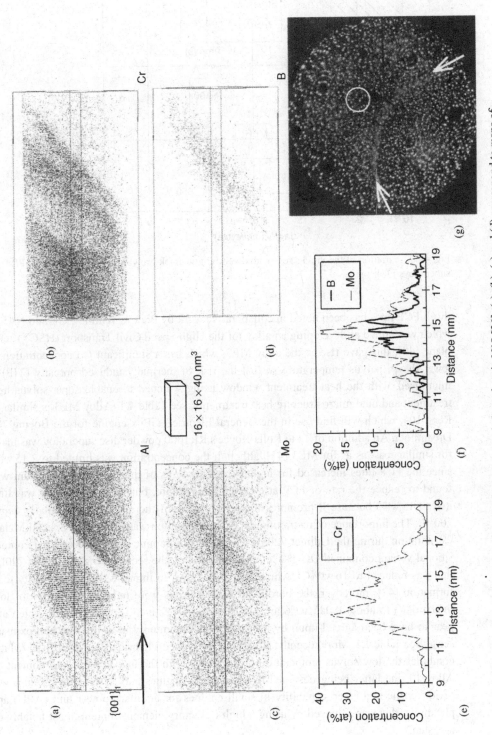

Fig. 4.38. Analysis using atom probe tomography of a γ / γ grain boundary in N18 [36]: (a), (b), (c) and (d) correspond to maps of Al, Cr, Mo and B; (e) and (f) are concentration profiles normal to the boundary and (g) is a field-ion image (FIM) of the grain boundary.

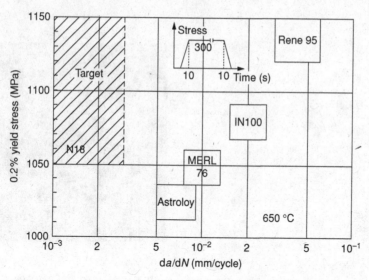

Fig. 4.39. Variation of the yield stress with dwell fatigue crack growth rate for a number of P/M superalloys [38].

dwell FCP rates has been noted in collaborative work by Pratt & Whitney, GE and NASA, which was aimed at developing an alloy for the High Speed Civil Transport (HSCT) aeroplane [40]; this gave rise to the alloy ME3, which has a significant Co concentration to lower the γ' solvus temperature so that the risk of thermally induced porosity (TIP) is minimised, with the heat treatment window being widened to enable super-solvus heat treatment and dual microstructure heat treatments; see Table 4.1. Alloy ME3 is similar to Rene 104, which will find use in the General Electric's GEnx engine for the Boeing 787 Dreamliner. An addition of Ta to Rolls-Royce's RR1000 powder disc superalloy was made for similar reasons as for ME3 [41], although the concentration was limited to 2.15 wt% since the FCP rates increased for higher concentrations of this element. Chromium was found to reduce the rate of FCP at elevated temperature, but its concentration was limited to 15 wt% because it promotes the precipitation of the σ phase, particularly above 700 °C. The importance of restricting the Cr concentration was learned by Special Metals Corporation during the Udimet 720 development programme [16] – the original compositional variant contained 18 wt%Cr, but in developing the second variant, Udimet 720Li, this was reduced to 16 wt%Cr, which has been shown to improve the susceptibility to σ formation [42]; moreover, the boron and carbon levels were reduced (Li refers to 'low interstitial') to about 0.025 wt% and 0.018 wt%, respectively, concentrations which appear to have been agreed upon by the superalloy community in at least an approximate sense; see Table 4.1. More recently, NASA has developed a hybrid disc alloy, LSHR, which combines the low solvus temperature of Rene 104 with the higher refractory content of Alloy 10, the latter being designed by Honeywell Engines and Systems to produce superior tensile and creep capability in small engines for auxiliary power unit (APU) applications; this was achieved by using a high refractory element content, most notably of tungsten.

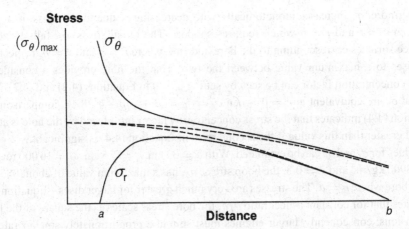

Fig. 4.40. Illustration of the stress field (solid lines) in a uniform disc of outer radius b, with a hole of radius a; σ_θ and σ_r denote the hoop and radial stress, respectively. Also shown (dashed lines) is the behaviour of σ_θ and σ_r for the case when $a = 0$, i.e. when no hole exists.

4.3 Service life estimation for turbine disc applications

4.3.1 Stress analysis of a turbine disc of simplified geometry

A geometry of a typical turbine disc is complicated by the requirement for fixturing to accommodate the turbine blading, the drive arms necessary to transmit loads to the shaft, cover plates to manage the flow of cooling air and flanges with holes for the bolting together of neighbouring discs. Moreover, the shape of the disc is generally tapered to restrict the stress field at the bore to manageable levels. However, to gain a first appreciation of the very large stresses to which the material is subjected, one can consider a cylindrical component of outer diameter b containing a hole of inner radius a, rotating at an angular velocity ω. The density is ρ. Clearly the stress field will depend upon whether the component is thin (such that it approximates a disc) or thick (in which case it is a cylinder). Considering the situation for a disc under plane stress conditions – and given the simplifying assumptions – the radial and hoop components of the stress field, denoted σ_r and σ_θ, respectively, are given by [43,44]

$$\sigma_r = \frac{3+\nu}{8}\rho\omega^2\left(b^2 + a^2 - \frac{a^2 b^2}{r^2} - r^2\right) \tag{4.2}$$

and

$$\sigma_\theta = \frac{3+\nu}{8}\rho\omega^2\left(b^2 + a^2 + \frac{a^2 b^2}{r^2} - \left[\frac{1+3\nu}{3+\nu}\right]r^2\right) \tag{4.3}$$

where the distance r is measured outwards from the axis of symmetry in a radial direction.

It is instructive to consider the stress field implied by these equations; see Figure 4.40. The hoop stress, σ_θ, is a maximum at the bore where $r = a$, where it takes a value of

$$(\sigma_\theta)_{max} = \left[\frac{3+\nu}{4}\right]\rho\omega^2 b^2 + \left[\frac{1-\nu}{4}\right]\rho\omega^2 a^2 \tag{4.4}$$

Furthermore, σ_θ increases monotonically with decreasing r, due to the mass of material between $r = a$ and $r = b$ which requires support. The radial stress, σ_r, falls to zero at the free surfaces corresponding to the bore and rim, where $r = a$ and $r = b$, respectively, but rises to a maximum value between the two. That the hole provides a considerable stress concentration factor can be seen by setting $a = 0$ in Equations (4.2) and (4.3). Then, σ_θ and σ_r are equivalent at $r = 0$, when $\sigma_r = \sigma_\theta = (3 + v)\rho\omega^2 b^2/8$. A comparison with Equation (4.4) indicates that the stress concentration factor introduced by the hole is at least 2, and greater than this value if the second term in Equation (4.4) is significant.

Values for σ_r and σ_θ can be estimated. With $b = 33$ cm, $a = 7.5$ cm, $\omega = 10\,000$ rev/min, $\rho = 8900$ kg/m^3 and $v = 0.3$, the hoop stress, σ_θ, has a maximum value of about 885 MPa at the bore where $r = a$. The stresses are very much greater for larger discs – Equation (4.4) indicates that for constant aspect ratio b/a, the bore stress scales as the square of the linear dimensions; consequently, larger engines must spin at a proportionately smaller rate. For example, if the linear dimensions are doubled at constant b/a, then ω must be halved for an equivalent stress field. These considerations confirm that materials with high-yield stresses are required for turbine-disc applications, and that the geometry needs to be chosen carefully such that the stress levels are restricted to tolerable levels. However, the level of conservatism in component design cannot be too great, otherwise excessive parasitic weight will be introduced.

4.3.2   Methods for lifing a turbine disc

The manufacture of a turbine disc calls for great care to be taken to avoid the introduction of defects. To see why this is the case, it is instructive to consider three common lifing methods.

A   The life-to-first-crack approach

In the life-to-first-crack approach, which is used in the United Kingdom, current requirements stipulate that fracture critical components such as an HP turbine disc are treated according to a safe-life policy in which component 'failure' is defined as the occurrence of an engineering crack of surface length 0.75 mm [45,46]. This figure is below the length required for fast catastrophic fracture, and is chosen because it is about the limit of common non-destructive inspection (NDI) methods. Reliance is placed on the testing of real discs in spin rigs, under stress and temperature conditions comparable to those experienced in service. Each test can cost between $250\,000 and $500\,000; consequently, the number of tests is small – it is rare for this to exceed five.

For the purposes of declaring a safe service life, it is assumed that (i) the number of cycles to failure is distributed according to a log-normal density function, and (ii) to a 95% confidence level, not more than 1 in 750 discs are expected to contain a crack of surface length greater than 0.75 mm [47,48]; thus one is working at the 1/750 quantile (-3σ) such that the ratio of the fatigue lives at the $+3\sigma$ and -3σ points is 6; see Figure 4.41. Given this assumed scatter factor, the life corresponding to the lower 1/750 quantile is located at a factor of $\sqrt{6}$ below the geometric mean obtained from the spin testing. This lower

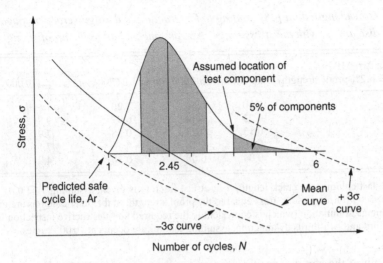

Fig. 4.41. Distribution of fatigue failure events superimposed on a $\sigma - N$ plot with the mean, $+3\sigma$ and -3σ values identified [47].

confidence bound translates to a factor of

$$6^{1.645/6\sqrt{n}} \tag{4.5}$$

where the 95% confidence limit corresponds to 1.645 standard deviations. This procedure allows a so-called 'safe life' to be calculated from the expression

$$\text{safe life} = \frac{\left(\prod_{i=1}^{n} N_i\right)^{1/n}}{2.449 \times 6^{\left(\frac{1.645}{6\sqrt{n}}\right)}} \tag{4.6}$$

The denominator in the above equation, i.e. the factor by which the 'safe life' is lower than the geometric mean of the results from spin testing, is 4.00 when $n = 1$, falling to 3.25 when $n = 3$ and 2.90 when $n = 8$ – thus, a safety factor is present, the magnitude dependent upon the number of components tested.

Note that the use of the 'first crack' as the basis for the calculation will be conservative, since the length of a surface crack which will burst the disc is usually likely to be in excess of 0.75 mm. This provides a further built-in margin of safety.

B   Damage-tolerant lifing

In practice, the 'extra' safety factor arising from the choice of a 0.75 mm long crack as constituting failure creates a dilemma. Precisely how large is it? If it is too large the procedure will be wildly conservative. It could, however, be very small or non-existent. Thus, in some ways the life-to-first-crack approach is unsatisfactory.

The 'damage-tolerant' approach to disc lifing (known alternatively as the 'retirement for cause' method) represents an alternative strategy [47,48]. One starts by estimating the length of crack required for 'burst', i.e. for fast catastrophic fracture, by using a fracture mechanics approach. Next, the number of cycles is estimated for the fatigue crack to grow

Table 4.4. *Crack length data [48] relating to fast fracture and fatigue crack propagation in turbine disc alloys, with the stress range,* $\Delta\sigma$*, assumed equal to the proof strength*

Stress range $\Delta\sigma$ /MPa (assumed = 0.2% proof strength)	a_{NDI}/mm	$a(3300)$/mm	$a(6700)$/mm	$a_{crit}(10\,000)$/mm
800	0.30	0.50	1.20	11.1
850	0.23	0.40	0.97	9.8
950	0.13	0.23	0.58	7.9
1000	0.10	0.18	0.46	7.1
1200	0.04	0.08	0.20	4.9

Note: In the last column, the crack length, a_{crit}, at fast fracture is given, assuming $Y = 0.67$ and K_{1c} equal to $100\,\mathrm{MPa\,m^{1/2}}$ and a stress equal to the proof strength. In the other columns are given the crack lengths at initiation (which is equivalent to the required non-destructive inspection limit a_{NDI}), one-third and two-thirds dysfunction, assuming that failure occurs at $10\,000$ cycles.

from the limit of the non-destructive inspection techniques, a_{NDI}, to the size required for burst, a_{crit}. Fatigue-crack propagation data in the form of da/dN data are required for this purpose, and this usually conforms to the usual Paris law expression

$$\frac{da}{dN} = A\Delta K^m \tag{4.7}$$

where ΔK is the range of stress intensity experienced by the crack. The exponent m is usually between 3 and 4 for superalloys. Upon inserting the expression for $\Delta K = Y\Delta\sigma\sqrt{\pi a}$, and after performing the appropriate integration, one has

$$N_0 = \frac{1}{A(m/2 - 1)Y^m \Delta\sigma^m \pi^{m/2}} \left[\frac{1}{a_{NDI}^{m/2-1}} - \frac{1}{a_{crit}^{m/2-1}} \right] \tag{4.8}$$

where N_0 is the number of cycles for failure.

It is instructive to plug in some numbers, following Knott [48]. For many of the polycrystalline nickel-based superalloys used for turbine disc applications, the Paris relationship is of the form

$$\frac{da}{dN} = 4 \times 10^{-12} \Delta K^{3.3} \tag{4.9}$$

where ΔK is expressed in $\mathrm{MPa\,m^{1/2}}$ and da/dN is the growth increment per cycle expressed in metres. Assuming a semicircular edge crack, $Y = 0.67$. For most nickel-based superalloys, the fracture toughness, K_{1c}, is about $100\,\mathrm{MPa\,m^{1/2}}$; if we assume $\Delta\sigma$ is equal to $800\,\mathrm{MPa}$, with failure occurring at the top of the cycle, one obtains $a_{crit} = 11\,\mathrm{mm}$. Next, one can determine an appropriate limit for a_{NDI} if one requires the turbine disc to last for at least $10\,000$ cycles – a figure of $0.30\,\mathrm{mm}$ is found. This is not very much smaller than the value assumed for the 'first crack' in the life-to-first-crack approach. The values of a at 3300 and 6700 cycles, corresponding to one-third and two-thirds dysfunction, respectively, are $0.50\,\mathrm{mm}$ and $1.20\,\mathrm{mm}$; see Table 4.4 [48]. Note that the majority of the fatigue crack growth occurs in the last few thousand cycles. In practice, one should note that estimates of the fatigue life made in this way are *usually* conservative, since no account has been taken of

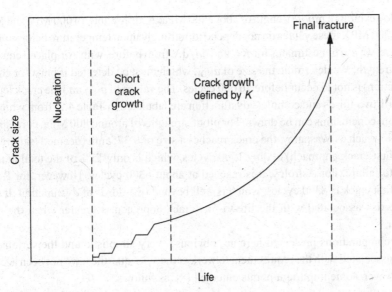

Fig. 4.42. Illustration of the different stages associated with fatigue crack failure: initiation, short crack growth, fatigue crack propagation and final fracture governed by the fracture toughness.

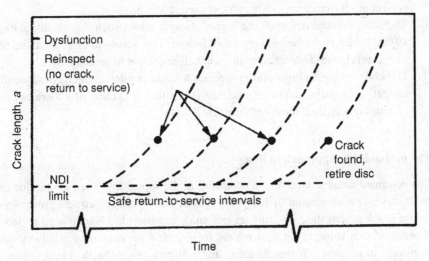

Fig. 4.43. Schematic representation of the damage-tolerant life prediction method [49].

the number of cycles required for fatigue crack growth to be initiated; see Figure 4.42. After the disc has operated for an appropriate number of cycles, it undergoes thorough inspection using the NDE techniques which would have been first employed before the disc entered service; provided that this reveals that no flaws are present, it is placed back into the engine for a further number of cycles. Figure 4.43 illustrates this so-called 'damage-tolerant' method [49].

It is helpful to make estimates of the safety margins to be expected from the use of nickel-based alloys of differing proof strengths, following the analysis presented in ref. [48].

Waspaloy exhibits a proof strength of about 800 MPa, Astroloy about 1000 MPa and Rene 95 about 1200 MPa; these values do not depend too strongly upon temperature below 600 °C. In each case, we use our estimates for K_{1c} and da/dN from before, with $\Delta\sigma$ placed equal to the proof strength. We determine the size of a_{NDI}, which must be detected if burst (or complete dysfunction) is not to occur before 10 000 cycles. The values of a_{NDI} and the crack lengths at one-third, two-thirds and complete dysfunction are tabulated in Table 4.2, from which some important conclusions can be drawn. For proof strengths of around 800 MPa, corresponding to an alloy such as Waspaloy, the crack reaches a size of 0.375 mm (deemed 'failure' in the life-to-first-crack approach) at only 1500 cycles, which is only 15% of the total number of cycles to failure. For Astroloy, it is reached at about 6300 cycles. However, for Rene 95, life-to-first crack is 8300 cycles, which is well beyond two-thirds of dysfunction. It is clear that the risk associated with the life-to-first-crack approach is greater when the alloy is stronger.

The computations presented here are obviously very simplistic and the various safety factors associated with the requirement to work at the -3σ quartile have not been accounted for. However, some important points emerge [48], as follows.

(i) Control of the manufacturing processes to prevent defects from being introduced is an absolute necessity. If pre-existing defects of length greater than 1 mm were to be present at service entry, catastrophic consequences might result.

(ii) The size of initial defect in the higher-strength alloys such that an adequate life of 10 000 cycles is reached is very small indeed. This serves to emphasise the need for very careful, non-destructive evaluation before entry into service.

(iii) The lower-strength alloys exhibit superior tolerance to defects, and consequently their reputation as fatigue-resistant materials is justified. It is clear why there can be some reluctance to replace these with the new alloys.

C   The probabilistic approach to lifing

The dominant damage mechanism occurring in the superalloys used for turbine disc applications is crack initiation by low-cycle fatigue (LCF) and subsequent growth by cyclic fatigue crack propagation. For this reason, much emphasis has been placed on laboratory experimentation using test pieces which, for practical reasons, have relatively small dimensions: of perhaps $\sim$10 mm diameter and $\sim$50 mm gauge length. The accuracy of this approach requires that the conditions experienced by the test pieces are representative of those in a turbine disc. This is expected to be so when the density of nucleation sites is large, which will be the case for LCF conditions with high levels of cyclic strain or for superalloys of low yield stress such as Waspaloy or IN718. For these alloys, crack initiation is found to be associated with intense slip bands which have a very high density, and the damage is representative of that occurring in turbine discs of much greater size. The lifing approach is then essentially deterministic, although scatter in the laboratory test data should, of course, be accounted for.

The situation is somewhat different for alloys such as Rene 95 or RR1000, which have significantly higher strengths and which are manufactured by powder-processing routes.

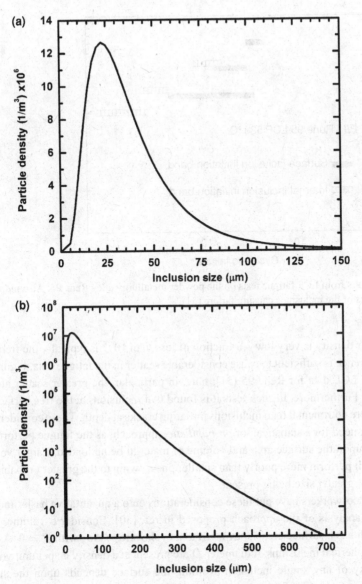

Fig. 4.44. Typical distribution of inclusions in a powder metallurgy alloy such as Rene 95 [50].
(a) Number of particles per unit volume against particle size, plotted on linear–linear axes;
(b) as for (a), but plotted on log–linear axes.

First, the higher strength means that a higher proportion of the fatigue life tends to be spent in the fully elastic regime, in which the density of slip bands is low. Thus, testing conditions are not necessarily representative of the service environment, and a significant size effect might be anticipated. Second, failure is found to be associated with ceramic inclusions – typically Al_2O_3 or SiO_2 – which are inherited from the processing route. A typical particle size distribution for the Rene 95 superalloy [50] is given in Figure 4.44.

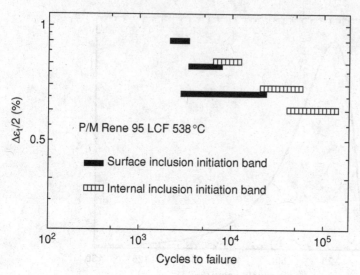

Fig. 4.45. Results from LCF fatigue tests on the powder metallurgy alloy Rene 95. Also indicated are the positions of the inclusions causing failure [51].

Although their density is very low – a fraction of less than 10^{-6} is typical – the frequency of their occurrence is sufficient to cause considerable scatter in the fatigue data. Figure 4.45 shows typical LCF data for Rene 95 [51]; note, in particular, the greater scatter at lower cyclic strains. Furthermore, in such tests it is found that inclusions at the free surfaces are very much more detrimental than inclusions initiating cracks at depth. These considerations emphasise the need for a statistical, or *probabilistic*, approach, as the fatigue performance now depends upon the surface area and volume of material being loaded. Larger volumes of material will perform more poorly than smaller ones, owing to the greater probability of an inclusion of greater size being present.

Pineau and co-workers have put these considerations onto a quantitative basis, and what follows is an analysis of the approach proposed in ref. [50]. Consider a volume, V, of material which has a surface area, S, exposed to a cyclic stress, $\Delta\sigma$, with $R = 0$. Assume that *uniform* spherical inclusions of diameter D are present at a density n_v per unit volume. The probability of any single inclusion avoiding the surface depends upon the surface to volume ratio, S/V; to a first approximation, it is given by the quantity $1 - DS/V$, if stereological effects associated with the intersection of the particle with the free surface are neglected. However, a total of $(n_v \times V)$ inclusions are present. It follows that the probability, p_1, of *any one* inclusion intersecting the free surface is given by

$$p_1\{D\} = 1 - \left(1 - \frac{DS}{V}\right)^{n_v V} \tag{4.10}$$

It is necessary also to acknowledge the manner in which any given inclusion interacts with the free surface; for example, a fully embedded inclusion is clearly more damaging than one which is only partially embedded to depth d, where $d < D$. If the flaw size, d, defines a fatigue life, N_0, consistent with the damage-tolerant approach described above,

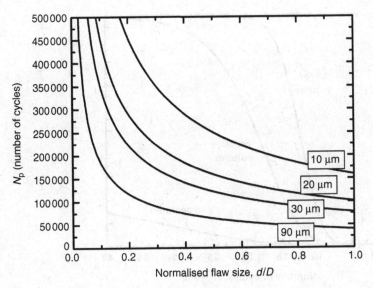

Fig. 4.46. Variation of the predicted life with the flaw size, d, normalised against the inclusion diameter, D, for various values of D.

the probability, p_2, that the *actual* life is N_p such that $N_p < N_0$ is

$$p_2\{N_p < N_0\} = \left[1 - \left(1 - \frac{DS}{V}\right)^{n_v V}\right]\left[\left(\frac{D-d}{D}\right)\right] \tag{4.11}$$

The term $(D - d)/D$ should be considered to be a 'harmfulness criterion', which acknowledges that, when $d = D$, the inclusion is at its most potent; then $N_p = N_0$ and $p_2 = 0$. At the other extreme, when the inclusion only just intersects the free surface, $d = 0$, and therefore it is not damaging; hence $p_2 = p_1$. In this limit, the N_p tends to infinity. Figure 4.46 illustrates the way in which N_p varies with d/D, for the various sizes of inclusion D implied by the particle size distribution in Figure 4.44. Note that as d/D tends to zero, an infinite life is expected since flaws are then absent.

In practice, a superalloy turbine disc material contains a *distribution* of particle sizes, and this must be accounted for. Suppose that the particle size distribution is discretised such that is contains n_{classes} classes, each of diameter D_i and number density n_v^i, such that $i = 1 \rightarrow n_{\text{classes}}$. The probability, p_3, that the life, N_p, is less than the potential life N_0 is then given by

$$p_3\{N_p < N_0\} = 1 - \prod_{i=1}^{n_{\text{classes}}} [1 - p_4\{N_p < N_0\}] \tag{4.12}$$

where, by appealing to the expression for p_2, one has

$$p_4\{N_p < N_0\} = \left[1 - \left(1 - \frac{D_i S}{V}\right)^{n_v^i V}\right]\left[\left(\frac{D_i - d}{D_i}\right)\right] \tag{4.13}$$

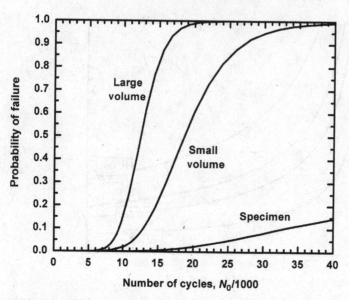

Fig. 4.47. Variation of the probability of failure, given by p_3, against the number of cycles, N_0, for the probabilistic model of Rene 95.

Note that the product in the expression for p_3 represents the probability that *every* class of inclusion, D_i, in the set of $n_{classes}$ is non-damaging; p_3 is then the probability that *any one* of the D_i provides the fatal flaw.

With this model, quantitative calculations can be carried out in the following way. First, the variation of N_0 with flaw size, d, is determined using the damage-tolerant approach described in the previous section. Data for K_{1c}, $\Delta\sigma$ and the constants in the Paris law are required. Second, the data for d are normalised against the D_i present in the particle size distribution, and a family of curves is produced which describes the variation of the life, N_p, with d/D_i, as in Figure 4.46. Third, a prescribed life, N_0, is chosen; this defines d, and hence d/D_i, for any given D_i so that a set of values for p_4 can be determined from the expression given, each for a corresponding D_i. This allows a value for p_3 to be calculated. The process can be repeated for various values of N_0 and $\Delta\sigma$; see Figure 4.47, in which N_0 is plotted against p_3, the probability that the life falls short of N_0.

The results in Figure 4.47 – calculated for the Rene 95 superalloy with $\Delta\sigma = 1000$ MPa, $Y = 0.67$, the constants for the Paris equation given in the previous section and the particle size distribution given in Figure 4.44 – demonstrate that the fatigue performance displays a considerable dependence upon the surface area of the volume under load. For a large disc of surface area $84\,000$ mm^2 and volume 10^{-2} m, the probability of failure varies from less than 0.01 for fewer than 5000 cycles, to greater than 0.99 beyond $20\,000$ cycles. A smaller disc of surface area 9300 mm^2 and volume 3×10^{-3} m, or else a standard fatigue test specimen of surface area 200 mm^2 and volume 3.5×10^{-6} m, exhibit considerably superior performance. At any given probability of failure, the model indicates that the inclusion which is the likeliest to cause failure has a size which depends upon the surface to volume ratio. For example, a test specimen is predicted to fail from an inclusion at

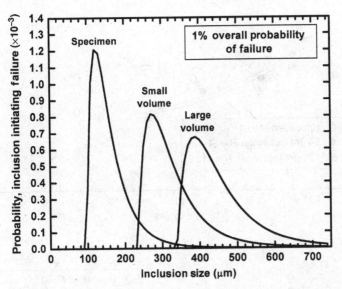

Fig. 4.48. Variation of the probability, p_4, with inclusion size, D_i, at a probability p_3 equal to 0.01, for the three geometries considered in the probabilistic model for Rene 95. The applied stress is 1000 MPa.

the lower end of the particle size distribution, whereas a turbine disc of reasonable size fails from an inclusion of greater size; see Figure 4.48. This is a direct consequence of a particle-sampling effect – a large particle (which has a low likelihood of occurrence) has a greater probability of causing failure when the surface area is significant. The sensitivity of the fatigue life to the particle size distribution has been confirmed [52] using low-cycle-fatigue (LCF) tests on powder-processed Udimet 720, and the results have been compared with those from the same material that had been intentionally doped with alumina particles of sizes 54 μm and 122 μm. The results (see Figure 4.49) indicate that the fatigue life is indeed impaired as the particle size increases; the reduction in life is greatest for lower total strain ranges and larger R-ratios. Figure 4.50 illustrates the results of fractography to identify the location of fatigue initiation in a typical turbine disc alloy. In principle, such experiments can be used to deduce the naturally occurring particle size distributions from the fatigue data of unseeded material – as required by the probabilistic modelling.

The above considerations justify the considerable emphasis which is placed by the superalloy industry on 'superclean melting technology' – the aim of which is to prevent non-metallic inclusions from becoming incorporated into superalloy forging stock and hence the turbine discs themselves. It is common nowadays for the rate of contamination to be less than one part per million on a weight per cent basis. Nevertheless, when using a probabilistic lifing method it is absolutely necessary to have a good estimate of the particle size distribution; the accuracy of the fatigue life depends upon it and, indeed, no estimate is possible if this is unavailable. To glean the necessary information, different approaches have been employed. Quantitative metallography has been used to examine plane, polished cross-sections. However, for 'superclean' material this can be problematical, since the number and

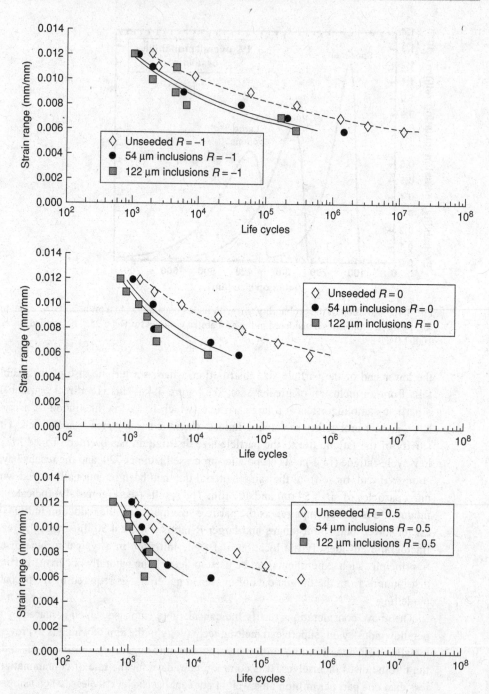

Fig. 4.49. Comparison of the LCF lives of powder metallurgy Udimet 720 in the unseeded and seeded conditions, for various R ratios [52].

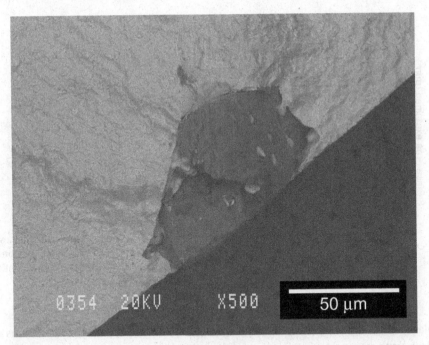

Fig. 4.50. Fatigue initiation site for powder metallurgy Udimet 720 seeded with 54 μm diameter inclusions, LCF tested at 427 °C under strain control with a range of 1.2%, $R = -1$ [52]. Surface initiation is apparent.

size of inclusions is very small, possibly to an extent which makes the work prohibitively time-consuming and expensive. This is confirmed by the data in Figure 4.51, which shows the cross-sectional area of material to be examined in order to obtain an estimate of the volume fraction of inclusions to an accuracy greater than 20%, with 80% confidence [53]; for a diameter of 100 μm and a volume fraction of 10^{-6}, the surface area required is $\sim 10^6$ mm², or one square metre. These difficulties have provided the incentive for the development of the electron beam button melting (EBBM) technique, in which about 1 kg of the alloy is drip-melted under vacuum by an electron beam; the inclusions rise to the surface of the pool (since they are lighter than the molten liquid), where they are subject to buoyancy and Marangoni forces. If the cooling is then properly controlled, the inclusions concentrate into the last portion to solidify, which then allows them to be analysed. The EBBM techique has shown some promise, although it has been criticised on the grounds that it does not provide an accurate measure of the *distribution* of particle sizes, since the flotation efficiency is a function of the particle size.

4.3.3 Non-destructive evaluation of turbine discs

It has been shown in Section 4.3.2 that the flaws in a turbine disc are of great significance; since they represent potential sites for the nucleation of cracks, they threaten the integrity of the component and reduce the fatigue life. For this reason, much emphasis is

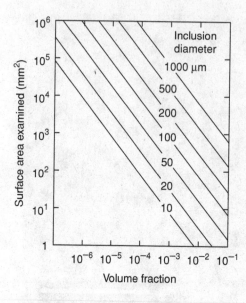

Fig. 4.51. Variation of the surface area which must be examined with volume fraction and diameter of inclusions, at the 80% confidence limit [53].

placed on using non-destructive evaluation (NDE) to characterise the dimensions of the flaws, both during the manufacturing of the turbine disc – i.e. before the turbine disc enters service for the first time – to ensure that it is of acceptable quality, and then whenever the disc is removed from the engine, prior to a decision being made to place it back in service. Obviously, the first situation arises regardless of the lifing method employed, the second only when the damage-tolerant method is being used. The various NDE methods which are suitable include (i) the liquid penetrant method, most usually using dyes which fluoresce under ultra-violet light, thus providing a measure of the length of each crack exposed to the surface; (ii) the eddy current method, which enables a quantitative estimate of the crack depth, but only for cracks located at or near the surface; (iii) X-ray radiography, which has the power to penetrate relatively thick sections, but which cannot usually resolve the finest flaws that are present; and (iv) ultrasonics – the use of high-frequency sound waves for the detection of inclusions, coarse grains, shrinkage pipe, tears, seams and laps. In practice, for turbine disc applications this final method is of the greatest importance.

To illustrate the use of ultrasonics [54] for NDE in this context, consider its application to a forged and heat-treated disc, which will have been machined to a pre-specified geometry known as the 'sonic shape' or 'condition of supply'. To enable ease of ultrasonic inspection, the features of the disc (for example, the blade root sockets, cover plate and flanges) are not introduced by the machining process as yet, so that, at this stage, it contains only a few, flat surfaces which intersect at 90° angles; see Figure 4.52. To detect the smallest flaws, care is taken also with the surface finish, which is typically better than 6 μm. Production and measurement of the ultrasonic waves is accomplished via the use of piezoelectric transducers and receivers – these rely upon crystals, such as quartz or barium titanate, which display a strong inverse piezoelectric effect. Since ultrasonic waves are attenuated strongly by air, the

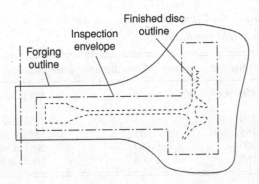

Fig. 4.52. Schematic illustration of the outline of a machined turbine disc, the sonic envelope used to test the material ultrasonically and the forging from which it is machined.

disc and transducer are either kept immersed in water during inspection, or else a thin layer of coupling fluid is maintained between the two; the waves are attenuated only very weakly by the material itself, so that very thick sections can be analysed. In practice, the waves are sent through the component in at least two different ways. First, the so-called 'straight-beam top inspection' is carried out to detect planar flaws lying parallel to surfaces whose normal lies parallel to the axis of the forging; for this purpose, a longitudinal ultrasonic beam is directed parallel to the normal to these same surfaces, using a single pulse-echo transducer, 13–40 mm in diameter, at a frequency in the range 1–5 MHz. Calibration is carried out using standard 50 mm diameter cylindrical reference blocks of varying heights, into which flat-bottomed holes are machined to simulate artificial flaws of different sizes located at different depths. Detection sensitivity is normally set to 10% of the size of the flaw machined into the calibration block. For example, if the #2 block in Series B of the ASTM standard E127 is found to be suitable for calibration [55], the expected sensitivity of flaw detection is then 0.08 mm since the flat-bottomed hole in that case has a diameter of 0.8 mm. A second set of tests are conducted using shear waves, to detect flaws that have an axial radial orientation; for this, an angle-beam transducer is used to introduce a beam at 45° to the surface normal, and inspection is performed by scanning in the circumferential direction, both clockwise and anti-clockwise, around the periphery of the forging. Calibration notches – typically 25 mm long and either V-shaped or rectangular – are cut axially on the inner and outer surfaces of the forging, with a width not exceeding twice the depth. The sensitivity of the detection is established by first adjusting the instrument controls to obtain a minimum 13 mm sweep-to-peak signal from the calibration notch on the outer surface, followed by a measurement of the response from the notch on the inside surface; the peaks corresponding to the two notches are connected thus establishing a reference line.

With regard to the inspection of different superalloys, the background noise is found to be significantly greater for coarse-grained material, and therefore the sensitivity of flaw detection is then much reduced; small flaws are then more likely to be undetected. The data given in Table 4.5 illustrate this [56]. Interestingly, for any given grain size, powder metallurgy (P/M) material exhibits ultrasonic inspectability which is very much better than for conventional cast-and-wrought product. This has been attributed to the

Table 4.5. *Data for the inspectability of the Udimet 720 superalloy for various grain sizes, in the powder and cast-and-wrought conditions [56]*

Form	Billet diameter /mm	Grain diameter /μm	Noise level
Powder metallurgy, extruded	150–300	3	6% at hole 1[a]
Powder metallurgy, sub-solidus			
HIP + cog	300	11	25% at hole 1
Powder metallurgy, HIP	165	11	15% at hole 1
Powder metallurgy, sub-solidus HIP	300	127	50% at hole 1
Cast and wrought	165	8	100% at hole 2

Note: [a] Hole 1 corresponds to a diameter of 0.40 mm.

more uniform grain size and the reduced levels of segregation when powder metallurgy is employed.

Questions

4.1   Refining of superalloys is often done with a combination of vacuum arc melting (VIM), electro-slag remelting (ESR) and vacuum arc remelting (VAR). Triple-melted stock is then processed by the combination of VIM/ESR/VAR, in that order. Explain why the induction-melted stock is subjected to electro-slag remelting prior to vacuum arc remelting, rather than the other way around.

According to Maurer, the maximum ingot diameter which can be supplied in the alloy IN718 to aerospace quality depends upon the secondary melting processes employed. For example, 17 inch is the limit for a VIM/ESR ingot, 20 inch for a VIM/VAR ingot and 24 inch for a triple-melted VIM/ESR/VAR ingot. Explain why this should be the case.

4.2   Careful control of electrode immersion is important during ESR refining, with a 1/4 inch immersion of the electrode being suggested for optimum processing. Why is control of the immersion depth important? What would be the consequences of (i) too great an immersion depth and (ii) too small a depth? What would be the influence on the voltage swing?

4.3   The ESR process gives rise to considerably less shrinkage pipe than the VAR process. Why is this?

4.4   During the secondary remelting of IN718 superalloy using the VAR process, freckling can occur. The most usual location is at mid-radius, implying that the material at centre and surface positions is not so freckle-prone. Explain.

4.5   During the solidification of the superalloys, the formation of TiN particles ahead of the growing solid/liquid interface has been implicated in the nucleation of stray grains. Explain how you would expect this columnar-to-equiaxed transition to depend upon (i) the growth velocity and (ii) the temperature gradient employed.

4.6   Cast ingots of the alloy IN718 are found to contain significant quantities of the Laves phase, $(Ni, Fe)_2Nb$. This can be found on the binary Fe–Nb phase diagram. This is despite the fact that IN718 contains only 5.35 wt% Nb. What is the reason for this? Would you expect powder IN718 to have more or less of the Laves phase?

4.7 Why, before the consolidation step in the processing of superalloys by powder-processing, is it usual to blend together particle distributions of different sizes?

4.8 Work aimed at optimising the compositions of polycrystalline superalloys for turbine disc applications has confirmed that additions of boron have a very beneficial effect on creep performance. Calculate the concentration of boron (in wt%) required to deposit a monolayer of boron on the grain boundaries of a superalloy with a grain size of (a) 1 μm, (b) 10 μm and (c) 100 μm. Do the concentrations calculated compare well with the values used in practice?

4.9 The microstructure of a polycrystalline turbine disc alloy can display the γ' phase in primary, secondary and tertiary forms. Give an explanation for this, invoking a description of possible heat-treatment schedules and the microstructural response expected due to nucleation, growth and coarsening theories.

4.10 A closed-die forging operation on a piece of a superalloy billet leaves a characteristic grain size distribution, which depends upon the thermal and strain history induced. Describe what this might look like. Will the grain size be greater at the rim or the bore region?

Explain why this grain size distribution is unfortunate, given the operating conditions experienced by the rim and bore in the gas turbine engine. How might a dual-heat-treatment process help in this regard?

4.11 Examine the low-cycle-fatigue data for various turbine disc alloys, which are given in Figure 4.30(a). At low total strain ranges, $\Delta\epsilon_t$, stronger alloys such as Rene 95 (each possessing a significant yield stress) have better fatigue lives than weaker ones such as Waspaloy. However, for large $\Delta\epsilon_t$, the converse is true. Give an explanation for this behaviour.

4.12 With regard to the lifing of a turbine disc, distinguish between the (a) life-to-first-crack and (b) damage-tolerant approaches to lifing. Which method would be the more appropriate for (i) a fatigue-resistant alloy such as that used for civil engines and (ii) a higher-strength alloy used for a military application?

4.13 The stress distribution in a rotating circular disc is of practical importance in the design of gas turbine engines. For a disc containing a central hole of radius a, the principal stresses in the plane of the disc depend upon its density, ρ, angular velocity, ω, and maximum radius, b, according to

$$\sigma_r = \frac{3+\nu}{8}\rho\omega^2\left(b^2 + a^2 - \frac{a^2 b^2}{r^2} - r^2\right)$$

$$\sigma_\theta = \frac{3+\nu}{8}\rho\omega^2\left(b^2 + a^2 + \frac{a^2 b^2}{r^2} - \left[\frac{1+3\nu}{3+\nu}\right]r^2\right)$$

For a disc containing a central hole with radius $a = b/10$, plot out the variation of σ_r and σ_θ with r. Take $\nu = 1/3$. Where are the maximum radial and hoop stresses? Compare the maximum value of the hoop stress with the maximum value for a disc with no hole ($a = 0$) and note the stress-concentrating effect of the hole. Hence explain why it is possible to spin a typical helicopter turbine disc very much faster than one in a civil aeroengine.

4.14   Derive an expression for the average stress, P_b, exerted by the blades on the rim of the turbine disc, which is taken to have a thickness, t, in the axial direction. Model each blade as a simple slab. Your expression should include the angular velocity, ω, the number of blades, N_b, the mass of each blade, M_b, and the typical dimensions. If the rotational speed of the HP shaft is 10 000 rev/min, estimate a value for P_b.

4.15   Lame's equation for the hoop stress at the bore of a penny-shaped disc of outer radius y and hole of radius x, loaded evenly around its outer circumference with a stress P_b, is given by

$$\sigma_{hoop,bore} = \left(\frac{1-\nu}{4}\right) x^2 \rho \omega^2 + \left(\frac{3+\nu}{4}\right) y^2 \rho \omega^2 + \frac{2 P_b y^2}{(y^2 - x^2)}$$

where ρ is the density and ν is Poisson's ratio. Use this expression to make a first estimate of the inner and outer diameters of the turbine disc of a large turbofan such as the GE90. How does (i) the weight and (ii) the hoop stress, $\sigma_{hoop,bore}$, vary as x and y are altered?

4.16   The operating conditions of a civil aeroengine are at their most extreme at take-off, when the shaft speed increases typically by about 5% over the value in steady-state flight at altitude. Using the equations given in previous questions, compare the stress distributions in the HP turbine disc (i) at take-off and (ii) during steady-state flight. Why might your considerations be overly simplistic?

4.17   Four turbine discs are tested in a spin-pit test rig, under conditions which are representative of those seen in service. Failure occurs at 8821, 9276, 9946 and 10 592 cycles. On the basis of the *life-to-first-crack* lifing philosophy, predict a safe working life for this component.

        After entry to service, one turbine disc is removed prematurely from the jet engine after 3000 cycles. The test rig is then used to spin it to destruction – this takes a further 5967 cycles. Given this information, calculate a revised predicted safe working life. State your assumptions. Has the *mean* life increased or decreased?

4.18   A gas turbine manufacturer is considering the use of Rene 95 for a high-pressure turbine disc, which is currently manufactured from Waspaloy. This will allow the range of cyclic stress, $\Delta\sigma$, to be increased from 800 MPa to 1200 MPa to reflect the increased yield stress of the new alloy. The fracture mechanics 'damage-tolerant' approach is used for turbine disc lifing.

        With regard to the limits set for surface flaw detection using non-destructive inspection, what are the ramifications of specifying Rene 95 for this application? Assume that fatigue crack growth in both alloys can be described by

$$\frac{da}{dN} = 4 \times 10^{-12} \Delta K^{3.3}$$

where ΔK is expressed in MPa m$^{1/2}$ and da/dN is the growth increment per cycle expressed in metres, and that $K_{1c} = 100$ MPa m$^{1/2}$ for most superalloys. Take $Y = 0.67$, corresponding to a semi-circular edge crack.

4.19 At ambient temperature, the fatigue crack propagation rate of many polycrystalline superalloys is found to depend upon the ratio $\Delta K/E$, where ΔK is cyclic stress intensity and E is Young's modulus. Relationships such as

$$\frac{da}{dN} \propto (\Delta K/E)^{3.5}$$

have been proposed [57], where it has been shown that growth rates are predicted within a factor of ± 2 provided that a suitable calibration constant is chosen.

How might a normalisation of ΔK by E be justified? Suggest a better normalisation, involving the product of the yield stress and modulus, based upon the general principles of fracture mechanics. Are there circumstances under which these approximations might be too crude?

4.20 In practice, the processing of a high-strength turbine disc superalloy such as Rene 95 superalloy cannot be achieved without introducing distributions of ceramic inclusions; see Figure 4.44.

Determine the *most likely* diameter of surface-breaking inclusion in the Rene 95 superalloy, given this distribution in Figure 4.44, for (i) a specimen of surface area 200 mm^2 and volume 3.5×10^{-6} m^3, (ii) a helicopter disc of surface area 9300 mm^2 and volume 2.9×10^{-4} m^3, and (iii) a disc for a civil turbofan engine of surface area 84 000 mm^2 and volume 1.0×10^{-2} m^3.

Given the fatigue performance and loading situations outlined in the previous question, and assuming that fatigue crack growth begins at the first cycle from a flaw of size equal to the diameter of the most likely surface-breaking inclusion, estimate the fatigue life for each of the situations (i), (ii) and (iii) described above. Hence demonstrate that the predicted fatigue life has a considerable dependence on component size.

4.21 The turbine disc material of an engine is to be replaced with a stronger grade of alloy. This will allow the engine to be run harder, although care will be taken to ensure that the hoop stresses at the bore do not exceed the same proportion of the yield stress. How would one expect the (i) fatigue initiation and (ii) the fatigue propagation lives to be altered? Does this have ramifications for the way in which one must life the turbine disc?

4.22 Consider the *damage-tolerant* (sometimes known as the *retirement for cause*) method of lifing a turbine disc. In practice, the number of cycles declared after inspection is assumed not to vary during repeated use of the disc – i.e. it is the same for the first and last service overhauls. Explain why this is not an optimum strategy in some ways.

4.23 You are promoted to Chief of Materials for a company specialising in the design and manufacture of large civil aeroengines. Unfortunately, your organisation has no experience with the use of powder metallurgy product for its turbine discs. Prepare a two-page memorandum for the Board of Directors, which summarises the challenges which need to be overcome before powder metallurgy product can be introduced into the product line.

4.24 During the inspection of forged superalloy billet using ultrasonic non-destructive testing, the probability of successfully detecting a defect depends strongly upon the grain size of the material. Why is this? And why does this give powder billet a big advantage in this respect?

4.25   The magnetic particle inspection (MPI) method is used as a non-destructive evaluation technique for the detection of flaws in combustor cans, but it is used much more rarely for turbine discs. Why is this? Can you suggest an alloy which is likely to respond favourably to MPI?

References

[1]   A. Choudhury, *Vacuum Metallurgy* (Materials Park, OH: ASM International, 1990).

[2]   G. E. Maurer, Primary and secondary melt processing – superalloys, in J. K. Tien and T. Caulfield, eds, *Superalloys, Supercomposites and Superceramics* (San Diego: Academic Press, 1989), pp. 49–97.

[3]   M. G. Benz, Preparation of clean superalloys, in C. L. Briant, ed., *Impurities in Engineering Materials: Impact, Reliability and Control* (New York: Marcel Dekker Inc., 1999), pp. 31–47.

[4]   P. P. Turillon, Evaporation of elements from 80/20 nickel-chromium during vacuum induction melting, in *Transactions of the Sixth International Vacuum Metallurgy Conference* (New York: American Vacuum Society, 1963), pp. 88–102.

[5]   J. W. Pridgeon, F. N. Darmara, J. S. Huntington and W. H. Sutton, Principles and practices of vacuum induction melting and vacuum arc remelting, in M. J. Donachie Jr, ed., *Superalloys: Source Book* (Metals Park, OH: American Society for Metals, 1984), pp. 205–214.

[6]   A. D. Patel, R. S. Minisandrum and D. G. Evans, Modeling of vacuum arc remelting of alloy 718 ingots, in K. A. Green, T. M. Pollock, H. Harada *et al.*, eds, *Superalloys 2004* (Warrendale, PA: The Minerals, Metals and Materials Society (TMS), 2004), pp. 917–924.

[7]   E. Samuelsson, J. A. Domingue and G. E. Maurer, Characterising solute-lean defects in superalloys, *Journal of Metals*, **40** (1990), 27–30.

[8]   L. A. Jackman, G. A. Maurer and S. Widge, New knowledge about white spots in superalloys, *Journal of Metals*, **43** (1993), 18–25.

[9]   W. Zhang, P. D. Lee, M. McLean and R. J. Siddell, Simulation of intrinsic inclusion motion and dissolution during the vacuum arc remelting of nickel-based superalloys, in T. M. Pollock, R. D. Kissinger, R. R. Bowman *et al.*, eds, *Superalloys 2000* (Warrendale, PA: The Minerals, Metals and Materials Society (TMS), 2000), pp. 29–37.

[10]   D. G. Evans and M. G. Fahrmann, A study of the effect of electro-slag remelting parameters on the structural integrity of large diameter Alloy 718 ESR Ingot, in K. A. Green, T. M. Pollock, H. Harada *et al.*, eds, *Superalloys 2004* (Warrendale, PA: The Minerals, Metals and Materials Society (TMS), 2004), pp. 507–516.

[11]   P. Aubertin, T. Wang, S. L. Cockcroft and A. Mitchell, Freckle formation and freckle criterion in superalloy castings, *Metallurgical and Materials Transactions*, **31B** (2000), 801–811.

[12] J. C. Borofka, J. K. Tien and R. D. Kissinger, Powder metallurgy and oxide dispersion processing of superalloys, in J. K. Tien and T. Caulfield, eds, *Superalloys, Supercomposites and Superceramics* (San Diego: Academic Press, 1989), pp. 237–284.

[13] B. L. Ferguson, Aerospace applications, in *ASM Handbook Volume 7: Powder Metallurgy*, 6th Edn (Materials Park, OH: ASM International, 1997), pp. 646–656.

[14] A. Banik, B. Lindsley, D. P. Mourer and W. H. Zimmer, Alternative processing for the production of powder metal superalloy billet, in G. E. Fuchs, A. W. James, T. Gabb, M. McLean and H. Harada, eds, *Advanced Materials and Processes for Gas Turbines* (Warrendale, PA: The Minerals, Metals and Materials Society (TMS), 2003), pp. 227–236.

[15] D. R. Chang, D. D. Krueger and R. A. Sprague, Superalloy powder processing, properties and turbine disc applications, in R. H. Bricknell, W. B. Kent, M. Gell, C. S. Kortovich and J. F. Radavich, eds, *Superalloys 1984* (Warrendale, PA: The Metallurgical Society of AIME, 1984), pp. 245–273.

[16] D. Furrer and H. Fecht, Ni-based superalloys for turbine discs, *Journal of Metals*, **51** (1999), 14–17.

[17] J. Jones and D. J. C. Mackay, Neural network modelling of the mechanical properties of nickel base superalloys, in R. D. Kissinger, D. J. Deye, D. L. Anton *et al.*, eds, *Superalloys 1996* (Warrendale, PA: The Minerals, Metals and Materials Society (TMS), 1996), pp. 417–424.

[18] D. Raynor and J. M. Silcock, Strengthening mechanisms in γ precipitating alloys, *Metal Science*, **4** (1970), 121–129.

[19] M. P. Jackson and R. C. Reed, Heat treatment of Udimet 720Li: the effect of microstructure on properties, *Materials Science and Engineering*, **A259** (1999), 85–97.

[20] J. Telesman, P. Kantzos, J. Gayle, P. J. Bonacuse and A. Prescenzi, Microstructural variables controlling time-dependent crack growth in a P/M superalloy, in K. A. Green, T. M. Pollock, H. Harada *et al.*, eds, *Superalloys 2004* (Warrendale, PA: The Minerals, Metals and Materials Society (TMS), 2004), pp. 215–224.

[21] J. C. Williams and E. A. Starke, Progress in structural materials for aerospace systems, *Acta Materialia*, **51** (2003), 5775–5799.

[22] K. R. Bain, M. L. Gambone, J. M. Hyzak and M. C. Thomas, Development of damage tolerant microstructures in Udimet 720, in S. Reichman, D. N. Duhl, G. Maurer, S. Antolovich and C. Lund, eds, *Superalloys 1988* (Warrendale, PA: The Metallurgical Society, 1988), pp. 13–22.

[23] R. V. Miner, Fatigue, in C. T. Sims, N. S. Stoloff and W. C. Hagel, eds, *Superalloys II* (New York: John Wiley and Sons, 1987), pp. 263–289.

[24] R. V. Miner, J. Gayda and R. D. Maier, Fatigue and creep-fatigue deformation of several nickel-base superalloys at 650°C, *Metallurgical Transactions*, **13A** (1982), 1755–1765.

[25] D. Locq, P. Caron, S. Raujol, F. Pettinari-Sturmel, A. Coujou and N. Clement, On the role of tertiary gamma prime precipitates in the creep behaviour at 700 °C of a powder metallurgy disk superalloy, in K. A. Green, T. M. Pollock, H. Harada *et al.*,

eds, *Superalloys 2004* (Warrendale, PA: The Minerals, Metals and Materials Society (TMS), 2004), pp. 179–188.

[26]   N. J. Hide, M. B. Henderson and P. A. S. Reed, Effects of grain and precipitate size variation on creep-fatigue behaviour of Udimet 720Li in both air and vacuum, in T. M. Pollock, R. D. Kissinger, R. R. Bowman *et al.*, eds, *Superalloys 2000* (Warrendale, PA: The Minerals, Metals and Materials Society (TMS), 2000), pp. 495–503.

[27]   K. Li, N. E. Ashbaugh and A. H. Rosenberger, Crystallographic initiation of nickel-base superalloy IN100 at RT and 538 deg C under low cycle fatigue conditions, in K. A. Green, T. M. Pollock, H. Harada *et al.*, eds, *Superalloys 2004* (Warrendale, PA: The Minerals, Metals and Materials Society (TMS), 2004), pp. 251–258.

[28]   A. H. Rosenberger, The effect of partial vacuum on the fatigue crack growth of nickel-base superalloys, in K. A. Green, T. M. Pollock, H. Harada *et al.*, eds, *Superalloys 2004* (Warrendale, PA: The Minerals, Metals and Materials Society (TMS), 2004), pp. 233–240.

[29]   C. W. Brown, J. E. King and M. A. Hicks, Effects of microstructure on long and small crack growth in nickel-base superalloys, *Metal Science*, **18** (1984), 374–380.

[30]   A. H. Rosenberger, E. Andrieu and H. Ghonem, Influence of high temperature elastic-plastic small crack growth behaviour in a nickel-base superalloy on the life prediction of structural components, in S. D. Antolovich, R. W. Stusrud, R. A. MacKay *et al.*, eds, *Superalloys 1992* (Warrendale, PA: The Minerals, Metals and Materials Society (TMS), 1992), pp. 737–746.

[31]   H. T. Pang and P. A. S. Reed, Fatigue crack initiation and short crack growth in nickel-base turbine disc alloys – the effects of microstructure and operating parameters, *International Journal of Fatigue*, **25** (2003), 1089–1099.

[32]   D. P. Mourer and J. L. Williams, Dual heat treatment process development for advanced disk applications, in K. A. Green, T. M. Pollock, H. Harada *et al.*, eds, *Superalloys 2004* (Warrendale, PA: The Minerals, Metals and Materials Society (TMS), 2004), pp. 401–408.

[33]   T. J. Garosshen, T. D. Tillman and G. P. McCarthy, Effects of B, C and Zr on the structure and properties of a P/M nickel-base superalloy, *Metallurgical Transactions*, **18A** (1987), 69–77.

[34]   L. Xiao, D. L. Chen and M. C. Chaturvedi, Effect of boron concentrations on fatigue crack propagation resistance and low-cycle fatigue properties of Inconel 718, in K. A. Green, T. M. Pollock, H. Harada *et al.*, eds, *Superalloys 2004* (Warrendale, PA: The Minerals, Metals and Materials Society (TMS), 2004), pp. 275–282.

[35]   S. K. Jain, B. A. Ewing and C. A. Yin, The development of improved performance P/M Udimet 720 turbine disks, in T. M. Pollock, R. D. Kissinger, R. R. Bowman *et al.*, eds, *Superalloys 2000* (Warrendale, PA: The Minerals, Metals and Materials Society (TMS), 2000), pp. 785–794.

[36]   D. Lemarchand, E. Cadel, S. Chambreland and D. Blavette, Investigation of grain-boundary structure-segregation relationship in N18 nickel-based superalloy, *Philosophical Magazine*, **82A** (2002), 1651–1669.

[37] D. Blavette, P. Duval, L. Letellier and M. Guttman, Atomic-scale APFIM and TEM investigation of grain boundary microchemistry in Astroloy nickel-base superalloys, *Acta Materialia*, **44** (1996), 4995–5005.

[38] C. Ducrocq, A. Lasalmonie and Y. Honnorat, N18: a new damage tolerant P/M superalloy for high temperature turbine discs, in S. Reichman, D. N. Duhl, G. Maurer, S. Antolovich and C. Lund, eds, *Superalloys 1988* (Warrendale, PA: The Metallurgical Society, 1988), pp. 63–72.

[39] D. D. Krueger, R. D. Kissinger and R. G. Menzies, Development and introduction of a damage tolerant high temperature nickel-base disk alloy Rene 88DT, in S. D. Antolovich, R. W. Stusrud, R. A. MacKay *et al.*, eds, *Superalloys 1992* (Warrendale, PA: The Minerals, Metals and Materials Society (TMS), 1992), pp. 277–286.

[40] E. S. Huron, K. R. Bain, D. P. Mourer, J. J. Schirra, P. L. Reynolds and E. E. Montero, The influence of grain boundary elements and microstructures of P/M nickel-base superalloys, in K. A. Green, T. M. Pollock, H. Harada *et al.*, eds, *Superalloys 2004* (Warrendale, PA: The Minerals, Metals and Materials Society (TMS), 2004), pp. 73–82.

[41] M. C. Hardy, B. Zirbel, G. Shen and R. Shankar, Developing damage tolerance and creep resistance in a high strength nickel alloy for disc applications, in K. A. Green, T. M. Pollock, H. Harada *et al.*, eds, *Superalloys 2004* (Warrendale, PA: The Minerals, Metals and Materials Society (TMS), 2004), pp. 83–90.

[42] R. C. Reed, M. P. Jackson and Y. S. Na, Characterisation and modelling of the precipitation of the sigma phase in Udimet 720 and Udimet 720Li, *Metallurgical and Materials Transactions*, **30A** (1999), 521–533.

[43] S. P. Timoshenko and J. N. Goodier, *Theory of Elasticity*, 3rd edn (New York: McGraw-Hill, 1970).

[44] E. Volterra and J. H. Gaines, *Advanced Strength of Materials* (Englewood Cliffs, NJ: Prentice-Hall, 1971).

[45] T. K. Goswami and G. F. Harrison, Gas turbine disk lifing philosophies: a review, *International Journal of Turbo and Jet Engines*, **12** (1995), 59–77.

[46] A. C. Pickard, Component lifing, *Materials Science and Technology*, **3** (1987), 743–751.

[47] G. F. Harrison and M. B. Henderson, Lifing strategies for high temperature critical components, in R. Townsend *et al.*, eds, *Life Assessment of Hot Section Gas Turbine Components* (London: The Institute of Materials, 2000), pp. 11–33.

[48] J. F. Knott, The durability of rotating components in gas turbines, in A. Strang *et al.*, eds, *Parsons 2000: Advanced Materials for 21st Century Turbines and Power Plant* (London: The Institute of Materials, 2000), pp. 950–960.

[49] A. K. Koul, P. Au, N. Bellinger, R. Thamburaj, W. Wallace and J. P. Immarigeon, Development of a damage tolerant microstructure for IN718 turbine disc material, in D. N. Duhl, G. Maurer, S. Antolovich, C. Lund and S. Reichman, eds, *Superalloys 1988* (Warrendale, PA: The Metallurgical Society, 1988), pp. 3–12.

[50] A. Pineau, Superalloy discs: durability and damage tolerance in relation to inclusions, in E. Bachelet *et al.*, eds, *High Temperature Materials for Power Engineering 1990* (Dordrecht: Kluwer Academic Publishers, 1990), pp. 913–934.

[51] C. E. Shamblen and D. R. Chang, Effect of inclusions on LCF life of HIP plus heat-treated powder metallurgy Rene 95, *Metallurgical Transactions*, **16B** (1985), 775–784.

[52] T. P. Gabb, J. Telesman, P. T. Kantzos, P. J. Bonacuse and R. L. Barrie, Initial assessment of the effects of non-metallic inclusions on fatigue life of powder-metallurgy processed Udimet 720, in G. Fuchs, A. James, T. Gabb, M. McLean and H. Harada, eds, *Advanced Materials and Processes for Gas Turbines* (Warrendale, PA: The Minerals, Metals and Materials Society (TMS), 2003), pp. 237–244.

[53] M. McLean, Nickel-based alloys: recent developments for the aero-gas turbine, in H. M. Flower, ed., *High Performance Materials in Aerospace* (London: Chapman & Hall, 1995), pp. 135–154.

[54] Y. Bar-Cohen and A. K. Mal, Ultrasonic inspection, in *ASM Handbook Volume 17: Nondestructive Evaluation and Quality Control* (Materials Park, OH: ASM International, 1992), pp. 231–277.

[55] A. Banik and K. A. Green, The mechanical property response of turbine discs using advanced powder metallurgy processing techniques, in R. D. Kissinger, D. J. Deye, D. L. Anton *et al.*, eds, *Superalloys 1996* (Warrendale, PA: The Minerals, Metals and Materials Society, 1996), pp. 69–78.

[56] G. E. Maurer and W. Castledine, Development of consolidated powder metallurgy superalloys for conventional forging to gas turbine components, in R. D. Kissinger, D. J. Deye, D. L. Anton *et al.*, eds, *Superalloys 1996* (Warrendale, PA: The Minerals, Metals and Materials Society, 1996), pp. 645–652.

[57] W. Hoeffelner, Fatigue crack growth in high temperature alloys, in R. H. Bricknell, W. B. Kent, M. Gell, C. S. Kortovich and J. F. Radavich, *Superalloys 1984* (Warrendale, PA: The Metallurgical Society of AIME, 1984), pp. 167–176.

5 Environmental degradation: the role of coatings

As with any material, the superalloys suffer chemical and mechanical degradation when the operating temperatures are too high. Obviously the incipient melting temperature of a superalloy represents an upper limit on the temperature that can be withstood; this is usually no greater than about 1600 K. Despite this, the turbine entry temperature (TET) of the modern gas turbine continues to increase, with a take-off value of 1750 K being typical at the turn of this century; see Figure 1.5. Such extreme operating conditions have become possible only because action is taken to protect the components using surface engineering. In fact, the provision of such coatings and measures to ensure that they remain in place during service has become the most critical issue in the gas turbine field; in a state-of-the-art engine the components in the combustor and turbine sections would degrade very quickly were it not for the protection afforded by the coatings placed on them [1]. Thus, whilst the primary role of the superalloy substrate is to bear the mechanical stresses developed, an additional requirement is for mechanical and chemical compatibility with the coatings required to protect them.

Figure 5.1 summarises the different coating technologies which have become available [2,3] and ranks coating life and the temperature enhancement conferred by them in a relative way. It also serves as an introduction to the terminology used. The so-called diffusion coatings remain the most common form of surface protection. For example, aluminium is often deposited onto the surface of the superalloys by chemical vapour deposition, which, in its crudest form, is known as 'pack-aluminisation'; subsequent heat treatment to promote adhesion encourages interdiffusion with the superalloy substrate such that an aluminium-rich layer is formed on the surface of the component – this is typically rich in the β–NiAl phase [4]. Due to their high surface concentration of aluminium, coated superalloys produced in this way are very efficient formers of an external, protective alumina layer or scale, and consequently resistance to oxidation is much improved. It has been demonstrated that the electrodeposition of a 5 to 10 μm-thick layer of platinum prior to the aluminisation process improves both high-temperature oxidation and hot corrosion resistance (for example, see ref. [5]). When treated in this way, superalloys are said to be protected by aluminide or platinum-aluminide coatings; indeed the platinum aluminide coatings have become an industrial standard against which other coating solutions are compared. For greater resistance to oxidation and corrosion, so-called overlay coatings are available, but at greater cost, since deposition must be carried out by air or vacuum plasma spraying (APS/VPS) [6] or else electron beam physical vapour deposition (EB-PVD) [7]. The MCrAlX-type materials in various compositional variants have become the standard overlay coatings for gas turbine

283

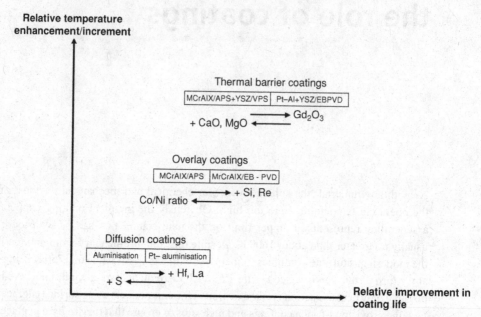

Fig. 5.1. Schematic illustration of the three common forms of protective coating used for turbine applications, and their relative coating lives and temperature enhancements which they afford.

applications [8,9]. Here, the 'M' refers to Ni or Co or combinations of these, and 'X' is a 'reactive' element (or mixture of them) at concentrations approaching 0.5 wt% – yttrium, hafnium and silicon are commonly used. The overlays have the advantage over the diffusion coatings of a greater flexibility of coating composition, with chemical and mechanical behaviour largely independent of the substrate on which they are placed; diffusion coatings on the other hand – by the nature of their formation – inherit a strong dependence on the substrate composition.

The thermal barrier coatings (TBCs) represent an alternative strategy – these are now finding very widespread usage in turbine engines [10]. A ceramic layer is deposited onto the superalloy, which, by virtue of its low thermal conductivity, provides thermal insulation and thus lowers the temperature of the metallic substrate. The effect can be very potent – a modern TBC of thickness 300 µm, if used in conjunction with a hollow component and cooling air, has the potential to lower metal surface temperatures by a few hundred degrees [11]. The first thermal barrier coatings emerged in the 1960s and were produced by the plasma spraying of calcia- or magnesia-stabilised zirconia. Whilst these were observed to perform well below ~1000 °C, above this temperature they were found to be unstable; formation of MgO and CaO occurred by diffusion of Mg^{2+} and Ca^{2+} ions, and the formation of the monoclinic form of zirconia was promoted causing a four-fold increase in the thermal conductivity. Modern TBCs that are now widely used to protect turbine blading, nozzle guide vanes and combustor sections in service are based upon zirconia containing about 7 wt% of yttria – so-called yttria-stabilised zirconia (YSZ) – a composition identified originally by NASA as giving the best thermal cycle life in burner rig tests. Unfortunately, TBCs based upon YSZ are unable to prevent ingress of oxygen due to rapid transport of oxide ions through them and, consequently, oxidation of the underlying substrate remains a possibility during operation. This fact, along with the mismatch in thermal expansion coefficients of superalloy substrate ($\sim$17 ppm/$^\circ$C) and ceramic top coat ($\sim$12 ppm/$^\circ$C), which causes the formation of thermal stresses during thermal cycling, leads to the eventual failure of the TBC during operation by spallation. To mitigate against this possibility, a so-called 'bond coat' is deposited on the superalloy substrate prior to the deposition of the ceramic top coat; some form of diffusion or overlay coating is used for this purpose [12,13]. The term TBC therefore strictly refers to the combination of ceramic layer and bond coat; these need to be chosen to ensure compatibility with the superalloy substrate. Even with a bond coat in place, spallation of the ceramic layer will eventually occur, albeit after a more prolonged period of time. Promoting TBC life and the design of procedures to estimate it are currently among the biggest challenges faced by materials scientists and engineers working in the gas turbine field [14].

5.1 Processes for the deposition of coatings on the superalloys

5.1.1 Electron beam physical vapour deposition

The electron beam physical vapour deposition (EB-PVD) method for the coating of gas turbine materials was introduced by Pratt & Whitney in the late 1960s, for the production of overlay coatings of the MCrAlX type [7]. Whilst the method is still used widely today for this purpose, the EB-PVD process has come into its own for the deposition of the ceramic TBC materials for turbine blade aerofoils [10]; for the overlays, EB-PVD has assumed decreased importance because many of these are now processed using plasma spraying (see Section 5.1.2). The characteristic feature of the EB-PVD process is the use of an electron beam to vaporise an ingot of the coating material, which is typically held in a water-cooled copper crucible; this causes a vapour cloud to be formed above the ingot in which the component for coating is manipulated; see Figure 5.2 [15]. As the coating material is evaporated, the feedstock is fed into the chamber to maintain a constant feedstock height – it is important to generate a completely molten evaporant surface free of splashing so that the coating process is stable. Thus no chemical reactions are involved. In order to increase the productivity of the process to a reasonable level (coating rates of between 5 and 10 µm per minute are typical), a powerful electron beam gun needs to be employed, for example, a 150 kW unit operating at 40 kV. Moreover, processing is carried out in a vacuum of better than a few Pa. Since there is also a requirement for the manipulation of the components within the vapour cloud, it turns out that the EB-PVD process is one of the most expensive of the coating processes used in the gas turbine industry. In practice, there are only around 50 establishments around the world which are capable of depositing EB-PVD TBCs to the required standard for engine service. Figure 5.3 shows an industrial-scale EB-PVD unit used for the deposition of coatings on turbine blading.

It has been found that superior TBC density, hardness, erosion resistance and spallation life are achieved only when the substrate temperature is raised to within the range 850 $^\circ$C to 1050 $^\circ$C during processing. By design, the resulting morphology then consists of a series of columnar colonies which grow competitively in a direction perpendicular to the surface

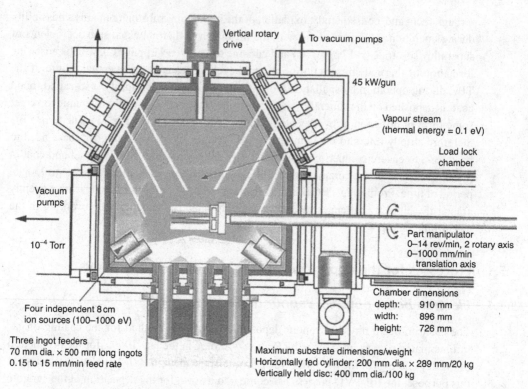

Fig. 5.2. Schematic illustration of the arrangement used for the electron beam physical vapour deposition (EB-PVD) method [15].

of the substrate. Figure 5.4 shows the microstructure of a typical 7 wt%Y_2O_3–ZrO_2 thermal barrier coating produced by the EB-PVD process, which has been shown to grow with a strong $\langle 111 \rangle$ texture [16]. The boundaries between the columnar deposits have been shown to be poorly bonded [17] – in effect, a bundle of columnar grains exists whose ends are attached to the substrate – so that a degree of strain tolerances is introduced into the coatings which would otherwise be prone to extreme brittleness and failure due to stresses developed during thermal cycling. Unfortunately, this also means that the coating quality is impaired if the surface to be coated is not perfectly clean; small imperfections are not covered up as they might be with other coating processes, but can result in a growth abnormality that is magnified as the coating thickens [7]. Experiments have shown that the size and morphology depends not just upon the substrate temperature, but also on the rate of component rotation which is required due to the complicated shape of the turbine aerofoils – EB-PVD being a line-of-sight process. At low temperatures and low rotational speeds, the columns vary not only in size from root to top, but also from one to another; see Figure 5.5. Increasing both the substrate temperature and the rotational speed improves the regularity and the parallelity of the microstructure and enlarges the column diameter. The need for component temperatures to be raised in this way prior to coating places further demands on the coating apparatus. For this purpose, either conventional radiation heating in the form of graphite furnaces placed in a preheating chamber, or else pre-heating by the electron beam itself is used.

Fig. 5.3. Industrial-scale electron beam physical vapour deposition (EB-PVD) unit for the deposition of ceramic coatings on gas turbine components. (Courtesy of Matt Mede, ALD Vacuum Technologies.)

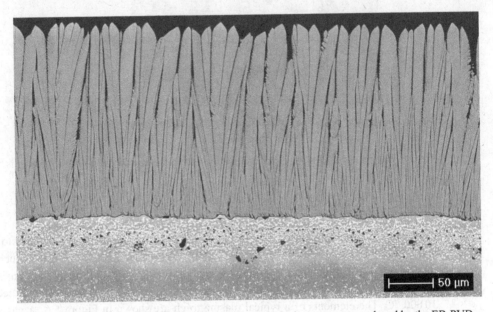

Fig. 5.4. Typical yttria-stabilised zirconia (YSZ) coating microstructure produced by the EB-PVD process. (Courtesy of Martijn Koolloos, NLR, the Netherlands.)

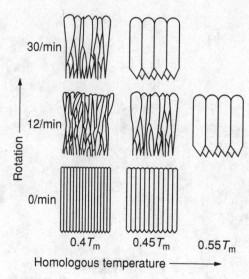

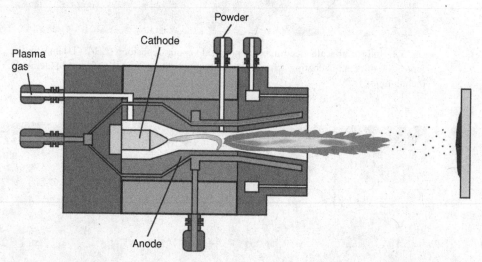

Fig. 5.5. Schematic illustration of the EB-PVD coating morphologies produced in YSZ material, as a function of substrate temperature and substrate revolution speed. (Courtesy of Uwe Schulz, DLR, Germany.)

Fig. 5.6. Schematic illustration of the cross-section of a plasma torch in which the powder is injected internally.

5.1.2 Plasma spraying

Plasma spray processes utilise the energy contained in a thermally ionised gas to melt and propel fine metal or oxide particles, initially in powder form, to a surface such that they adhere and agglomerate to produce a coating [18–20]. The plasma itself consists of gaseous ions, free electrons and neutral atoms – with its temperature in the torch exceeding 10 000 °C. The elements of a typical plasma torch are shown in Figure 5.6. A gas, most typically argon or nitrogen, is made to flow around a tungsten cathode in an annular space housed by a water-cooled copper anode of complex shape. A high-frequency electrical

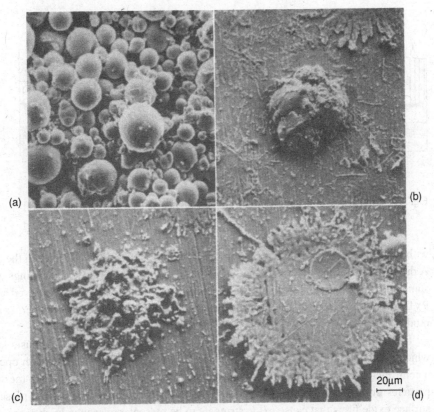

Fig. 5.7. Solidified NiCoCrAlY particles produced by plasma spraying: (a) initial condition, prior to spraying (b) and (c) partially melted, and (d) liquid condition upon impact [21].

discharge is then used to strike a direct current arc between the electrodes which is carried by the ionised plasma – causing a region of very high temperature into which particles of the powder to be sprayed are introduced. These are rapidly melted and then propelled towards the substrate to be coated. The entry of the powder should be introduced uniformily, either where the nozzle diverges or just beyond the exit of the torch. The powder velocities which are produced are very high – several hundred metres per second is common. The plasma-sprayed coatings are then built upon particle-by-particle, each producing a characteristic 'splat' morphology [21]; see Figure 5.7. Due to the large differences in thermal mass of particles and substrate, high cooling rates are achieved in the range 10^6 to 10^8 K/s. Most powder used for plasma spraying has a diameter between 5 and 60 μm; to achieve uniform heating and acceleration of it such that a high coating quality is produced, it should also be introduced into the plasma at a uniform rate. A narrow size range is preferred since this improves deposition efficiency; finer particles give denser deposits with less porosity, but with higher levels of residual stresses and oxide inclusions formed by oxidation of metallic particles during flight. A significant advantage of plasma spraying is that the composition of the powder usually closely matches that of the coating – which is not always the case

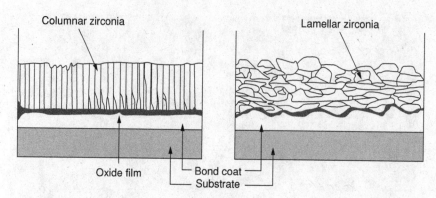

Fig. 5.8. Comparison of the microstructure of EB-PVD and plasma-sprayed TBCs [22].

with the EB-PVD process, due to different evaporation rates of the elements in the charge. A disadvantage of the process is that it is difficult to produce fully dense coatings without some porosity. Figure 5.8 compares the coating microstructure of plasma-sprayed and EB-PVD coatings, which display very different characteristics by virtue of the differing growth modes [22].

There are many types of plasma spraying apparatus available, from the basic systems which operate in air to the more sophisticated, fully automated systems which operate in a controlled environment or else in vacuum. With air plasma spraying (APS), the sprayed particles impact and deform, but perfect bonding is not achieved due to the presence of surface oxide films that cause oxide stringers to be present; consequently, they are characterised by their relatively low bond strengths and appreciable porosity. The low-pressure plasma spraying (LPPS) or vacuum plasma spraying (VPS) process circumvents some of these difficulties, because clean coatings are produced with virtually no oxide inclusions; see Figure 5.9. Since the plasma jet exhausts into a enclosed chamber held between 0.04 and 0.4 atm, significant expansions of the temperature isotherms and velocity contours of the plsama jet are achieved relative to APS; this is due to the increased mean free path between ions and electrons in the plasma – leading to very long spray distances. For example, spray distances for MCrAlX-type overlay coatings using the APS process are in the range 75 to 100 mm, whereas for VPS these are extended to greater than 400 mm [6]. At these so-called stand-off distances, the plume diameters are about 10 mm for APS and 50 mm for VPS. These characteristics lead to more uniform and consistent coatings, high densities close to the theoretical values, low residual stresses and the attainment of increased deposition thicknesses. Nevertheless, the use of VPS does come with some additional complications. The enclosed environment of the chamber necessitates robotic manipulation of the spray gun, vacuum interlocks, exhaust cooling, dust filtration and chamber-wall cooling, all of which add to the cost of the process. Consequently, the use of VPS is often used only where absolutely necessary; for combustor applications, for example, it is common to deposit an MCrAlX-type overlay bond coat using VPS prior to a YSZ-based thermal barrier coating using APS, to take advantage of lower cost and higher coating porosity of the latter process.

Fig. 5.9. Commercial vacuum plasma spraying (VPS) equipment for the production of thermally sprayed coatings. (Courtesy of Sulzer Metco.)

Recently, attention has been paid to techniques by which 'segmentation cracks' – running perpendicular to the coating surface and penetrating a significant fraction of the coating thickness – can be deliberately introduced into plasma-sprayed TBCs of YSZ [23]. The microstructure then resembles that of a TBC produced by the EB-PVD method; see Figure 5.10. It has been demonstrated that coatings fabricated in this way have improved resistance to spallation during thermal cycling, so that thicker TBCs can be produced which provide greater thermal insulation. Control of the substrate interpass temperature during spraying has been shown to be vital if segmentation cracks are to be introduced successfully – the range 500 to 800 °C is appropriate, with higher temperatures producing a greater crack density. The use of solid powder (fused and crushed) feedstock rather than the hollow-sphere (HOSP) variety is also helpful [24], although the latter gives rise to a greater porosity levels, which promote lower thermal conductivities. Although more research is required to elucidate the formation mechanism of segmentation cracks, it seems probable that the higher interpass temperatures cause successive splat layers to be present in the partially molten state, as evidenced by a common columnar solidification structure which then runs across them; the thermal strains induced by cooling are then able to promote segmentation cracking upon cooling, with improved intersplat bonding possibly playing a role. Unfortunately, the

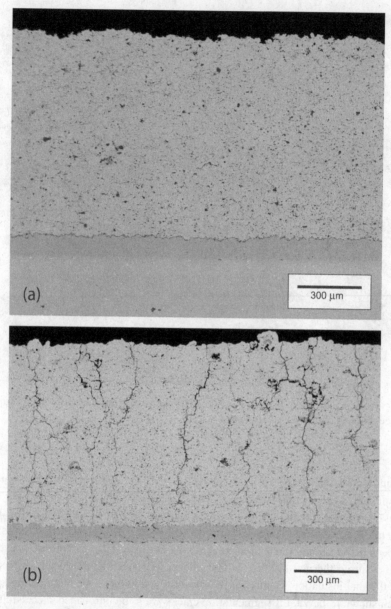

(a)

300 μm

(b)

300 μm

Fig. 5.10. Typical coating structures produced by plasma spraying of yttria-stabilised zirconia: (a) conventional processing conditions and (b) conditions designed to produce segmentation cracking. (Courtesy of Hongbo Guo.)

high porosity levels required for coatings of the lowest thermal conductivity are produced with low rather than high substrate temperatures – this indicates that, if segmentation cracks are to be introduced into any given coating material to improve its strain tolerance, then a greater thermal conductivity must be accepted.

5.1.3 Pack cementation and chemical vapour deposition methods

In the pack cementation process, the components to be coated are immersed in a powder mixture within a sealed or semi-sealed retort – this is then heated in a protective atmosphere until a coating of sufficient thickness is formed [25–28]. The pack consists of (i) the coating element source, (ii) an activator (such as NaF, NaCl or NH$_4$Cl) and (iii) an inert filler material, often alumina, which prevents the source from sintering [25]. The pack processes include chromising and siliconising, but aluminising is probably the most important and widely used in the gas turbine industry; for example, the vast majority of aeroengine turbine blade aerofoils are aluminised in some way to improve their resistance to high-temperature oxidation or as a surface preparation prior to the deposition of a TBC. In this case, the source can be either pure aluminium or an aluminium alloy (for example, Cr–30 wt%Al) depending upon the aluminium activity required; it is the difference between the activity of Al in the source and the substrate surface which represents the driving force for coating deposition. Broadly speaking, the activator reacts with the aluminium source to form aluminium halides; these are transported to the surface of the substrate where they are deposited, allowing them to react with the alloy and thus release aluminium.

The flexibility arising from the choice of aluminium activity in the source allows two distinct types of coating microstructure to be built up [4,29]. When the Al activity is high, an external brittle layer of δ–Ni$_2$Al$_3$ forms at the substrate surface (see Figure 5.11); aluminium is then the main diffusing species and the coating grows inwards. The diffusion coefficient of Al in δ is such that the process can then be carried out at relatively low temperatures, for example, 870 °C for 20 h – hence the term 'low-temperature, high-activity' (LTHA) is applied in this case. Subsequent heat treatment – for example, 1100 °C for 1 h – is used to promote the diffusion of Al inwards and Ni outwards such that a surface layer of β–NiAl is formed on the γ/γ' substrate, with an 'interdiffusion zone' (IDZ) forming between the two (see Figure 5.12); microanalysis using energy dispersive spectroscopy confirms that the IDZ is rich in slow-diffusing elements such as Re, W and Ta between the two (see Figure 5.13). Alternatively, if a low-activity pack is used then an external β–NiAl layer forms immediately during aluminisation, predominantly by the outward diffusion of nickel [4]. Since a relatively high temperature of 1100 °C is required in this case to promote reasonable diffusion rates, it is referred to as the 'high-temperature, low-activity' (HTLA) version of the process. Often, platinum-modified aluminide coatings are produced by electroplating a thin layer of Pt onto the surface of the superalloy; to encourage its adherence, a grit-blasting treatment is often applied prior to Pt electrodeposition. The incorporation of Pt into the aluminide layer in this way has been shown to improve the long-term oxidation resistance [30,31]. Traditionally, this has been attributed [32–34] to the exclusion of refractory elements from the outer aluminide layer caused by the presence of Pt; however, recent work [35] has disputed this view, the beneficial effect of platinum being linked to the minimisation of void formation at the coating/oxide interface and a retention of the oxide scale.

One drawback of the pack-cementation process is that it can only be used effectively to coat those blade surfaces which may be brought into direct contact with the pack, so

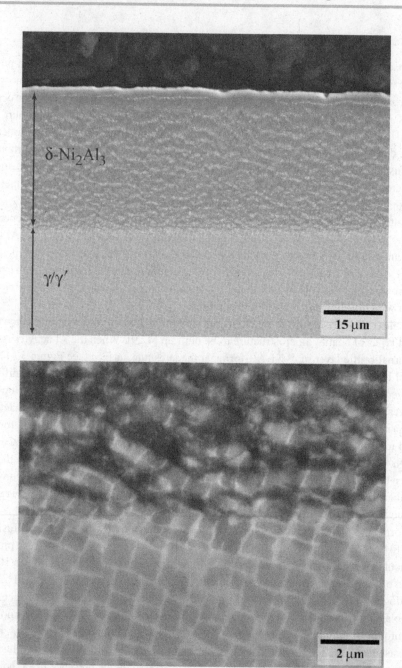

Fig. 5.11. Microstructures of the CMSX-10 single-crystal superalloy, following the low-temperature, high-activity, pack-cementation aluminisation process of 870 °C for 20 h. (Courtesy of Matthew Hook.)

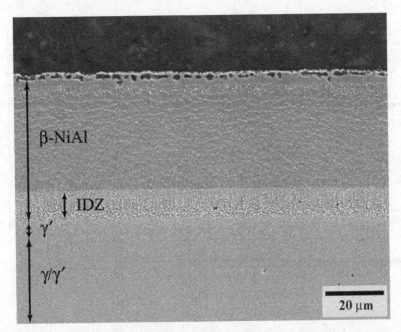

Fig. 5.12. The microstructure of the CMSX-10 superalloy following the low-temperature, high-activity pack-cementation aluminisation process, and subsequent heat treatment at 1100 °C for 1 h and further ageing at 875 °C for 16 h. (Courtesy of Matthew Hook.)

that the intricate cooling channels and internal surfaces found in many turbine blades are left uncoated [36,37]. To circumvent this difficulty, chemical vapour deposition (CVD) methods can be used such that aluminisation takes places exclusively in the gas phase [38]. Two distinct variants of gas-phase aluminisation can be identified. In the 'above-pack' process, the components are suspended within a reactor above beds of aluminium–chromium chips, upon which a powder of $AlCl_3$ is distributed. The reactor is filled with argon by means of a continuous gas feed, heated to a temperature of ~1100 °C, at which point the aluminisation reaction is initiated by the introduction of hydrogen gas into the Ar feed. The reactions which occur are beyond the scope of this book [36], but hydrogen plays a role in forming the reduced form of aluminium chloride, aluminium monochloride ($AlCl$), which reacts at the superalloy surface to produce the coating layer. The resulting reactants are transported to the surface of superalloy components by the thermal eddy currents which are set up within the reactor. This 'above-pack' process differs from the 'true' chemical vapour deposition method [36,37], in which the trichloride gas is generated outside of the reactor prior to its introduction into the reaction vessel – this allows its flow rate and activity to be accurately controlled, so that consistent and uniform aluminide layers are produced. A potential advantage of this approach, which needs further technological development, is that the generation and supply of reactant species outside of the reactor permits a possible 'reactive element co-deposition' so that additional elemental species may be incorporated into the coating to improve its effectiveness [39]. Schematic diagrams of the 'above-pack' and 'true' vapour deposition processes are given in Figure 5.14.

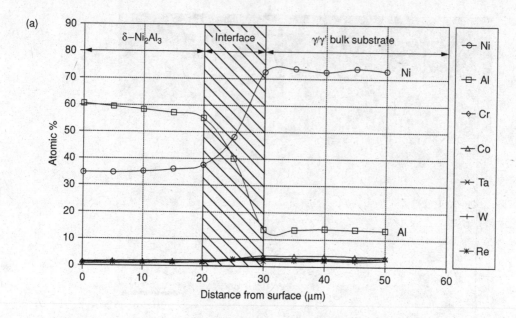

(a)

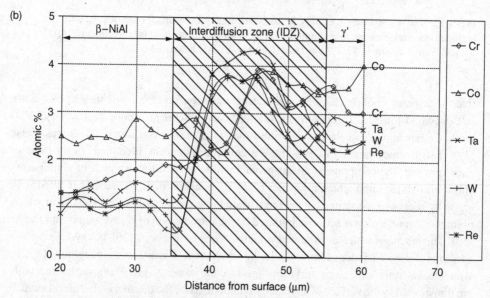

(b)

Fig. 5.13. Surface concentration profiles of the CMSX-10 superalloy (a) after the LTHA aluminisation process of 870 °C for 20 h, and (b) after subsequent heat treatment 1100 °C for 1 h and further ageing at 875 °C for 16 h. (Courtesy of Matthew Hook.)

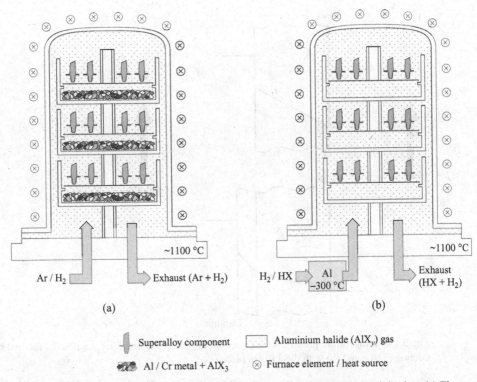

Fig. 5.14. Schematic diagram of the two reactor types employed for CVD aluminisation. (a) The 'above-pack' technique relies on the *in situ* evolution of the aluminium halide species, whereas the 'true' CVD process, (b), generates the aluminide halides externally and prior to introduction into the reactor.

5.2 Thermal barrier coatings

5.2.1 *Quantification of the insulating effect*

The potency of the insulation effect produced by a thermal barrier coating is best illustrated with the following quantitative example.

Example calculation

Consider a hollow, high-pressure turbine blade operating in a gas stream at 1450 °C, such that the outer and inner surfaces are at 1000 °C and 800 °C, respectively – the mean metal temperature is then 900 °C. The wall thickness can be taken as 5 mm; see Figure 5.15(a). Determine the decrease in metal temperature and steady-state heat flux arising from the addition of a 1 mm-thick thermal barrier coating (TBC); see Figure 5.15(b). The superalloy and TBC have thermal conductivities of 100 and 1 W/(m K), respectively. Assume that the heat transfer coefficient at the outer surface of the blade is unaltered by the addition of the TBC.

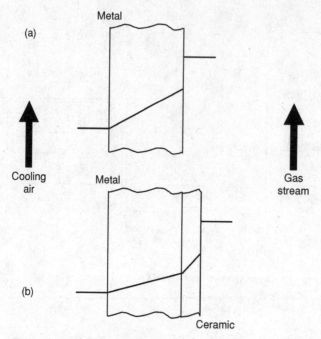

Fig. 5.15. Temperature profiles across a cooled turbine blade aerofoil (a) without a thermal barrier coating (TBC) and (b) with the TBC in place.

Solution   Without the TBC, the temperature gradient across the superalloy is $200/5 \times 10^{-3}$ or 4×10^4 K/m. Multiplying by the thermal conductivity gives a value of the heat flux, q, of 4×10^6 W/m^2; this must equal the product of the temperature drop across the gas stream/metal interface of 450 °C and the film transfer coefficient, h. An estimate of this last quantity is therefore $4 \times 10^6/450$ or 8.9×10^3 W/(m^2 K).

With the TBC added, let x and y be the new temperatures at the metal/ceramic interface and ceramic/gas stream interfaces, respectively. The equivalence of the new heat flux, q', through the coated system means that

$$q' = 100 \times (x - 800)/5 \times 10^{-3} = 1 \times (y - x)/1 \times 10^{-3} = 8.9 \times 10^3(1450 - y)$$

On solving these equations one finds that the temperature of the outer surface of the superalloy is now 828 °C, so that the mean metal temperature is reduced to 814 °C. The new flux, q', is then 5.6×10^5 W/m^2, i.e. 14% of the value without the TBC in place. The analysis can be used to confirm that both the flux and mean metal temperature decrease as the thickness of the ceramic layer is increased and its thermal conductivity is reduced.

5.2.2   The choice of ceramic material for a TBC

The above considerations confirm the usefulness of a thermal barrier coating and its dependence upon the thermal conductivity of the material employed. It is well known that ceramic compounds have low values of thermal conductivity, but which one is best for this

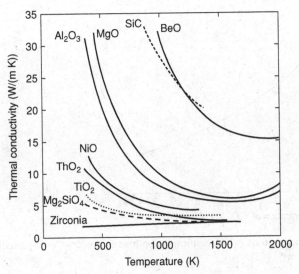

Fig. 5.16. The thermal conductivity of various polycrystalline oxides as a function of temperature [11,40].

application? Figure 5.16 presents some data [11,40] for thermally insulating oxide materials, in polycrystalline form. The thermal conductivities are in the range 1 to 30 W/(m K), the values decreasing with increasing temperature – the reasons for this are discussed in Section 5.2.3. Of the ceramic materials shown, zirconia (ZrO_2) displays the lowest conductivity, at about 2 W/(m K); moreover, no strong temperature dependence is found, so that this material does not suffer from the drawback of a steeply increasing conductivity at lower temperatures. These findings explain why ZrO_2-based ceramics have emerged as the materials of choice for TBC applications.

However, modern state-of-the-art TBCs are not fabricated from pure ZrO_2. Instead, they contain yttria (Y_2O_3) in the range 6 to 11% by weight, with a composition of 7 wt% of yttria being very commonly used [41]. This composition was originally identified by NASA during the 1970s as giving the best thermal cycle life in burner rig tests; see Figure 5.17 [42]. It turns out that this is the limiting composition at which the metastable 'non-transformable' tetragonal t′ phase is formed when the ceramic is quenched from the cubic phase field (see Figure 5.18). This avoids the tetragonal to monoclinic (t → m) phase transformation, which is associated with a 4% volume change and thus the poor thermal cycling resistance due to cracking and spalling at low Y_2O_3 contents; it also reduces the thermodynamic driving force for the dissociation at temperature into the equilibrium tetragonal t and cubic phases which inevitably occurs over time. The tetragonal t′ phase is thought to be the same tetragonal polymorph as the equilibrium tetragonal phase but with a very different microstructure, consisting of anti-phase domain boundaries and numerous twins [43]. When Y_2O_3 is combined with ZrO_2 in this way so that the t′ structure is formed, the resulting solution is known as 'partially stabilised zirconia'; the use of the adjective *partially* distinguishes it from 'fully stabilised zirconia' which arises at higher Y_2O_3 contents when the cubic phase is stable from ambient conditions to the melting point. For this reason, partially stabilised zirconia

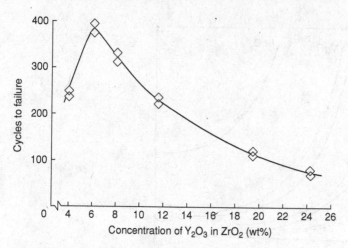

Fig. 5.17. Results from burner rig tests, which confirm the optimum yttria content of partially stabilised zirconia at about 7 wt% [42].

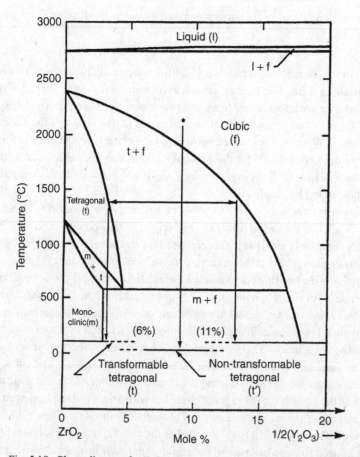

Fig. 5.18. Phase diagram for ZrO_2–Y_2O_3, showing the phase fields corresponding to the monoclinic (m), tetragonal (t) and cubic (f) phases, and the composition range for which the non-transformable tetragonal t′ form is found [44]. Note that sintering at 1400 °C of a composition corresponding to t′ must nonetheless lead to separation into the cubic and transformable tetragonal phases.

has emerged as the standard for TBC applications, being deposited in production onto combustor sections since the 1970s using thermal spraying and onto turbine nozzle guide vanes and blading since the 1980s using EB-VPD. It shows excellent thermal stability up to temperatures of 1200 °C, but unfortunately beyond this point it suffers from two drawbacks. First, partial reversion to the equilibrium t and cubic phases occurs with the tetragonal to monoclinic transformation then occurring upon subsequent thermal cycling [44], the sluggishness of this reverse transformation being due to the need for diffusion of Y through the zirconia. Second, extended exposure to high temperatures causes the sintering of the YSZ, with an associated increase in the elastic modulus [45]. The combination of these effects causes a loss of thermal cycling resistance.

Because of the importance of highly stable, low-thermal-conductivity ceramic materials for TBC applications, research has been carried out on compositions distinct from the 7YSZ standard. For example, CeO_2 rather than Y_2O_3 has been recommended (at a level of 25 wt%) for the stabilisation of ZrO_2 since no formation of the monoclinic phase was observed at 1400 °C and during limited exposures to 1600 °C; unfortunately, the erosion resistance is not improved by CeO_2 [46]. Similarly, 90% substitution of Sc_2O_3 for Y_2O_3 in so-called scandia/yttria-stabilised zirconia has been found to have significantly better tetragonal t' phase stability at 1400 °C [47,48]. Alternatively, ZrO_2-based ceramics can be dispensed with entirely. For example, hafnia (HfO_2) is a good candidate for an alternative ceramic material; HfO_2–Y_2O_3 TBCs have been found to display thermal cycling resistance comparable to 7YSZ, but at higher Y_2O_3 compositions of up to 27 wt% when the crystal structure is fully cubic [49]. Furthermore, lanthanum zirconate ($La_2Zr_2O_7$), which displays the pyrochlore crystal structure, a thermal conductivity of about 1.6 W/(m K) and excellent thermal stability and thermal shock resistance, has been proposed as a TBC material [50,51], as have other rare-earth zirconates based upon Gd and Sm [52]; interestingly, none of these compounds contains structural vacancies, so there is some uncertainty as to the reason for the low thermal conductivity of these materials. Finally, lanthanum hexaaluminate, which displays the magnetoplumbite structure, is considered by some to be a promising competitor to partially stabilised zirconia for operation above 1300 °C due to its resistance to sintering [53]. The science of these emerging ceramic materials for TBC applications has been reviewed recently in ref. [54]; at the time of writing, these new ceramics are not being used widely in service – although Pratt & Whitney claim to be using a $Gd_2Zr_2O_7$-based TBC for niche applications in military turbines. A significant difficulty is that many of the new ceramic materials have inadequate resistance to erosion which can occur due to sand ingestion and foreign object damage (FOD) accumulation. This is one further reason why partially stabilised rather than fully stabilised zirconia is preferred; in this system, the erosion resistance falls off monotonically with increasing yttria content.

5.2.3 *Factors controlling the thermal conductivity of a ceramic coating*

The choice of ceramic provides an intrinsic resistance to heat flow, but there are other features on the scale of the microstructure which provide a further *extrinsic* contribution. To appreciate this one should recognise that in crystalline solids in general, heat is transferred by

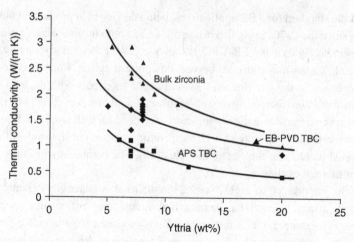

Fig. 5.19. The thermal conductivity of zirconia-based TBC materials as a function of yttria content and processing method [56].

three mechanisms [11]: (i) electrons, which are expected to provide the major contribution at very low temperatures, (ii) lattice vibrations, i.e. anharmonic phonon–phonon scattering, which will dominate at intermediate temperatures and (iii) radiation, i.e. photons, which will contribute strongly at very high temperatures. Ceramic materials in general are reasonable electronic insulators; indeed, zirconia is a very good one, with oxygen ion diffusion being responsible for electrical conductivity at elevated temperatures, so that any electronic contribution can be neglected. The data in Figure 5.16 support this conclusion, since the thermal conductivities decrease with increasing temperature, displaying in general the $1/T$ dependence characteristic of the Umklapp phonon–phonon process [55] – there is no increase with increasing temperature which would be expected from the electronic contribution. At very high temperatures the curves reach plateaux, becoming independent of temperature; as noted by Clarke [11], if not blocked, the radiative heat transport during thermal conductivity measurement can lead to an apparent increase in κ at high temperatures – there is evidence of this in Figure 5.16. Beyond 1250 °C, partially stabilised zirconia becomes significantly transparent to thermal radiation at wavelengths between 0.3 and 2.8 μm, with 90% of the incident radiation lying in this range with about 10% of the heat flux through the zirconia being due to radiation [56].

The above arguments indicate that further extrinsic reductions in thermal conductivity can be won by introducing inhomogeneities on the scale of the crystal lattice and grain structure. These considerations explain the significant decrease in the κ of yttria-stabilised zirconia (YSZ) observed with increasing yttria content and the potent influence of processing method used for deposition [56]; see Figure 5.19. Increasing the yttria content from 5 wt% to 20 wt% decreases the thermal conductivity by a factor of 2. However, the introduction of microstructural features such as intersplat boundaries and columnar grains in PS and EB-PVD coatings, respectively, has a comparable effect, with the former being more effective because they lie perpendicular to the direction of heat flow. Thus a coating has a lower

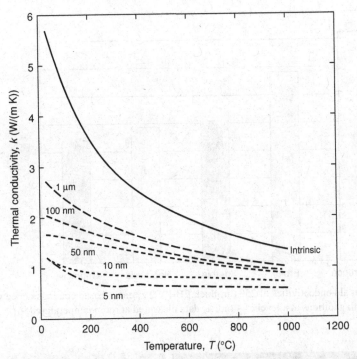

Fig. 5.20. Thermal conductivity of yttria-stabilised zirconia (7YSZ) as a function of temperature and grain size [56].

κ when deposited by PS than by EB-PVD. In general, porosity has been shown to have a profound influence on the thermal conductivity of TBCs fabricated by plasma spraying [57]. Reducing the grain size has also been found to reduce the thermal conductivity [56] (see Figure 5.20), although this effect is much less pronounced at elevated temperatures where only the point defects are effective. Alternatively, dopants have been introduced to YSZ-based ceramics to encourage phonon dispersion – rare-earth elements added in their oxide form, for example, erbia, ytterbia, neodymia and gadolinia, have been shown to be effective at reducing the thermal conductivity [56] (see Figure 5.21) because they act as phonon scatterers. A further novel strategy proposed recently has been a 'layered' form of TBC produced by sequential evaporation of two compositional variants of ceramic using a single electron beam, made to jump between two different ceramic ingots. An advantage of this approach is that the layer/layer boundaries lie perpendicular to the TBC growth direction; see Figure 5.22. Columnar, mixed ceria–zirconia and alumina–zirconia TBCs have been produced in this way, with layers exhibiting periodicity as small as 0.1 µm [58]. Similar structures have been produced using a glow discharge plasma to vary the density of the coating deposited, the layering being produced in this case by oscillating the d.c. bias applied to the substrate during the EB-PVD process, with reductions of about 40% in κ being claimed relative to the state-of-the-art TBCs being used in service [56]. Further microstructural modifications in the form of 'herringbone' structures have been produced

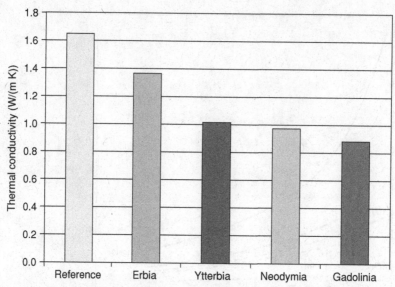

Fig. 5.21. Thermal conductivities of 250 μm-thick EB-PVD zirconia-based coatings doped with various rare-earth additions at a level of 4 mol%; data measured at room temperature [56].

(a)   (b)

Fig. 5.22. (a) Microstructure of multilayered EB-PVD ceria-stabilised zirconia TBC, illustrating (b) the formation of micro-columns formed from the vapour phase using the two-source jumping beam method [58].

by manipulation of the substrate during deposition, giving symmetrical zig-zag or wavy columns [59]; see Figure 5.23 – it has been claimed that the thermal conductivity of the coatings produced in this way can be up to 25% lower than so-called straight-columned TBCs produced ordinarily. Similar structures have been produced using high-velocity gas flows to focus the vapour plume, a process which has been christened 'directed vapour deposition' – conductivities of the 7YSZ coatings produced in this way were shown to be equivalent to those produced by APS processing [60].

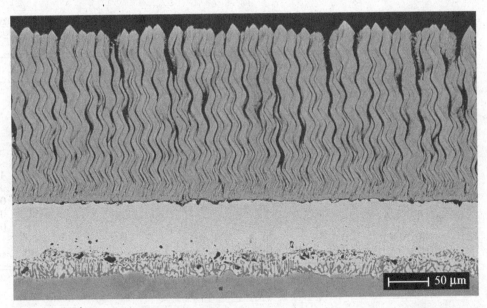

Fig. 5.23. Thermal barrier coating produced with a 'herringbone' microstructure [59]. (Courtesy of Martijn Koolloos, NLR, the Netherlands.)

5.3 Overlay coatings

Overlay coatings placed on the superalloys are generally of MCrAlX type, where M is either nickel (Ni) or cobalt (Co) or a combination of the two, and X is a reactive element added in minor proportions – yttrium (Y) is very commonly used. The coatings are thus metallic in form. By their nature, the composition of the coating is independent of that of the substrate (although some interdiffusion inevitably occurs) – hence, in principle, they allow the desired surface properties to be attained for any given application, for example with an optimum blend of oxidation and corrosion resistances, strength and ductility. The first overlay coatings appeared in the late 1960s and were deposited using the emerging EB-PVD technology. These were based upon cobalt alloyed with 20 to 40 wt%Cr, 12 to 20 wt%Al and yttrium levels around 0.5 wt% – with the composition Co–25Cr–14Al–0.5Y being very successful [3]. Such Co-based overlay coatings are still used extensively and are recognised for their superior hot-corrosion resistance. However, following much subsequent research effort, a wide range of coating compositions are now available with additions such as Si, Ta, Hf [61,62] or precious metals such as Pt, Pd, Ru or Re [63] – Table 5.1 lists the compositions of some common overlay coatings, including some which featured in patent applications during this early period of alloy development [8]. As of 1998, it has been stated [64] that there were at least 40 patented variations of the original MCrAlY concept; the exact compositions are sometimes of a proprietary nature, and it should be noted that the compositions used in practice are those which circumvent existing patents and are not always those which display optimum properties. As an example, Figure 5.24 illustrates the improvement in cyclic oxidation performance of plasma-sprayed NiCoCrAlY-type coatings modified with additions of Si and Hf at optimised concentrations of 0.4 and 0.25 wt%,

Table 5.1. *Compositions of some commonly used and prototype [8] MCrAlY overlay coatings, in weight %*

	Ni	Co	Cr	Al	Y	Ti	Si	Hf	Others
NiCrAlY	Bal		25	6	0.4				
NiCrAlY	Bal		22	10	1.0				
NiCrAlY	Bal		31	11	0.6				
NiCrAlY	Bal		35	6	0.5				
CoNiCrAlY	32	Bal	21	8	0.5				
CoCrAlY	—	Bal	25	14	0.5				
NiCoCrAlTaY	Bal	23	20	8.5	0.6				4 Ta
NiCoCrAlYSi	Bal	0–40	12.5–20	2–8	0–0.25	0–10	2–10		0–4 Nb
									0–4 Mo
									0–20 Fe
									0–5 Mn
NiCrAlTi	Bal		30–40	1–10		1–5			
NiCoCrAlHf	Bal	0–40	10–45	6–25				0–10	

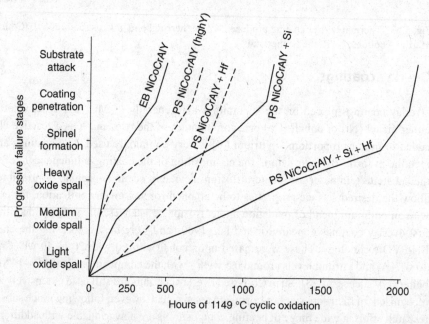

Fig. 5.24. Progressive degradation of EB-PVD deposited (EB) and PS sprayed overlay coatings showing synergistic benefit of silicon and hafnium additions [61].

respectively [61]. Much of this work became possible because of the development of plasma-spraying technology, since it was found that the EB-PVD process was not amenable to the transfer of low-vapour-pressure elements (for example, Si, Ta, Hf) along with the required Ni, Cr, Al and Y in the desired proportions. From this work emerged the nickel-based overlay coatings which were designed with oxidation rather than corrosion resistance in mind, such that an adherent Al_2O_3-based oxide scale is formed during exposure to high

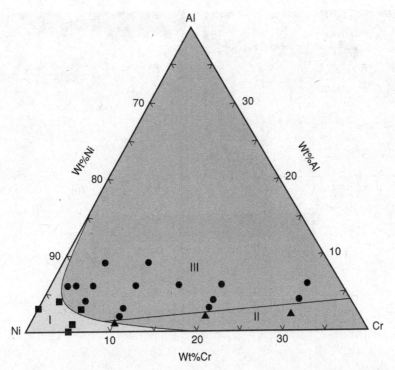

Fig. 5.25. Classification of Ni–Cr–Al alloys based on oxide and oxidation mechanism at 1000 °C [67].

temperatures. Compositions such as Ni–25Cr–6Al–0.4Y and Ni–22Cr–10Al–1Y (wt%) now find frequent use as overlay coatings, for example, for the bond coats in TBC systems. Recently, the Siemens Company has introduced Re into their MCrAlY overlay coatings used for industrial gas turbines, and demonstrated that the oxidation and thermomechanical properties are much improved [63]. The microstructure of these coatings consists of a two-phase mixture of the β–NiAl phase and the solid-solution of γ–Ni [65].

5.3.1 Oxidation behaviour of Ni-based overlay coatings

Given the emergence of the NiCrAlY-type compositions as nickel-based overlay coatings, one should identify the physical reasons for their suitability for this application. Some early papers of Pettit and co-workers [66,67] provide considerable insight. In a series of careful and methodical experiments, it was confirmed that in Ni–Cr–Al ternary alloys, and depending upon the precise mean composition, either an alumina (Al_2O_3) or a chromia (Cr_2O_3) oxide scale can form *exclusively* during exposure to high temperatures, this despite nickel being the majority component in these alloys so that nickel oxide might have been expected under all circumstances. This effect has been christened 'selective oxidation'. The ternary diagram in Figure 5.25 illustrates the regions in which the different scales were found to form – termed Group I, Group II and Group III behaviour – at 1000 °C,

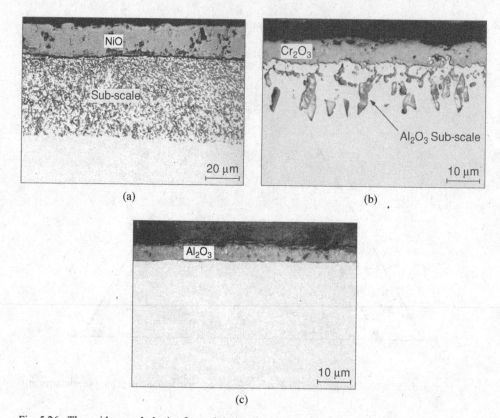

Fig. 5.26. The oxide morphologies formed during the high-temperature oxidation of Ni–Cr–Al alloys: (a) Group I, (b) Group II and (c) Group III [67].

although it should be noted that the I/III, II/III and I/II boundaries can alter with temperature, heating rate and grain size. The features of the three different scales are illustrated in Figure 5.26. Group I alloys, which are rich in Ni and thus have only low levels of Al and Ti, are characterised by a continuous, external scale of NiO accompanied by a sub-scale of Cr_2O_3, Al_2O_3 and/or the spinel $Ni(Al, Cr)_2O_4$. Oxidation begins with the formation of $Ni(Al, Cr)_2O_4$, which is rapidly overtaken by a brittle, outward-growing NiO scale; Cr_2O_3, and then Al_2O_3, are found at further increasing depths into the substrate due to their low equilibrium partial pressures of oxygen. The oxidation of Group II alloys, which contain high levels of Cr, causes an external scale of Cr_2O_3 and internal precipitation of Al_2O_3 – these alloys contain relatively high and low concentrations of chromium and aluminium, respectively. Initially, a transient $NiO/Ni(Al, Cr)_2O_4$ mixed external scale and Cr_2O_3/Al_2O_3 sub-scale form as before; however, unlike Group I alloys, there is sufficient chromium to support the flux required to form a continuous Cr_2O_3 sub-scale, which, after time, replaces the $NiO/Ni(Al, Cr)_2O_4$ as the external oxide. Finally, oxidation of Group III alloys – which possess appreciable concentrations of both chromium and aluminium – result in the exclusive formation of a compact external Al_2O_3 scale; and, unlike Group I and Group II alloys, no evidence of internal sub-scale precipitation is shown. Once again, a $NiO/Ni(Al, Cr)_2O_4$ external scale arises first, but an internal and continuous sub-scale of

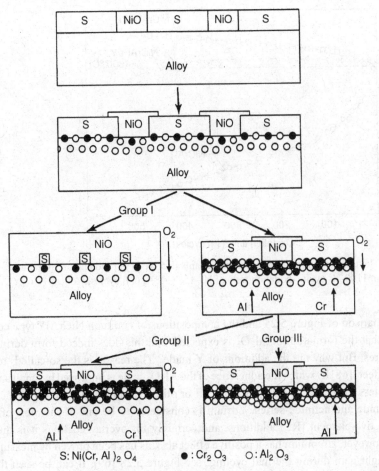

Fig. 5.27. Schematic illustration of the oxidation mechanisms occurring in Ni–Cr–Al alloys [67].

Al_2O_3 forms soon afterwards; the $NiO/Ni(Al, Cr)_2O_4$ eventually spalls away, leaving the Al_2O_3 as the external scale. Thus the basis of the Pettit theory is that nuclei of all oxide constituents arise at temperature, but that Al_2O_3 or Cr_2O_3 can predominate because of their greater thermodynamic stability, despite their growth kinetics being much slower than for NiO. A schematic diagram of the oxidation processes occurring in Groups I–III is given in Figure 5.27. One very important conclusion from this work is that it is possible to form Group III morphologies in the ternary system at aluminium concentrations as little as 5 wt% – this is in contrast to the Ni–Al binary system, which requires aluminium concentrations in the vicinity of 10 wt%. The ability of chromium to encourage the premature external oxidation of aluminium has been referred to as a 'gettering effect', but it may be more appropriate to refer to it as a 'third-element effect' since it is likely that Cr is having an effect on the Al activity. For example, Figure 5.25 indicates that 20 wt%Al is required in a binary Ni–Al alloy for exclusive Al_2O_3 formation, but that this figure falls to 5 wt%Al when 5 wt%Cr is added.

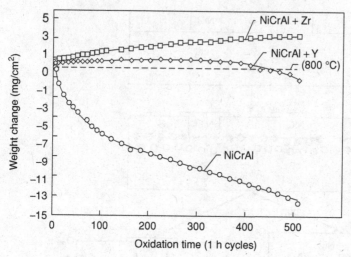

Fig. 5.28. The effect of yttrium or zirconium dopants (0.5 wt%) in producing Al_2O_3 scale adherence in 1100 °C cyclic oxidation tests of Ni–15Cr–13Al (1 h cycles) [69].

A comparison of Figure 5.25 and the compositions of common NiCrAlY-type coatings confirms that the formation of Al_2O_3 is expected, and this does indeed form during high temperatures. But why are the additions of Y made? The reason is the so-called 'reactive element effect' [68] by which the adherence of the Al_2O_3 scale is improved by small amounts (typically less than 1 wt%) of Y, La, Zr or Hf or other Group 3 and 4 elements, including the lanthanide and actinide series. Yttrium is considered to be one of the most effective of the reactive-element (RE) additions, and certainly for overlay coatings it is the most popular; however, zirconium has a positive effect also, as has been proven by measurement of the weight gain during thermal cycling; see Figure 5.28 [69]. It can be seen that the adherent scale on doped Ni–Cr–Al exhibits parabolic growth and only minimal weight gain; the undoped material, on the other hand, shows a mass loss which is very significant and approximately linear with time, consistent with a high rate of re-oxidation after spalling. It is generally accepted that the amount of the reactive-element dopant is critical if scale adhesion is to be optimal; see Figure 5.29 [70–72]. Low concentrations (less than 100 ppm) have been reported to reduce the growth stresses and scale-buckling effects arising from the higher specific volume of the scale. Levels around 0.1 wt% appear to be optimal for scale adhesion and low Al_2O_3 growth rates. However, at levels greater than about 0.5 wt%, the growth rate increases again – this can be attributed to the copious formation of oxides of the RE dopants, for example, from the oxidation of Ni_5Y and Ni_2AlZr precipitates, since the solubility limit of the dopant is then exceeded. One should note that the level of yttrium in the raw coating powder needs to be different from that in the deposited coating; for example, if air plasma spraying is used, the coating material needs to contain about 1 wt% yttrium to be most effective, since yttrium is oxidised by the plasma torch [69] – although this effect is less pronounced when vacuum plasma spraying is employed.

The reactive-element effect (sometimes known as the rare-earth effect) is very well known (see, for example, refs. [68] and [73]), and much research has been carried out in an attempt

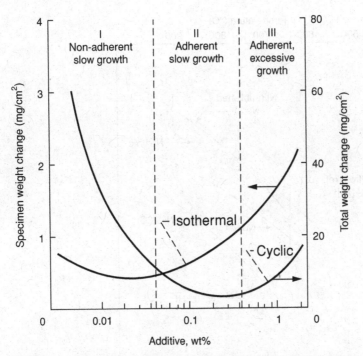

Fig. 5.29. Schematic diagram of the effects of dopant level on growth rate and degree of spallation in 100 h tests of MCrAlX alloys at 1100 °C [70–72]. (Courtesy of Jim Smialek, NASA.)

to explain the improved adherence which is conferred. For example, mechanisms proposed to account for the beneficial effects of the additions include (i) the formation of protective pegs which 'anchor' the scale to the substrate metal alloy [68], (ii) the prevention of vacancy coalescence at the metal/scale interface by providing alternative sites for coalescence [74], (iii) the enhancement of scale plasticity [75] and (iv) the improvement of the cohesive energy between metal and scale [68,76]. Most of these ideas have not been developed in a quantitative fashion; much controversy still exists concerning the relative importance of these mechanisms, their plausibilities and possible interaction. However, in recent years, the most widely accepted explanation emphasises the role played by sulphur in promoting Al_2O_3 spallation [77]. It has been known for some time that sulphur embrittles nickel and its alloys due to its equilibrium segregation to grain boundaries, which is Gibbsian in nature (see, for example, refs. [78] and [79]); its solubility in γ–Ni is strongly temperature-dependent, decreasing from about 500 ppm by atom at 1200 K to less than 10 ppm at 700 K, with elements such as Cr and Hf reducing the sulphur solubility and also its degree of grain-boundary segregation; see Figure 5.30. Consequently, it is not surprising that studies using Auger spectroscopy have also confirmed that sulphur segregation is present at the metal/scale interfaces formed during oxidation [80]; however, these also confirmed a role of yttrium in interacting with indigenous sulphur to form refractory sulphides such as YS and Y_2S_3, so that the amount available to concentrate at the scale/metal interface is then much reduced, or even eliminated [73]. These findings are supported by recent research on the PWA1480 single-crystal superalloy, hydrogen annealed in a gettered $5\%H_2/Ar$ flowing

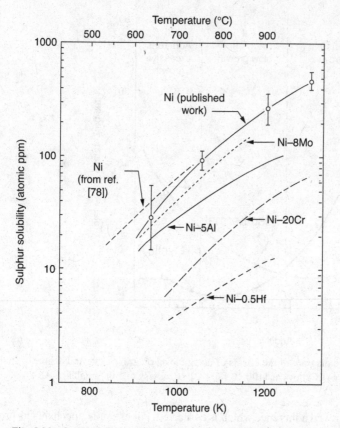

Fig. 5.30. Comparison between the sulphur solubility in Ni and its alloys, computed from grain-boundary segregation data, and that measured during an independent assessment [78].

gas mixture – this reduced sulphur content from 6.7 to less than 0.12 ppm by weight, with very much improved scale adherence; see Figure 5.31. By considering alloys with various levels of yttrium doping, the critical Y/S ratio for adhesion was found to be ~1:1 atomic, or 3:1 by weight. These findings justify the recent efforts of the superalloy melting community to implement melt desulphurisation methods into the procedures used for the production of superclean casting stock. However, it is not yet confirmed whether the addition of reactive elements to a sulphur-free coating material leaves the scale adherence unaffected – as would be expected if the sole influence is via the 'gettering' of sulphur; in fact, the results presented in ref. [77] indicate that the addition of a reactive element to a low-sulphur containing systems is beneficial to scale adhesion, presumably because the scale–metal interface is then strengthened.

5.3.2 *Mechanical properties of superalloys coated with overlay coatings*

The deposition of an overlay coating alters the mechanical performance of a superalloy turbine blade component [81]. In general, the high-temperature strength of the coating is lower than that of the superalloy on which it is placed; consequently, upon loading, the yield stress of the coating is exceeded, so that the stresses in it are quickly relieved and

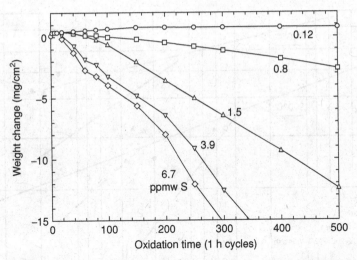

Fig. 5.31. Effect of sulphur content on the 1100 °C cyclic oxidation behaviour of the single-crystal superalloy PWA1480. Note ppmw = parts per million by weight. (Courtesy of Jim Smialek.)

redistributed across the cross-section. In practice, the extent to which this is beneficial or detrimental depends upon a number of factors – the corrosiveness of the environment, the operating conditions which dictate the temperatures experienced, the nature of the loading (i.e. static or cyclic) since crack nucleation may, in fact, be promoted and the degree to which the surface microstructure and stress state is altered. For example, during isothermal creep conditions which are more representative of those experienced by blading in an industrial (land-based) gas turbine, Tamarin and co-workers [1] have shown that the stress rupture life of the JS6U superalloy at 800 °C and 900 °C is severely compromised when testing is carried out with a gas turbine fuel (GZT) synthetic ash applied to the specimens; however, the use of a Ni–20Co–20Cr–12Al–Y overlay coating restored the properties to levels equivalent to those of the uncoated material, even with the ash present; see Figure 5.32. Under thermal-mechanical fatigue (TMF) loading, overlay coatings are found to perform very much better than the aluminide coatings, which under isothermal LCF conditions have been shown to impair the fatigue life at 600 °C, i.e. below the ductile-to-brittle transition temperature (DBTT) of the aluminide coating, although the detrimental effect disappears at higher temperatures [82].

When fatigue loading is severe – which is more representative of the aeroengine situation – property debits have been observed in isothermal laboratory tests when overlays are employed; thus, the behaviour can be compromised by the addition of this form of coating. Porosity inherited from plasma spraying, the increased surface roughness and the inherent brittleness of the coating contribute to this effect. For example, during isothermal low-cycle-fatigue (LCF) testing of the VJL12U superalloy, application of a Ni–20Cr–12Al–Y overlay coating by the EB-PVD method was found to be detrimental (see Figure 5.33), with the reduction in life being more pronounced when testing was carried out at 20 °C rather than 800 °C [1]. To explain this kind of observation, it has been noted that the overlay coatings are prone to cracking below a DBTT – the magnitude of which is dependent upon the coating composition. Traditionally, a recognition of the importance of the

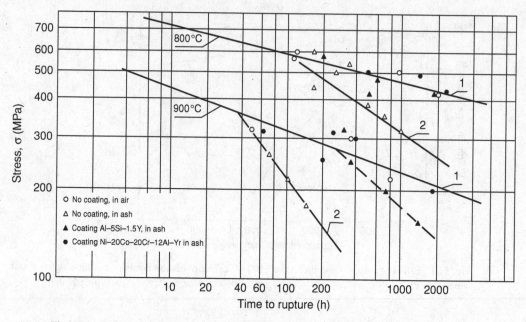

Fig. 5.32. Creep rupture curves for the JS6U superalloy, tested in the coated and uncoated conditions and with and without the application of a coating of synthetic ash [1].

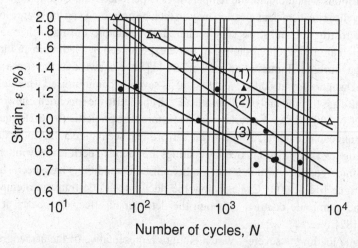

Fig. 5.33. Low-cycle-fatigue (LCF) test results [1] for the VJL12U superalloy tested at 20 °C, in the uncoated condition (1), with an aluminised coating (2) and with a Ni–20Cr–12Al–Y overlay coating (3).

DBTT in controlling cracking tendency has been used with success where in-service fatigue problems are identified – for example, the replacement of an aluminide diffusion coating on the blading of Rolls-Royce's Pegasus engine (used on the Harrier, which pioneered the vertical take-off/landing (VTOL) capability), where a CoCrAlY overlay eliminated the problem of cracking, but only after the Al content had been reduced to 7.5 wt% to reduce

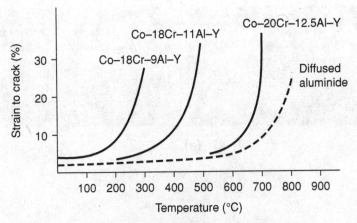

Fig. 5.34. Strain to cracking as a function of temperature for a number of CoCrAlY overlay coatings, illustrating the strong composition-dependence of the ductile-to-brittle transformation temperature [83].

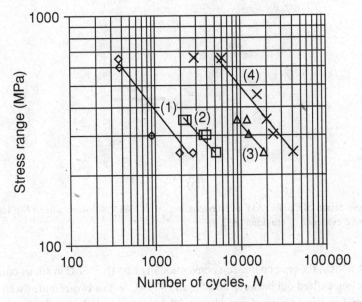

Fig. 5.35. Thermal mechanical fatigue (TMF) results (load control, out-of-phase temperature and loading cycles) showing the cyclic life of the CMSX-4 superalloy in the uncoated condition (4) and with Ni–7Co–12Cr–17Al–Y (3), Ni–8Co–12Cr–7Al–Y (2) and Co–32Ni–21Cr–8Al–Y (LCO-22) (1) overlay-type coatings [1,84].

the DBTT by several hundred degrees; see Figure 5.34 [83]. During testing under thermal cycling conditions, the effect of an overlay coating is mixed. Tamarin and co-workers [1,84] have reported that when CMSX-4 is coated with overlay coatings and tested in a load-controlled thermal-mechanical fatigue (TMF) rig, the coating reduces specimen life very considerably, with cobalt-based overlays being more detrimental than nickel-based ones, consistent with their greater brittleness; see Figure 5.35. On the other hand, Wood et al. [85]

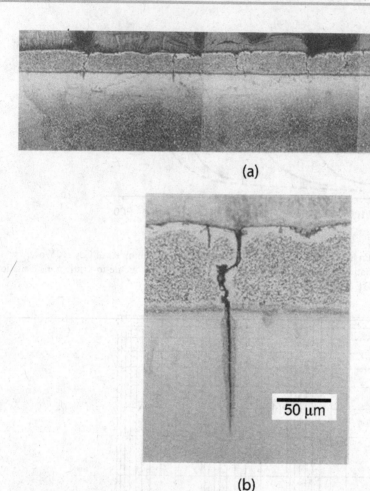

(a)

(b)

Fig. 5.36. Cross-section of CoNiCrAlY overlay coating on IN738LC substrate after TMF testing, showing extensive evidence of cracking [85].

have tested a CoNiCrAlY-type coating on conventionally cast IN738LC in strain-controlled TMF, with cycling carried out between 300 °C and 900 °C with a two-minute dwell period after heating and cooling. Extensive cracking of the coating was observed prior to failure (see Figure 5.36); however, the fatigue lives of coated and uncoated specimens were not substantially different.

When a thermal barrier coating (TBC) is employed, the resistance to thermal cycling (for example, in a burner rig test) is dependent upon the composition of the overlay coating onto which it is placed; see Figure 5.37 [86]. The TBC life has been found to correlate strongly with the resistance to oxidation displayed by the overlay, which is now acting as a bond coat [87,88]. For instance, during thermal cycling tests of TBCs fabricated from ZrO_2–6.1 wt%Y_2O_3 and NiCrAlY overlay coatings of various compositions which were placed on the Mar-M200+Hf superalloy, the spallation life is found to correlate strongly with the specific weight gain due to oxidation; see Figure 5.38. TBC spallation life is lowest

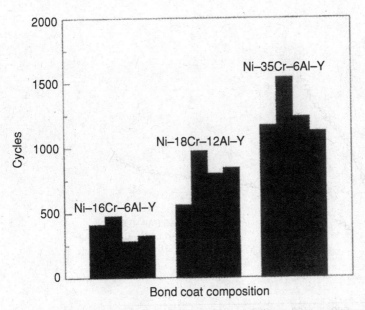

Fig. 5.37. Results from burner rig thermal cycle tests showing lives of TBCs using three different compositions of NiCrAlY bond coat, placed on Waspaloy substrates, with ZrO$_2$–7 wt%Y$_2$O$_3$ top coat [86].

when the rate of oxidation is greatest, and vice versa. These effects are discussed further in Section 5.5.

5.4   Diffusion coatings

Aluminide coatings are prone to microstructural degradation during service. Broadly speaking, two distinct modes arise. The first involves interdiffusion of the aluminide coating – which is initially rich in the β–NiAl phase – with the superalloy substrate causing concentration gradients set up by the coating deposition to become less pronounced. In particular, the aluminium concentration becomes reduced over time such that the γ' phase is promoted at the expense of β, which itself becomes increasingly susceptible to transformation to the martensitic form β' via the reaction $\beta \rightarrow \beta'$ as its stoichiometry is lost [89]. Due to these interdiffusional fluxes, Kirkendall voiding can also be found in some coating/substrate combinations [90]. Particularly for the third-generation single-crystal superalloys which are rich in Re, interdiffusion promotes the formation of a 'secondary reaction zone' (SRZ) beneath the interdiffusion zone (IDZ) [91,92]; see Figure 5.39. This occurs by a discontinuous phase transformation, in which a γ' matrix forms containing needles of the γ and the topologically close-packed phase P; thus, SRZ formation causes recrystallisation of the γ/γ' structure, but it has been demonstrated that there is no significant texture induced [92]. It is interesting to note that a single-crystal superalloy can be prone to the SRZ effect even though TCP precipitation does not occur within the bulk, even after extended periods of time at temperature – an anomaly which occurs, for example, with the MC-NG superalloy [92].

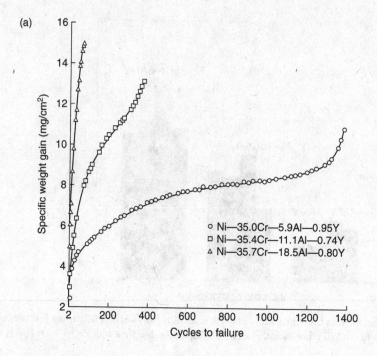

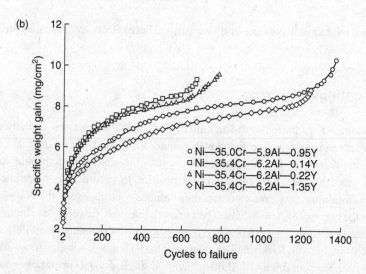

Fig. 5.38. Effect of bond coat concentration on the oxidation behaviour and TBC life of Ni–Cr–Al–Y/ZrO$_2$–6.1 wt%Y$_2$O$_3$ system for cyclic testing to a maximum temperature of 1100 °C: (a) aluminium concentration and (b) yttrium concentration [87,88].

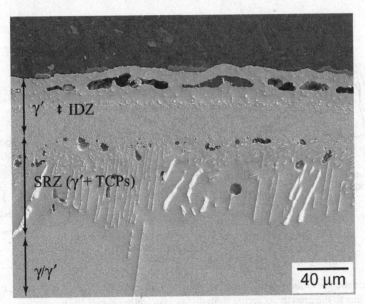

Fig. 5.39. Scanning electron micrograph of a Pt–aluminide diffusion coating on the CMSX-10 single-crystal superalloy, heat treated at 1100 °C for 1000 h; note the formation of the secondary reaction zone (SRZ) beneath the interdiffusion zone. (Courtesy of Matthew Hook.)

When a SRZ forms, it is usually already present in the as-coated condition but it then thickens parabolically with time, consistent with growth being diffusion-controlled. The nucleation of the SRZ reaction has, however, been shown to depend upon the surface preparation of the superalloy prior to coating – electropolishing prevents the reaction in cases where shot peening promotes it, with grit-blasting and grinding giving behaviour between these two limits [91]. The second mode of degradation is due to the formation of an alumina scale by the process of oxidation. Both the first and second modes of degradation cause depletion of Al from the coating such that its useful working life is rendered finite.

In view of the high fraction of the β–NiAl phase in aluminide coatings, emphasis has been placed on elucidating the mechanism of Al_2O_3 formation which occurs on 'model' NiAl-based alloys and the effects of doping them with different elements [93]. It has become clear that pure β–NiAl is prone to the formation of metastable cubic (otherwise known as 'transition') aluminas such as γ, δ or θ–Al_2O_3, particularly at temperatures below 1050 °C, when water vapour is present and/or when the oxygen partial pressure is low [94]. Characteristics of the cubic aluminas include (i) an unusual morphology consisting of whiskers, which is indicative of a growth mechanism controlled by outward migration of the Al^{3+} cation, as confirmed by $^{18}O_2$ tracer studies [95] and (ii) an epitaxial orientation relationship with the superalloy substrate, which is absent for the α–Al_2O_3 compound [96]. However, transformation to the equilibrium α–Al_2O_3 is always observed to occur rapidly thereafter, beginning at the metal–oxide interface. The polycrystalline scale which forms always displays (at least for undoped NiAl) a characteristic 'ridged' morphology at the high angle boundaries (see Figure 5.40) [93,97], probably as a consequence of outward diffusion of the Al^{3+} cation along them [98]; low-angle grain boundaries, cracks and porosity are also

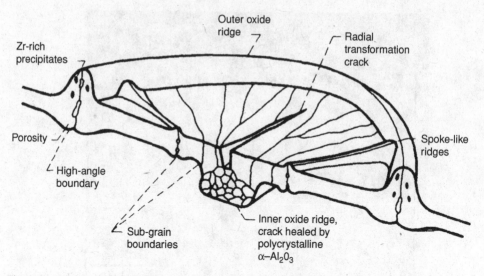

Fig. 5.40. Schematic illustration of the ridged morphology of α–Al_2O_3 formed on β–NiAl, the ridges forming due to the growth occurring from distinct nucleation events [97].

Fig. 5.41. Scanning electron micrograph images of the surface of a CVD (Ni,Pt)Al coating oxidised at 1150 °C for 200 h. Note that no scale spallation is observed, but thin plates of alumina had apparently flaked off some parts of the surface, leaving the underlying scale intact [99].

present, see Figure 5.41 [99]. In their analysis of Zr-doped NiAl, which was oxidised first in $^{16}O_2$ then $^{18}O_2$ at 1200 °C, Prescott *et al.* [100] measured secondary ion mass spectrometry (SIMS) depth profiles on untransformed θ–Al_2O_3 and transformed α–Al_2O_3 (see Figure 5.42), and demonstrated that (i) one has ^{18}O and ^{16}O enrichment, respectively, in these two cases, (ii) a ^{18}O signal, which is greater than that for ^{16}O at depth, and (iii) a Zr

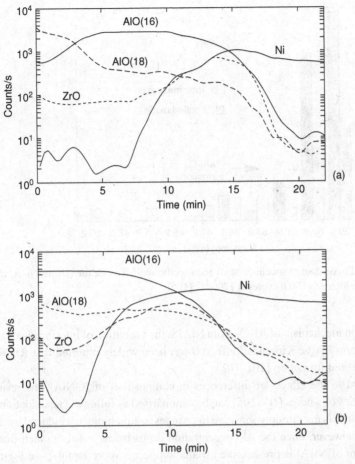

Fig. 5.42. Secondary ion mass spectrometry (SIMS) depth profiles obtained from Zr-doped NiAl oxidised for 12 min at 1200 °C [100]: (a) from untransformed, metastable θ–Al$_2$O$_3$ and (b) from transformed, stable α–Al$_2$O$_3$.

signal, which is greatest at the oxide/scale interface in the two cases. These observations imply that during growth of α–Al$_2$O$_3$ there is a strong component of O^{2-} anion transport inwards; taken with the observation of the ridged morphology, it seems likely that growth occurs by mixed Al^{3+} cation and O^{2-} anion grain-boundary transport. However, doping with reactive elements (REs) such as Y, Hf, La and Zr has some significant effects. There is a tendency for thick oxide scales to exhibit a columnar morphology, although the effects of the REs are less pronounced when scale formation occurs at higher temperatures [94]. It has been postulated in a 'dynamic segregation theory' that the REs, which have a strong tendency to segregate to the grain boundaries in α–Al$_2$O$_3$, suppress the outward diffusion of Al^{3+} [73]; while there is, as yet, no direct evidence to confirm this, it seems plausible on account of the significant affinity of the REs for oxygen, which should promote their migration outwards along the grain boundaries such that rare-earth oxides are formed at the scale/gas interface – for which there is considerable evidence [73]. The dominant

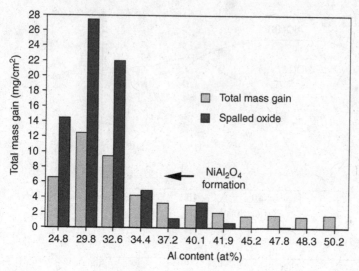

Fig. 5.43. Total mass gain of specimen/spall and specific spalled oxide for various Ni–Al alloys (+0.05 at% Hf) after ten 100 h cycles at 1200 °C [105].

scale-spallation mechanism of Al_2O_3 from NiAl is the formation of interfacial voids, which presumably form by the Kirkendall effect; it has been widely reported that RE additions suppress such void formation [101–103].

With regard to the effects of differences in composition of β–NiAl, the results of a systematic series of studies [104,105] can be summarised as follows. The oxidation performance of β–NiAl is not strongly sensitive to changes in stoichiometry induced by differing Al/Ni ratios; however, once the Al concentration drops below $\sim$40 at% such that a substantial amount of Ni_3Al is present, then spallation occurs more rapidly; see Figure 5.43. Hafnium-doped NiAl has been found to form an extremely slow-growing, adherent scale [105]; in fact, when doped with 0.05 at%Hf, NiAl has been found to be the most oxidation resistant substrate in a wide-ranging comparison of all types of α–Al_2O_3 formers and RE dopants [94]. Zirconium is also best suited at a dopant concentration of 0.05 at%, but Y is required at higher levels of 0.1 at% and La at lower levels of a few 100 ppm; there is some indication of a synergistic effect by which co-doping of two or more REs allows the total content to be reduced [106]. The solubility of the RE in the coating is considered to be an important issue, which explains why La and Y work less well than Zr in β–NiAl diffusion coatings, but adequately in the MCrAlY overlay coatings. The Pt effect has been studied, but no strong evidence has been found to support the widely held view that it acts as a diffusion barrier, or that it reduces substantially the α–Al_2O_3 scale growth rate [12,107]. Moreover, it appears that Pt does not act as a RE; 100 times more Pt than Hf is required to improve the scale adherence of near-stoichiometric NiAl (see Figure 5.44), and the scale morphology is not altered substantially. However, recent studies of the oxidation behaviour of ternary Ni–Al–Pt alloys have been revealing [105] – for pure Ni–Al unalloyed with Pt, the concentration below which a Ni-rich rather than an α–Al_2O_3 scale was found to form during 500 h cyclic oxidation between ambient and 1000 °C was found to lie between 40.0

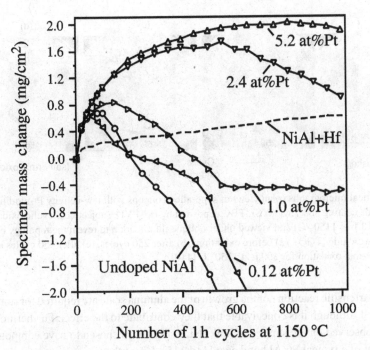

Fig. 5.44. Specimen mass change during oxidation (1 h cycles at 1150 °C in dry, flowing O_2) for various near-stiochiometric NiAl alloys alloyed with Pt [105].

and 42.5 at%Al. However, when 5.6 at%Pt was added to Ni–38.7 at%Al, no Ni-rich scale was formed and low rates of mass gain were observed. Thus, the addition of Pt was found to suppress the formation of Ni-rich oxides and improved the selective oxidation of Al to α–Al_2O_3, by lowering the Al concentration required to achieve it. The strong influence of Pt on the behaviour of Al in Ni–Al–Pt ternary alloys has recently been confirmed by Gleeson and co-workers [107], who have reported up-hill diffusion of Al in the γ' phase in the presence of a concentration gradient of platinum, implying that Pt reduces the Al activity; it is possible that similar effects would occur in the β-phase, but this needs to be tested. Platinum modification has been shown to reduce the sensitivity to sulphur and to suppress the formation of voids at the oxide/metal interface [108].

Recently, much attention has been paid to the gradual surface roughening of the aluminide bond coats during thermal cycling, primarily because of its perceived influence on TBC life [109]. The effect occurs predominantly during cyclic oxidation (see Figure 5.45), being much less pronounced after isothermal oxidation [110]. The effect, which has also been termed 'rumpling', has been postulated to occur due to growth strains in the oxide combined with the oxide/metal thermal expansion mismatch [111,112]. Transformation strains associated with martensitic reaction of the β-phase have also been suggested to have an influence [113]. However, recent experiments have eliminated some of the possible reasons for the rumpling effect; it has been found to persist even when thermal cycling is carried out in vacuum and when cycling is carried out between 1150 °C and 750 °C, the lower temperature being above the martensite start temperature [114,115]. These observations imply that

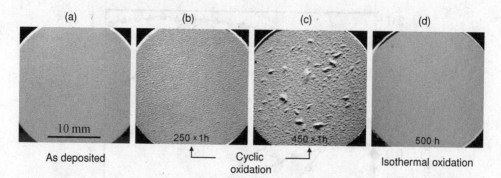

(a)　　　　　(b)　　　　　(c)　　　　　(d)

10 mm

250 × 1h　　　　450 × 1h　　　　500 h

As deposited　　　　Cyclic oxidation　　　　Isothermal oxidation

Fig. 5.45. Optical micrographs (plan view) of superalloy coupons with low-activity Pt-modified CVD aluminide coating and 7 wt%YSZ TBC deposited by EB-PVD, subjected to cyclic oxidation (dwell time of 1 h at 1150 °C) and viewed under oblique illumination to reveal the topology of the thermally grown oxide (TGO): (a) before oxidation, (b) after 250 cycles, (c) after 450 cycles and (d) after isothermal oxidation for 500 h at 1150 °C [110].

neither the martensitic reaction nor the growth of the alumina scale are required for rumpling to be observed, although it is conceivable that they contribute to the effect. Furthermore, no rumpling is observed on the surface of bulk (Ni, Pt)Al alloys chosen to have compositions similar to that of a typical Pt–Al bond coat [114,115]. Given these circumstances, it seems likely that interactions between the bond coat and the superalloy substrate are responsible for the rumpling effect. Moreover, since rumpling occurs with a characteristic wavelength close to the β-grain size of the aluminide coating [116,117], the $\beta \rightarrow \gamma'$ transformation is implicated since γ' precipitation occurs preferentially on the β-grain boundaries contained within the coating.

5.5　Failure mechanisms in thermal barrier coating systems

5.5.1　Introduction

A thermal barrier coating (TBC) system is prone to failure under the action of thermal cycling between ambient temperature (as at engine start-up) and the working conditions experienced during service. Most usually, this manifests itself as delamination at (or near to) the interface between the bond coat and the ceramic top coat; see Figure 5.46 [118]. It occurs during cooling, or very soon after ambient conditions are attained; however, it is rare for the coating to fail completely – some pieces of it remain intact, and significant scatter is observed in the time to failure if testing is repeated under identical conditions. The driving force for delamination arises from the temperature gradients set up during thermal cycling coupled with the differing thermal expansion coefficients of the various constituents present (for example, superalloy substrate, bond coat, ceramic top coat), but, as will be seen, various degradation mechanisms arise [14,119]; some or all of these must operate before failure finally occurs. It is emphasised that, at this point, controversy exists concerning the circumstances leading to TBC failure – possibly due to the difficulty in reproducing the

Fig. 5.46. Failure of an EB-PVD YSZ/NiCoCrAlY thermal barrier coating on a service-exposed high-pressure vane of a commercial civil aeroengine [118].

failure conditions (for example, temperature gradients, cycling times) pertinent to the gas turbine in the laboratory; although much research is still being carried out on this topic, some common characteristics have been identified. It has also become clear that TBC life is very dependent upon the combination of materials (substrate, bond coat, ceramic) and processes (EB-PVD, plasma spraying) used. The possibility of TBC failure represents a major threat to the integrity of a turbine engine – consequently, much attention is being paid to the factors which exacerbate it. From an engineering standpoint, a predictive capability is required so that full advantage can be taken of the extra temperature capability warranted by the presence of the TBC; this will allow them to be used in 'prime-reliant' rather than 'band-aid' mode. However, at the time of writing, this has not yet been achieved – there is a need to identify and quantify the chemical and mechanical factors which are the precursors to failure.

5.5.2 Observations of failure mechanisms in TBC systems

It is helpful to summarise the various observations which have been made concerning the microstructural degradation effects and failure mechanisms occurring in TBC systems. They are as follows.

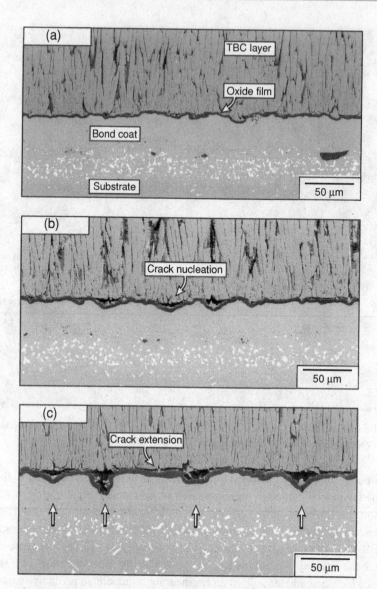

Fig. 5.47. Microstructural and morphological evolution at various stages of coating life in a Pt-modified aluminide/EB-PVD YSZ thermal barrier coating [120]. (a) After 20 cycles, corresponding to 8% of cyclic life; (b) 80 cycles, corresponding to 34% of life; (c) 180 cycles, corresponding to 76% of life. Note that imperfections and cracks are present after only 34% of cyclic life.

(1) Since the ceramic top coat is unable to prevent ingress of oxygen, a 'thermally grown oxide' (TGO) forms during thermal cycling due to oxidation at its interface with the underlying bond coat; see Figure 5.47 [120]. The thickness of the TGO layer increases parabolically with time (see Figure 5.48) [121]; this is strong evidence that its formation is diffusion-controlled. A TGO is already present after deposition of the ceramic; by

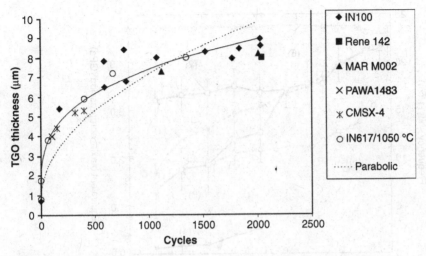

Fig. 5.48. Growth kinetics of thermally grown oxide (TGO) in a number of EB-PVD NiCoCrAlY/YSZ TBC systems [121].

the time that the thickness reaches 10 μm, TBC failure has usually occurred [122]. Thus there is a correlation between the thickness of the TGO and the TBC life which has been consumed.

(2) The bond coat becomes deleted of aluminium over time; this is due to (i) formation of the TGO, which has been shown to be predominantly α-alumina [123] and (ii) interdiffusion with the substrate, which causes the β-phase to dissolve and the formation of γ' to be promoted [12,124]. Interdiffusion with the substrate provides an opportunity for elements not present in the coating to reach the bond coat/TGO interface, and to influence TBC life [121].

(3) Due to the thermal stresses arising from a mismatch in thermal expansion coefficients and growth stresses due to the mechanism of its formation, the TGO layer forms with a significant biaxial in-plane compressive stress in it; photo-luminescence studies indicate that these can approach 4 GPa, with the strain increasing with thickness [125]. The action of thermal cycling reduces the average compressive stress measured in the TGO; see Figure 5.49 [110].

(4) As TBC life is consumed during cycling, imperfections at or near the TGO (due to processing-related variations in thickness, the mechanism of formation of the α-alumina, the formation of γ' and/or further chemical reactions which produce further phases such as spinels and garnets) become amplified [120,126,127]. Particularly for TBCs processed by EB-PVD, the TGO is initially smooth but develops undulations over time as it thickens [120]; see Figure 5.47. However, when PS is used, the TGO inherits considerable roughness from the processing step and does not roughen further to any appreciable extent before failure; see Figure 5.50 [128].

(5) Incipient cracks in the form of separations nucleate at the imperfections – these grow and coalesce to reduce the elastic strain energy in the system [129]. Final failure occurs once the cracks have grown to a size to support buckle propagation (see Figure 5.51),

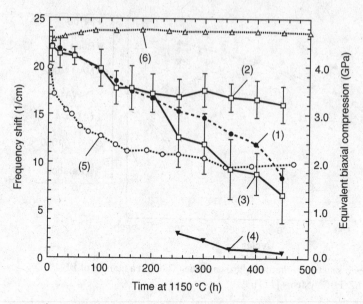

Fig. 5.49. Frequency shift of Cr^{3+} luminescence and equivalent biaxial compression in Al$_2$O$_3$ scales during cyclic oxidation at 1150 °C (1 h cycles) [110]. (1) Average stress underneath 60 μm TBC; local stress values for the (2) visually intact and (3) separated regions on the same sample; (4) stress in fractured or spalled fragments of the TGO; (5) average stress on similar samples without the TBC; and (b) with 130 μm TBC and fully intact TBC–TGO interface.

usually upon cooling when the ceramic layer gets placed in compression [130]; however, delaminations of the TBC in the form of incipient buckles are present before final failure, implying that they can grow in a stable fashion [131].

(6) The lifetimes of different combinations of substrate, bond coat and top coat vary considerably [121]; the processing method used for the deposition of the ceramic top coat (for example, EB-PVD, plasma spraying) is also found to be influential. TBCs with ceramic top coats produced by EB-PVD are generally found to perform better than those produced by plasma spraying (see Figure 5.52) [132], on account of their superior 'strain tolerance' arising from the microcracks inherited from the coating process. The best method for producing the bond coat (for example, aluminide by CVD, MCrAlY by EB-PVD or an alternative) is not yet apparent.

(7) Various other complicating factors can arise during engine operation. For example, TBCs can be subject to 'foreign object damage' (FOD) arising from the impact of airborne matter ingested by the engine [133]. Alternatively, calcia, magnesia, alumina and silica (CMAS) can deposit upon the surface of the TBC, causing partial dissolution of the TBC and exacerbating failure [134] since its strain tolerance is reduced and the stiffness increased. Vanadium impurities are found in kerosene, and these can leach Y out of the YSZ coating over time. Differences in substrate geometry, changes in its local radius of curvature and the testing method influence TBC life [135].

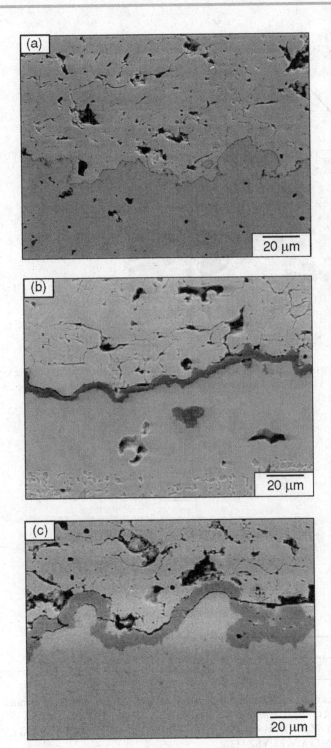

Fig. 5.50. Evolution of the ceramic/bond coat interface in an APS-sprayed YSZ/LPPS-sprayed CoNiCrAlY system: (a) as-deposited, (b) after exposure for 216 h at 1010 °C, showing the formation of a 3-μm-thick TGO and (c) after exposure for 1944 h at 1010 °C, showing the presence of cracks penetrating the TGO next to imperfections in the coating [128].

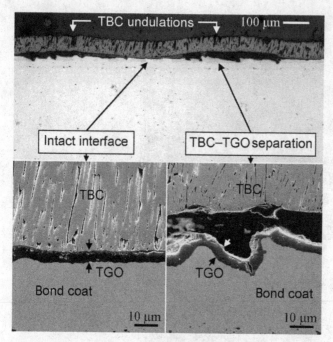

Fig. 5.51. Microstructure of a Pt-modified aluminide/EB-PVD YSZ thermal barrier coating system after 450 × 1 h cycles to 1150 °C showing separation of the ceramic/TGO interface and the onset of small-scale buckling [110].

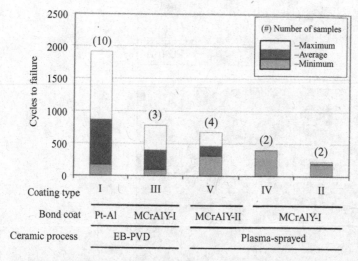

Fig. 5.52. Spallation lives of five production TBC systems tested in a cyclic furnace-controlled thermal rig, from ambient to 1121 °C with 1 h cycles: (I) Rene N5/Pt–Al/EB-PVD YSZ; (II) IN939/MCrAlY/PS YSZ; (III) CMSX-4/MCrAlY/EB-PVD YSZ (IV) CMSX-4/MCrAlY/PS YSZ and (V) MarM509/MCrAlY/PS YSZ [132].

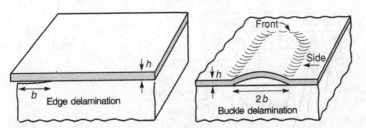

Fig. 5.53. Delamination modes for compressed films: (a) edge delamination and (b) buckling delamination for the straight-sided blister [137].

5.5.3 *Lifetime estimation models*

Existing methods for estimating the number of cycles before a TBC spalls away from the surface of a superalloy are highly empirical; for example, use is made of cylindrical testpieces exposed to burner rigs or else experienced gained from full-scale engine 'type' tests. Early work [136] made use of the notion of a 'critical thickness' of the TGO for TBC failure, termed δ_c, for TBC failure, which was then incorporated into a modified version of Miner's law – such that the ratio δ/δ_c (where δ is the sub-critical TGO thickness) is then a quantity that can be related (by a suitable functional form) to the life consumed. Unfortunately, this kind of approach is not fundamentally based upon the principles of fracture mechanics; for example, the driving force for interfacial crack growth and the interfacial toughness remain unaccounted for. Clearly, there is a need for physically based laws which model the failure mechanisms occurring and which, ideally, account for the microstructural effects causing the substantially different behaviour of varying combinations of substrate, bond coat and ceramic types.

Whilst much work remains to be done, progress in this regard has been made by Evans and co-workers (for example, see ref. [129]), who have produced 'failure maps' via the mathematical analysis of the micromechanics of TBC systems. Emphasis is placed on the competition (see Figure 5.53) between (i) delamination from edges, such as the trailing edge of the turbine blade, which is resisted by the mode II interfacial fracture toughness, Γ_{II}, and (ii) buckling of the TGO from pre-existing interfacial separations which have already nucleated. Figure 5.54 shows examples of failure maps arising from these calculations [137]; these provide insight into the factors controlling durability. The thickness of the ceramic top coat, h, is plotted against its modulus anisotropy, defined as the ratio of the Young's modulus in the in-plane and out-of-plane directions, i.e. $E_\parallel/E_\perp$; for YSZ produced by EB-PVD, this quantity is less than unity on account of the columnar microstructure present for this processing method. In the upper right-hand corners of the maps, the thickness and in-plane modulus are high such that the residual stresses and elastic strain energy density are large – this provides the driving force for edge delamination. The extent of this region is reduced as the mode II interfacial toughness increases. In the lower left-hand corners, the low thickness and in-plane modulus are unable to resist buckling; the extent of this second region increases with the characteristic width, b. Between these two limits lies a 'fail-safe' regime where failure is predicted not to occur – implying that there is an optimum thickness and in-plane modulus for a TBC, consistent with experience. The extent of the

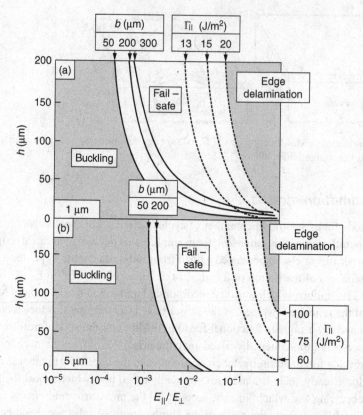

Fig. 5.54. Examples of TBC failure maps identifying the top coating thickness, t, and modulus ratio such that neither buckling nor edge delamination occurs in a TBC system of typical characteristics [137]: (a) with a 1 μm-thick TGO and (b) with a 5 μm-thick TGO. The term Γ_{II} is the mode II interfacial fracture toughness and b is the characteristic width of the delamination (see Figure 5.53).

fail-safe region also depends upon the thickness of the TGO; compare Figures 5.54(a) and 5.54(b).

Although the calculation of the failure maps requires consideration of multiple layers of the TBC system (i.e. substrate, bond coat, TGO and ceramic) the relative positions of the regimes on the failure map can be rationalised by considering a single layer film of thickness h – approximating the ceramic top coat – on a thick substrate. For the case of edge delamination, the energy release rate, G_0, approaches a steady-state, becoming independent of the crack length, b, once its length exceeds several layer thicknesses; thus [138]

$$G_0 = \frac{1}{2}\left(\frac{1 - v^2}{E_\parallel}\right)\sigma_0^2 h \qquad (5.1)$$

where E and v are the Young's modulus and Poisson's ratio of the TBC, respectively, and σ_0 is the compressive, biaxial misfit stress given by

$$\sigma_0 = E_\parallel \epsilon_T / (1 - v^2) \qquad (5.2)$$

where the misfit strain, ϵ_T, is equal to $\Delta\alpha \times \Delta T$, where $\Delta\alpha$ is the difference in thermal expansion coefficients of TBC and substrate and ΔT is the cooling range of approximately $1300\,^\circ$C. Inserting reasonable values for the variables, for example, $E_\| = 20\,$GPa, $\nu = 0.3$, $h = 100\,\mu$m and $\epsilon_T = 5 \times 10^{-3}$, one finds that G_0 is about $25\,$J/m^2. This is substantially lower than the mode II fracture toughness, Γ_{II}, which is in the range 60 to $100\,$J/m^2 [139,140]; thus in the as-deposited state there is insufficient energy density to delaminate the TBC. However, should h be increased by a factor of about 2 – so that one moves from the fail-safe to the edge-delamination regime of Figure 5.54 – then failure is predicted to occur. Alternatively, the same effect is caused by larger misfit strains arising from temperature gradients or else sintering of the TBC, which has been shown to lead to a three-fold increase in $E_\|$ [141]. In the case of a buckle delamination (see Figure 5.53) consider the case of a straight-side blister of width $2b$ and coating thickness, h. If it is clamped along its straight edges, it undergoes buckling when the compressive stress, σ_c, in the coating attains a value

$$\sigma_c = \frac{\pi^2}{12}\left(\frac{E_\|}{1 - \nu^2}\right)\left(\frac{h}{b}\right)^2 \tag{5.3}$$

which is a standard result from buckling theory, for example, for the Euler column. Note that when σ_0 is in excess of σ_c, buckling is expected. This occurs when h is small, b is large or for low in-plane stiffness, $E_\|$, i.e. on the bottom left-hand side of the failure map.

5.5.4 The role of imperfections near the TGO

At the present time, the major lack of understanding concerning the durability of TBCs relates to the physical and chemical factors controlling the growth of imperfections at or near the TGO, which are the precursors to delamination. Since these are a prerequisite for spallation, it follows that any useful model for lifetime prediction requires an accurate estimate of their rate of nucleation. Currently, quantitative theory for this purpose is unavailable; this situation is hampering the development of models for TBC lifetime estimation which are physically based, such that the temporal evolution of defects is accounted for. Unfortunately, at the present time there is no strong consensus on the origin of the imperfections and the mechanism controlling their development – and it is very probable that these differ with the physical and chemical properties of the substrate/bond coat/ceramic combination and the processing methods employed.

When the damage mechanisms arising in thermally cycled TBC systems are examined, the pervasive observations concern the gradual roughening of the bond coat/TGO and TGO/ceramic interfaces. Since the TBC is transparent to light, the undulations which form are detectable optically under oblique illumination (see Figure 5.45 [110]), and their amplitude has been shown to increase as cycling proceeds. Such findings are consistent with the 'rumpling' or 'scalloping' identified during the cyclic oxidation of both aluminide [142–144] and MCrAlY-type coatings [143] when the TBC is absent. Nevertheless, it remains to be proven that such roughening can be attributed solely to the oxidation process which causes TGO thickening. However, there is little doubt, as shown by the Cr^{3+}

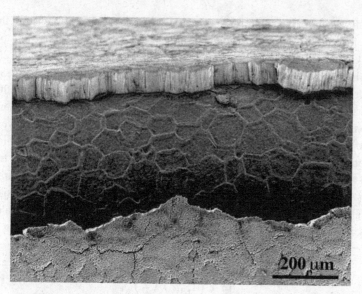

Fig. 5.55. Example of edge spallation in an platinum-aluminised/EB-PVD YSZ TBC system subjected to cyclic oxidation testing consisting of 330 × 1 h cycles from ambient to 1150 °C [145]. Note that the β-grain structure is apparent on the areas which have spalled.

luminescence studies [110], that the α–Al_2O_3 forms with a compressive stress of several GPa due to contributions from the thermal misfit and growth stresses, the former being considered to be the larger of the two [14]; thus there is a considerable driving force for buckling of the TGO. Many observations confirm that the TGO does indeed become more serpentine such that its overall length increases with time; see Figure 5.51. The TGO's stress is almost completely relieved once debonding occurs but, even when bonding remains intact, it is reduced by the action of thermal cycling – possibly due to sub-critical cracking, creep relaxation or else its morphological evolution. Once the ceramic layer has spalled, the 'ridged' morphology characteristic of α-alumina grown on polycrystalline β–NiAl is visible (see Figure 5.55 [145]) so that the bond coat/TGO and TGO/top coat interfaces are not smooth, and thus the ridges are possible sites for damage initiation, as emphasised in ref. [138]. The TBC life has been shown to be influenced by the surface condition of the aluminised coating prior to ceramic deposition; for example, grit blasting is detrimental and the polishing away of the ridged morphology inherited from the CVD process is beneficial.

Most observations on platinum-aluminised EB-PVD coatings indicate that imperfections in the TGO form and grow preferentially into the bond coat [120]; see Figure 5.47. This is consistent with an inward-growing scale controlled by O^{2+} diffusion along the grain boundaries in the α-alumina TGO; the formation of the oxide preferentially at the β-grain boundaries intersecting the TGO/bond coat interface once the oxygen partial pressure reaches a high enough value would yield a density of imperfections consistent with all observations so far, and indeed the postulate that the β-grain boundaries in the bond coat are potent sites for crack initiation [146]. This picture of the microstructural damage events occurring is consistent with recent observations [116,117] that the TGO/bond coat interface moves preferentially into β rather than the γ' phase forming by aluminium depletion due to oxidation and interdiffusion with the substrate – presumably because aluminium diffusion in the former is about an order of magnitude faster than in the latter phase; thus, the γ' grains gradually decorate the TGO/bond coat interface over time; see Figure 5.56 [116,117] with their aspect ratio being modified accordingly. Once again, the density of nucleation events is consistent with the coating's β-grain size since γ' nucleation is found to occur on the prior β-grain boundaries. If this microstructural model of the damage evolution is correct, then it does not seem necessary to invoke the ratcheting mechanism which is founded on the principles of mechanics [111]. There are however, many complicating factors which need to be explained. It is found, for example, that other oxide phases form at or near to the α-alumina; spinels have been observed particularly when NiCrAlY-type bond coats are employed, probably because of the presence of Cr, and garnets can form in TBC systems employing aluminide coatings. Moreover, there are substantial differences between the TBC lives of systems with diffusion-based and overlay-type bond coats, and indeed which of the two bond coat systems is the better one has still not been resolved. Furthermore, when the same ceramic/bond coat combination are employed, TBC life is a function of the substrate chemistry [121]. This is particularly the case when platinum-diffused 'low-cost' bond coats are employed, which avoids the use of the aluminisation treatment [147].

5.6 Summary

Due to the extreme operating conditions of the modern gas turbine, great reliance is now placed on coating technologies for the protection of components from oxidation and corrosion. This is particularly the case for the turbine blading, nozzle guide vanes and combustor sections which reside directly in the gas stream. Processes available for this purpose include electron beam physical vapour deposition (EB-PVD), plasma spraying (PS) and the chemical vapour deposition (CVD) method and its variants; the choice of process used is critical in determining the coating performance exhibited. Overlay coatings (of composition NiCoCrAlY) fabricated by PS and diffusion-type coatings (for example, the aluminides and platinum-aluminides) produced by CVD are very common. Thermal barrier coatings (TBCs) – in which a combination of a ceramic top coat produced by EB-PVD is deposited on either a CVD diffusion-type aluminide or alternatively an overlay coating – are an increasingly critical technology for gas turbine applications due to the thermal insulation they provide. Unfortunately, they are prone to failure by spallation during thermal cycling. A challenge for the future is the optimisation of TBC systems – by which which one means the ceramic top coat, the bond coat and superalloy substrate – so that coating life is maximised. Further research is needed to clarify the mechanisms of failure of TBC systems, which depend upon the combination of processes and coating materials used. Methods for TBC lifetime estimation are still immature, and these need to be developed further if full advantage is to be taken of the extra temperature capability conferred by the use of these coatings.

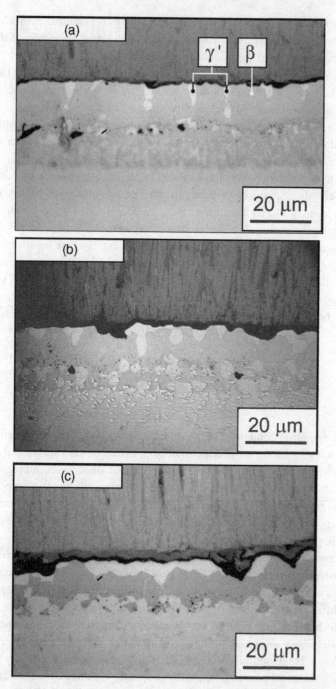

Fig. 5.56. Optical micrographs of furnace-cycled TBC-coated Rene N5/Pt–Al testpieces, at (a) 8%, (b) 34% and (c) 76% of TBC life [116,117]. Note that decomposition of the β phase to γ', which then decorates the bond coat/TGO interface at the later stages. The testpieces are the same as those in Figure 5.46 [120] – the etchant used here draws out contrast from the β and γ' phases more adequately.

Questions

5.1 The components of a thermal barrier coating system include (i) the bond coat and (ii) the ceramic top coat. When designing a thermal barrier coating system, what are the requirements of these two layers?

5.2 What are the characteristics of an air plasma-sprayed (APS) thermal barrier coating which make it unsuitable in some ways for application on a turbine aerofoil? Why might APS be used in place of EB-PVD in some instances?

5.3 Thermal barrier coatings are fabricated from ceramic materials; they are required to prevent the build up of heat in components such as turbine blading and combustor sections. What physical properties must the ceramic material possess for this application? By conducting a survey of the properties of the available ceramic materials, choose a suitable one for this application.

5.4 After the engine testing of a thermal barrier coating on the high-pressure (HP) turbine blading, spallation is observed on the leading edge of the blade. Draw a cross section of the blading, identifying the leading and trailing edges. Which positions are likely to experience the hottest temperatures during operation? What could be done to the geometry of the blade in order to prevent spallation of the TBC at these locations?

Hence rationalise Rolls-Royce's experience with the HP turbine blade of the Trent 500 engine, which necessitated increasing the radius of curvature at the leading edge to ensure TBC adherence.

5.5 The high-pressure turbine blading in an aeroengine is working in a gas stream of temperature 1500 °C. The metal temperatures at the outer and inner (cooled) surfaces are 1000 °C and 800 °C, respectively; the wall thickness is 5 mm. If the single-crystal superalloy from which the blading is cast has a thermal conductivity of 100 W/(m K), estimate (i) the steady-state heat flux entering the blading and (ii) the heat transfer coefficient at the blade/gas interface.

It is proposed to add a 1.2 mm-thick thermal barrier coating (TBC) to the blading, of thermal conductivity 1 W/(m K). Assuming that the heat transfer coefficient at the outer surface remains unaltered, estimate (i) the new value of the steady-state flux and (ii) the reduction in the mean metal temperature which the TBC confers.

Over time, the TBC is expected to sinter, causing its thermal conductivity to increase by 50%. What influence will this have?

5.6 The deposition of an aluminide coating on a superalloy by pack cementation can be carried out with either a high or low activity of aluminium in the pack. In which of these does one expect to see the δ–Ni_2Al_3 phase at the outer surface of the coating? Given that aluminium is the main diffusing element in δ and hyperstoichiometric β–NiAl of greater than 51 at%Al, rationalise the observation that, in this case, the coating grows inwards.

Why in the case of a low-activity pack is there a possibility that pack particles become embedded in the outer surface of the coating?

5.7 The resistance of a coated superalloy to cyclic oxidation is usually much inferior to its performance in an isothermal test; for example, spallation is much less likely in an isothermal 1000 h test at 1150 °C than it is in 1000 cycles consisting of heating/cooling between ambient and 1150 °C with a 1 h hold at that temperature. Why is this?

5.8 List the advantages and disadvantages of the electron beam physical vapour deposition (EB-PVD) process for the deposition of ceramic coatings.

5.9 Coatings produced by pack aluminisation offer satisfactory performance for many aviation, industrial and marine engine applications. However, under severe hot corrosion conditions, or at temperatures above 1050 °C, only limited protection is afforded. Which coatings are then more appropriate, and how might these be deposited?

5.10 The thermal conductivity of yttria-stabilised zirconia is strongly influenced by the presence of point defects. An estimate of the point defect spacing can be made by noting that an oxygen vacancy is introduced into the zirconia lattice for every two Y^{3+} that substitute for each pair of Zr^{2+} ions. For partially stabilised zirconia containing 7 wt% of yttria, estimate the average distance between oxygen vacancies. How is the mean free path for phonon dispersion, on which the thermal conductivity depends, likely to be altered by this effect?

5.11 Overlay coatings are often based on the NiCrAlY system, but many compositional variants are available. For example, proprietary coatings can contain significant proportions of Si, Re, Hf, or combinations of these. Make a literature review summarising the research which has been carried out to identify these 'modified' overlay coatings, and critique the advantages which are claimed in each case. For what type of turbine applications might these coatings be best employed?

5.12 During the oxidation of the Ni-rich, Group I-type Ni–Cr–Al alloys, an external scale of NiO forms. However, the internal scale consists of chromia and alumina, with the latter being found at greater depths than the former. Rationalise these observations.

5.13 Explain why the optimum level of a reactive element (such as Y) in the raw coating powder to be used for the plasma spraying of an NiCrAlY-type overlay coating depends upon whether the processing is to be carried out in air or under vacuum.

5.14 Examine the cracks in the overlay coating in Figure 5.36. One can see that the cracks are more or less uniformly spaced. Give a physical explanation for this.

5.15 Depletion of aluminium from the bond coat is a crucial precursor to the failure of a thermal barrier coating. This occurs due to (i) formation of a thermally grown oxide (TGO) of α-alumina and (ii) interdiffusion with the superalloy substrate. Which is the more important of these in practice? It is claimed in ref. [117] that it is the former, but in another, [148], that it is the latter. Perform an analysis, using data from the literature where necessary, to decide.

5.16 Prove that the biaxial compressive in-plane stress, σ_0, which develops in a thin coating on a stiff substrate during cooling through a temperature, ΔT, is given by

$$\sigma_0 = \frac{E_0 \Delta \alpha_0 \Delta T}{1 - \nu_0}$$

where $\Delta \alpha_0$ is the difference between the thermal expansion coefficients of the substrate and coating, and E_0 and ν_0 are the Young's modulus and Poisson's ratio of the coating, respectively.

Hence estimate the stress forming in the TGO of a TBC system. What effects might occur that would lead your simple estimate to be in error? (The thermal expansion coefficients for bond coat and alumina TGO are in the ranges 13–16 and 8–9 ppm/K, respectively. The Young's modulus and Poisson's ratio of the TGO are 200 GPa and 0.33, respectively.)

5.17 Estimate the transformation 'growth' stresses associated with the formation of a TGO on a β-NiAl substrate, given that the densities of α-Al$_2$O$_3$, β-NiAl and γ-Ni$_3$Al are 3.9, 5.6 and 7.5 g/cm^3, respectively.

5.18 Is the failure of a TBC nucleation- or propagation-controlled? Discuss.

5.19 Watanabe et al. [149] have used the wafer curvature method for the determination of the residual stress level in a Pt–aluminide bond coat deposited on the Rene N5 single-crystal superalloy. The bond coat was found to have a residual stress level of about 140 MPa (tensile) – varying only slightly with thickness. Rationalise this observation by considering the microstructure of the bond coat and the thermal expansion coefficients of the various phases present.

5.20 Piezospectroscopy, i.e. the strain-induced shift of luminescence and Raman spectra, can be used for the non-destructive evaluation of the TGO formed in a TBC system [114]. When illuminated with a laser of appropriate wavelength, luminescence from the Al$_2$O$_3$ formed by the oxidation of the bond coat can be detected through the thickness of the TBC. What are the prerequisites which make this possible?

5.21 The sintering resistance of ceramic materials scales with the melting temperature. Carry out a data-mining exercise to estimate the melting temperatures of candidate TBC materials, and thus show that a proposal for a hafnia-based TBC is sensible.

5.22 Evans and co-workers made concerted efforts in the late 1990s to model the failure mechanisms relevant to thermal barrier coatings; see, for example, ref. [111]. A ratcheting mechanism was emphasised strongly. Critique these early papers, in the light of later observations (for example, ref. [117]), which emphasise that diffusional reactions occur at and beneath the TGO. Is it likely that a ratcheting mechanism is operative?

5.23 The concentrations of slow-diffusing elements such as Re, W and Mo near the surface of an aluminised coating is expected to depend upon whether a high- or low-activity pack is used. Why? How would one expect the (i) mechanical and (ii) environmental properties of the two types of coating to differ?

5.24 The presence of a ridged surface morphology of an as-processed diffusion coating influences the life of a TBC system. Explain. Why does the ridged morphology form?

5.25 A NiCrAlY-type overlay coating will interdiffuse with a single-crystal superalloy substrate during service, such that microstructural changes occurring at the interface will depend upon the remanent life of the component. Describe the microstructural evolution which is expected under typical circumstances.

5.26 Walston and colleagues have measured the increase in thickness of the secondary reaction zone (SRZ) with time in a aluminide-coated third-generation single-crystal superalloy at 1090 °C; growth was found to be parabolic with time, with an effective diffusion constant equal to 6.7×10^{-11} cm^2/s. Is this finding consistent with the diffusion data of Figure 2.34? What element is rate-controlling?

5.27 The reasons why the addition of Pt to an aluminide coating, as in the so-called platinum–aluminide coating, are beneficial are still not completely understood. List some postulates which might be the cause and then critique them for their validity by appealing to observations in the literature. Which mechanism is best supported by the available experimental data?

5.28 Will the rate of TGO formation on an aluminised single-crystal superalloy during *isothermal* oxidation be altered by the addition of a ceramic YSZ layer? If so, why?

5.29 The root-mean-square roughness of a turbine blade aerofoil is important since it influences aerodynamic performance. The roughness of an uncoated blade is close to 1 μm; how will this figure be altered by the addition of a (i) plasma-sprayed coating and (ii) an EB-PVD coating?

5.30 An estimate for the TGO thickness, h, at which decohesion occurs by edge spallation can be made by equating its strain energy per unit area to the adhesion energy per unit area (i.e. the interfacial fracture toughness) which is designated Γ_i. Hence demonstrate that

$$h = \frac{E\Gamma_i}{\sigma_0^2(1 - v^2)}$$

where E and v are the Young's modulus and Poisson's ratio of the TGO, respectively.

The property Γ_i has been found to lie in the range 5 to 20 J/m^2. Hence, using typical property values, demonstrate that the value for h at decohesion determined in this way is appreciably smaller than 1 μm. What is the significance of your result in the light of experimentally observed thicknesses of the TGO at failure?

5.31 A turbine blade aerofoil is to be cooled during service via the addition of drilled holes which connect the outer surface to the internal cooling channels. Thereafter, a thermal barrier coating is to be applied.

What manufacturing methods are available for the production of the holes? What are the advantages and disadvantages of the available processing methods? Which coating method is the more suitable, if blocking of the holes is to be avoided? Justify your answer by appealing to the physical principles involved.

References

[1] Y. Tamarin, *Protective Coatings for Turbine Blades* (Materials Park, OH: ASM International, 2002).

[2] J. H. Wood and E. Goldman, Protective coatings, in C. T. Sims, N. S. Stoloff and W. C. Hagel, eds, *Superalloys II* (New York: John Wiley and Sons, 1987), pp. 359–384.

[3] J. R. Nicholls, Advances in coating design for high performance gas turbines, *MRS Bulletin*, **28** (2003), 659–670.

[4] G. W. Goward and D. H. Boone, Mechanisms of formation of diffusion aluminide coatings on nickel-base superalloys, *Oxidation of Metals*, **3** (1971), 475–495.

[5] H. W. Tawancy, N. M. Abbas and T. N. Rhys-Jones, The role of platinum in aluminide coatings, *Surface Coatings Technology*, **49** (1991), 1–7.

[6]   J. R. Davis, ed., *Handbook of Thermal Spray Technology* (Materials Park, OH: The ASM Thermal Spray Society, 2004).

[7]   D. H. Boone, Physical vapour deposition processes, *Materials Science and Technology*, **2** (1986), 220–224.

[8]   A. R. Nicholl, H. Gruner, G. Wuest and S. Keller, Future developments in plasma spray coating, *Materials Science and Technology*, **2** (1986), 214–219.

[9]   A. R. Nicholl and G. Wahl, Oxidation and high temperature corrosion behaviour of modified MCrAlY cast materials, in R. H. Bricknell, W. B. Kent, M. Gell, C. S. Kortovich and J. F. Radavich, eds, *Superalloys 1984* (Warrendale, PA: The Metallurgical Society of AIME, 1984), pp. 805–814.

[10] N. P. Padture, M. Gell and E. H. Jordan, Thermal barrier coatings for gas-turbine applications, *Science*, **296** (2002), 280–284.

[11] D. R. Clarke, Materials selection guidelines for low thermal conductivity thermal barrier coatings, *Surface and Coatings Technology*, **163–164** (2003), 67–74.

[12] B. A. Pint, I. G. Wright, W. Y. Lee, Y. Zhang, K. Prussner and K. B. Alexander, Substrate and bond coat compositions: factors affecting alumina scale adhesion, *Materials Science and Engineering*, **A245** (1998), 201–211.

[13] J. S. Wang and A. G. Evans, Effects of strain cycling on buckling, cracking and spalling of a thermally grown alumina on a nickel-based bond coat, *Acta Materialia*, **47** (1999), 699–710.

[14] A. G. Evans, D. R. Mumm, J. W. Hutchinson, G. H. Meier and F. S. Pettit, Mechanisms controlling the durability of thermal barrier coatings, *Progress in Materials Science*, **46** (2001), 505–553.

[15] D. E. Wolfe, J. Singh, R. A. Miller, J. I. Eldridge and D. M. Zhu, Tailored microstructure of EB-PVD 8YSZ thermal barrier coatings with low thermal conductivity and high thermal reflectivity for turbine applications, *Surface and Coatings Technology*, **190** (2004), 132–149.

[16] K. Wada, N. Yamaguchi and H. Matsubara, Crystallographic texture evolution in $ZrO_2–Y_2O_3$ layers produced by electron beam physical vapour deposition, *Surface and Coatings Technology*, **184** (2004), 55–62.

[17] D. H. Boone, T. E. Strangman and L. W. Wilson, Some effects of structure and composition on the properties of electron beam vapor deposited coatings for gas turbine applications, *Journal of Vacuum Science and Technology*, **11** (1974), 641–646.

[18] C. C. Berndt and H. Herman, Failure during thermal cycling of plasma-sprayed thermal barrier coatings, *Thin Solid Films*, **108** (1983), 427–437.

[19] C. C. Berndt and R. A. Miller, Failure analysis of plasma-sprayed thermal barrier coatings, *Thin Solid Films*, **119** (1984), 173–184.

[20] B. J. Gill and R. C. Tucker, Plasma spray coating processes, *Materials Science and Technology*, **2** (1986), 207–213.

[21] H. Gruner, Vacuum plasma spray quality control, *Thin Solid Films*, **118** (1984), 409–420.

[22] National Materials Advisory Board, *Coatings for High-Temperature Structural Materials: Trends and Opportunities* (Washington D. C.: Academy Press, 1996).

[23] H. B. Guo, R. Vassen and D. Stover, Atmospheric plasma sprayed thick thermal barrier coatings with high segmentation crack density, *Surface and Coatings Technology*, **186** (2004), 353–363.

[24] H. B. Guo, H. Murakami and S. Kuroda, Effects of heat treatment on microstructures and physical properties of segmented thermal barrier coatings, *Materials Transactions*, **46** (2005), 1775–1778.

[25] R. Mevrel, C. Duret and R. Pichoir, Pack cementation processes, *Materials Science and Technology*, **2** (1986), 201–206.

[26] R. Bianco, R. A. Rapp and N. S. Jacobson, Volatile species in halide-activated diffusion coating packs, *Oxidation of Metals*, **38** (1992), 33–43.

[27] R. Bianco and R. A. Rapp, Pack cementation aluminide coatings on superalloys: codeposition of Cr and reactive elements, *Journal of the Electrochemical Society*, **140** (1993), 1181–1190.

[28] S. C. Kung and R. A. Rapp, Analyses of the gaseous species in halide-activated cementation coating packs, *Oxidation of Metals*, **32** (1989), 89–109.

[29] G. Fisher, P. K. Data, J. S. Burnell-Gray, W. Y. Chan and R. Wing, An investigation of the oxidation resistance of platinum aluminide coatings produced by either high or low activity processes, in J. Nicholls and D. Rickerby, eds, *High Temperature Surface Engineering* (London: The Institute of Materials, 2000), pp. 1–12.

[30] K. Bungardt, G. Lehnert and H. W. Meinhardt, Protective diffusion layer on nickel and/or cobalt-based alloys, United States Patent 3819338 (1974).

[31] R. G. Wing and I. R. McGill, The protection of gas turbine blades: a platinum aluminide diffusion coating, *Platinum Metals Review*, **25** (1981), 94–105.

[32] M. R. Jackson and J. R. Rairden, The aluminisation of platinum and platinum-coated IN-738, *Metallurgical Transactions*, **8A** (1977), 1697–1707.

[33] A. L. Purvis and B. W. Warnes, The effects of platinum concentration on the oxidation resistance of superalloys coated with single-phase platinum aluminide, *Surface and Coatings Technology*, **146–147** (2001), 1–6.

[34] G. R. Krishna, D. K. Das, V. Singh and S. V. Joshi, The role of Pt content in the microstructural development and oxidation performance of Pt-aluminide coatings produced using a high activity aluminizing process, *Materials Science and Engineering*, **251A** (1998), 40–47.

[35] Y. Zhang, J. A. Haynes, W. Y. Lee *et al.*, Effect of Pt incorporation on the isothermal oxidation behaviour of chemical vapour deposition aluminide coatings, *Metallurgical Transactions*, **32A** (2001), 1727–1741.

[36] A. Squillace, R. Bonetti, N. J. Archer and J. A. Yeatman, The control of the composition and structure of aluminide layers formed by vapour aluminising, *Surface and Coatings Technology*, **120–121** (1999), 118–123.

[37] A. B. Smith, A. Kempster and J. Smith, Vapour aluminide coating of internal cooling channels in turbine blades and vanes, *Surface and Coatings Technology*, **120–121** (1999), 112–117.

[38] B. M. Warnes and D. C. Punola, Clean diffusion coatings by chemical vapour deposition, *Surface Coatings Technology*, **94–95** (1997), 1–6.

[39] B. M. Warnes, Reactive element modified chemical vapour deposition low activity platinum aluminide coatings, *Surface and Coatings Technology*, **146–147** (2001), 7–12.

[40] W. D. Kingery, Thermal conductivity: temperature dependence of conductivity for single-phase ceramics, *Journal of the American Ceramic Society*, **38** (1955), 251–255.

[41] D. R. Clarke and C. G. Levi, Materials design for the next generation of thermal barrier coatings, *Annual Review of Materials Research*, **33** (2003), 383–417.

[42] S. Stecura, Advanced thermal barrier system bond coatings for use on nickel-, cobalt- and iron-base alloy substrates, *Thin Solid Films*, **136** (1986), 241–256.

[43] S. Alperine and L. Lelait, Microstructural investigations of plasma-sprayed yttria partially stabilised zirconia TBC, *Transactions ASME: Journal of Engineering for Gas Turbines and Flows*, **116** (1994), 258–265.

[44] R. A. Miller, J. L. Smialek and R. G. Garlik, Phase stability in plasma-sprayed partially stabilised zirconia-yttria, in A. H. Heuer and L. W. Hobbs, eds, *Science and Technology of Zirconia, Advances in Ceramics: Volume 3* (Columbus, OH: American Ceramic Society, 1981), 241–253.

[45] J. A. Thompson and T. W. Clyne, The effect of heat treatment on the stiffness of zirconia top coats in plasma-sprayed TBCs, *Acta Materialia*, **49** (2001), 1565–1575.

[46] J. R. Brandon and R. Taylor, The phase stability of zirconia-based thermal barrier coatings: 1. Zirconia-yttria alloys, *Surface Coatings Technology*, **46** (1991), 75–90.

[47] R. L. Jones and D. Mess, Improved tetragonal phase stability at 1400 °C with scandia, yttria-stabilised zirconia, *Surface and Coatings Technology*, **86–87** (1996), 94–101.

[48] M. Leoni, R. L. Jones and P. Scardi, Phase stability of scandia-yttria-stabilised zirconia TBCs, *Surface and Coatings Technology* **108–109** (1998), 107–113.

[49] K. Matsumoto, Y. Itoh and T. Kameda, EB-PVD process and thermal properties of hafnia-based thermal barrier coating, *Science and Technology of Advanced Materials*, **4** (2003), 153–158.

[50] R. Vassen, X. Q. Cao, F. Tietz, D. Basu and D. Stover, Zirconates as new materials for thermal barrier coatings, *Journal of the American Ceramic Society*, **83** (2000), 2023–2028.

[51] B. Saruhan, P. Francois, K. Kritscher and U. Schulz, EB-PVD processing of pyrochlore-structured La$_2$Zr$_2$O$_7$-based TBCs, *Surface and Coatings Technology*, **182** (2004), 175–183.

[52] J. Wu *et al.*, Low thermal conductivity rare-earth zirconate for potential thermal-barrier coating applications, *Journal of the American Ceramic Society*, **85** (2002), 3031–3035.

[53] R. Gadow and M. Lischka, Lanthanum hexaaluminate – novel thermal barrier coatings for gas turbine applications – materials and process development, *Surface and Coatings Technology*, **151–152** (2002), 392–399.

[54] C. G. Levi, Emerging materials and processes for thermal barrier systems, *Current Opinion in Solid State and Materials Science*, **8** (2004), 77–91.

[55]   J. M. Ziman, *Principles of the Theory of Solids* (Cambridge: Cambridge University Press, 1972).

[56]   J. R. Nicholls, K. J. Lawson, A. Johnstone and D. S. Rickerby, Methods to reduce the thermal conductivity of EB-PVD TBCs, *Surface and Coatings Technology*, **151–152** (2002), 383–391.

[57]   I. O. Golosnoy, S. A. Tsipas and T. W. Clyne, An analytical model for simulation of heat flow in plasma-sprayed thermal barrier coatings, *Journal of Thermal Spray Technology*, **14** (2005), 205–214.

[58]   U. Schulz, K. Fritscher and C. Leyens, Two-source jumping beam evaporation for advanced EB-PVD TBC Systems, *Surface and Coatings Technology*, **133–134** (2000), 40–48.

[59]   M. F. J. Koolloos and G. Marijnissen, Burner rig testing of herringbone EB-PVD thermal barrier coatings, *Proceedings of TurboMat: International Symposium on Thermal Barrier Coatings and Titanium Aluminides* (Cologne: German Aerospace Center, Institute of Materials Research, 2002), pp. 18–21.

[60]   D. D. Hass, A. J. Slifka and H. N. G. Wadley, Low thermal conductivity vapor-deposited zirconia microstructures, *Acta Materialia*, **49** (2001), 973–983.

[61]   D. K. Gupta and D. S. Duvall, A silicon and hafnium modified plasma sprayed MCrAlY coating for single crystal superalloys, in R. H. Bricknell, W. B. Kent, M. Gell, C. S. Kortovich and J. F. Radavich, eds, *Superalloys 1984* (Warrendale, PA: The Metallurgical Society of AIME, 1984), pp. 711–720.

[62]   T. A. Taylor and D. F. Bettridge, Development of alloyed and dispersion-strengthened MCrAlY Coatings, *Surface and Coatings Technology*, **86–87** (1996), 9–14.

[63]   N. Czech and W. Stamm, Optimisation of MCrAlY-type coatings for single crystal and conventional cast gas turbine blades, in J. Nicholls and D. Rickerby, eds, *High Temperature Surface Engineering* (London: The Institute of Materials, 2000), pp. 61–65.

[64]   G. W. Goward, Progress in coatings for gas turbine aerofoils, *Surface and Coatings Technology*, **108–109** (1998), 73–79.

[65]   D. R. G. Archer, R. Munoz-Arroyo, L. Singheiser and W. J. Quadakkers, Modelling of phase equilibria in MCrAlY coating systems, *Surface and Coatings Technology*, **187** (2004), 272–283.

[66]   F. S. Pettit, Oxidation mechanisms of nickel-aluminium alloys at temperatures between 900 °C and 1300 °C, *Transactions AIME*, **239** (1967), 1297–1305.

[67]   C. S. Giggins and F. S. Pettit, Oxidation of Ni-Cr-Al alloys between 1000 °C and 1200 °C, *Journal of the Electrochemical Society*, **118** (1971), 1782–1790.

[68]   D. P. Whittle and J. Stringer, Improvements in high-temperature oxidation resistance by additions of reactive elements or oxide dispersions, *Philosophical Transactions of the Royal Society of London*, **295A** (1980), 309–329.

[69]   J. L. Smialek, Maintaining adhesion of protective Al_2O_3 scales, *Journal of Metals*, **52** (2000), 22–25.

[70] I. M. Allam, D. P. Whittle and J. Stringer, The oxidation behaviour of CoCrAl systems containing active element additions, *Oxidation of Metals*, **12** (1978), 35–66.

[71] A. S. Khan, C. E. Lowell and C. A. Barrett, The effect of zirconium on the isothermal oxidation of nominal Ni-14Cr-24Al alloys, *Journal of the Electrochemical Society*, **127** (1980), 670–679.

[72] C. A. Barrett, A. S. Khan and C. E. Lowell, The effect of zirconium on the cyclic oxidation of NiCrAl alloys, *Journal of the Electrochemical Society*, **128** (1981), 25–32.

[73] B. A. Pint, Experimental observations in support of the dynamic-segregation theory to explain the reactive-element effect, *Oxidation of Metals*, **45** (1996), 1–38.

[74] J. K. Tien and F. S. Pettit, Mechanism of oxide adherence on Fe-25Cr-4Al (Y or Sc) alloys, *Metallurgical Transactions*, **3** (1972), 1587–1599.

[75] J. E. Antill and K. A. Peakall, Influence of an alloy addition of yttrium on the oxidation behaviour of an austenitic and a ferritic stainless steel in carbon dioxide, *Journal of the Iron and Steel Institute*, **205** (1967), 1136–1142.

[76] J. E. McDonald and J. G. Eberhart, Adhesion in aluminum oxide-metal systems, *Transactions of the Metallurgical Society of AIME*, **233** (1965), 512–517.

[77] P. Y. Hou, Beyond the sulphur effect, *Oxidation of Metals*, **52** (1999), 337–351.

[78] R. A. Mulford, Grain boundary segregation in Ni and binary Ni alloys doped with sulphur, *Metallurgical Transactions*, **14A** (1983), 865–870.

[79] D. McLean, *Grain Boundaries in Metals* (Oxford: Clarendon Press, 1957).

[80] J. G. Smeggil, A. W. Funkenbusch and N. S. Bornstein, A relationship between indigenous impurity elements and protective oxide scale adherence characteristics, *Metallurgical Transactions*, **17A** (1986), 923–932.

[81] T. N. Rhys-Jones and T. P. Cunningham, The influence of surface coatings on the fatigue behaviour of aeroengine materials, *Surface and Coatings Technology*, **42** (1990), 13–19.

[82] T. C. Totemeier and J. E. King, Isothermal fatigue of an aluminide-coated single-crystal superalloy: part I, *Metallurgical and Materials Transactions*, **27A** (1996), 353–361.

[83] G. W. Meetham, Use of protective coatings in aero gas turbine engines, *Materials Science and Technology*, **2** (1986), 290–294.

[84] Y. A. Tamarin, V. G. Sundyrin and N. G. Bychkov, Thermo-mechanical fatigue tests of coatings for turbine blades, in J. Nicholls and D. Rickerby, eds, *High Temperature Surface Engineering* (London: The Institute of Materials, 2000), pp. 157–169.

[85] M. I. Wood, D. Raynor and R. M. Cotgrove, Thermomechanical fatigue of coated superalloys, in R. Townsend *et al.*, eds, *Life Assessment of Hot Section Gas Turbine Components* (London: The Institute of Materials, 2000), pp. 193–207.

[86] W. J. Brindley and R. A. Miller, Thermal barrier coating life and isothermal oxidation of low-pressure plasma-sprayed bond coat alloys, *Surface and Coatings Technology*, **43/44** (1990), 446–457.

[87] S. Stecura, Effects of yttrium, aluminum and chromium concentrations in bond coatings on the performance of zirconia-yttria thermal barriers, *Thin Solid Films*, **73** (1980), 481–489.

[88] S. Stecura, Two-layer thermal barrier coatings i: Effects of composition and temperature on oxidation behaviour and failure, *Thin Solid Films*, **182** (1989), 121–139.

[89] M. W. Chen, K. J. T. Livi, P. K. Wright and K. J. Hemker, Microstructural characterisation of a platinum-modified diffusion bond coat for thermal barrier coatings, *Metallurgical and Materials Transactions*, **34A** (2003), 2289–2299.

[90] J. Angenete, K. Stiller and E. Bakchinova, Microstructural and microchemical development of simple and Pt-modified aluminide diffusion coatings during long term oxidation at 1050 °C, *Surface and Coatings Technology*, **176** (2004), 272–283.

[91] W. S. Walston, J. C. Schaeffer and W. H. Murphy, A new type of microstructural instability in superalloys – SRZ, in R. D. Kissinger, D. J. Deye, D. L. Anton *et al.*, eds, *Superalloys 1996* (Warrendale, PA: The Minerals, Metals and Materials Society, 1996), pp. 9–18.

[92] O. Lavigne, C. Ramusat, S. Drawin, P. Caron, D. Boivin and J.-L. Pouchou, Relationships between microstructural instabilities and mechanical behaviour in new generation nickel-based single crystal superalloys, in K. A. Green, T. M. Pollock, H. Harada *et al.*, eds, *Superalloys 2004* (Warrendale, PA: The Minerals, Metals and Materials Society (TMS), 2004), pp. 667–675.

[93] J. Doychak, in J. H. Westbrook and R. L. Fleischer, eds, *Intermetallic Compounds: Principles and Practices, Volume 1* (New York: John Wiley and Sons, 1994), pp. 977–1016.

[94] M. P. Brady, B. A. Pint, P. F. Tortorelli, I. G. Wright and R. J. Hanrahan, High temperature oxidation and corrosion of intermetallics, in M. Schutze, ed., *Materials Science and Technology: A Comprehensive Treatment, Vol. II: Corrosion and Environmental Degradation* (Weinham, Germany: Wiley-VCH Verlag, 2000), pp. 229–325.

[95] B. A. Pint, J. R. Martin and L. W. Hobbs, The oxidation mechanism of θ-Al_2O_3 scales, *Solid State Ionics*, **78** (1995), 99–107.

[96] J. C. Yang, E. Schumann, I. Levin and M. Ruhle, Transient oxidation of NiAl, *Acta Materialia*, **46** (1998), 2195–2201.

[97] G. C. Rybicki and J. L. Smialek, Effect of theta-alpha Al_2O_3 transformation on the oxidation behaviour of β-NiAl+Zr, *Oxidation of Metals*, **31** (1989), 275–304.

[98] B. A. Pint, M. Treska and L. W. Hobbs, The effect of various oxide dispersions on the phase composition and morphology of Al_2O_3 scales grown on β-NiAl, *Oxidation of Metals*, **47** (1997), 1–20.

[99] J. A. Haynes, B. A. Pint, K. L. More, Y. Zhang and I. G. Wright, Influence of sulphur, platinum and hafnium on the oxidation behaviour of CVD NiAl bond coatings, *Oxidation of Metals*, **58** (2002), 513–544.

[100] R. Prescott, D. F. Mitchell, M. J. Graham and J. Doychak, Oxidation mechanisms of β-NiAl + Zr determined by SIMS, *Corrosion Science*, **37** (1995), 1341–1364.

[101] C. S. Giggins, B. H. Kear, F. S. Pettit and J. K. Tien, Factors affecting adhesion of oxide scales on alloys, *Metallurgical Transactions*, **5** (1974), 1685–1688.

[102] J. L. Smialek, Oxide morphology and spalling model for NiAl, *Metallurgical Transactions*, **9A** (1978), 309–320.

[103] B. A. Pint, On the formation of interfacial and internal voids in α-Al_2O_3 scales, *Oxidation of Metals*, **48** (1997), 303–328.

[104] B. A. Pint, J. A. Haynes, K. L. More, I. G. Wright and C. Leyens, Compositional effects on aluminide oxidation performance: objectives for improved bond coats, in T. M. Pollock, R. D. Kissinger, R. R. Bowman *et al.*, eds, *Superalloys 2000* (Warrendale, PA: The Minerals, Metals and Materials Society (TMS), 2000), pp. 629–638.

[105] B. A. Pint, J. A. Haynes, K. L. More and I. G. Wright, The use of model alloys to understand and improve the performance of Pt-modified aluminide coatings in K. A. Green, T. M. Pollock, H. Harada *et al.*, eds, *Superalloys 2004* (Warrendale, PA: The Minerals, Metals and Materials Society (TMS), 2004), pp. 597–606.

[106] B. A. Pint, K. L. More and I. G. Wright, The use of two reactive elements to optimise oxidation performance of alumina-forming alloys, *Materials at High Temperatures*, **20** (2003), 375–386.

[107] S. Hayashi, W. Wang, D. J. Sordelet and B. Gleeson, Interdiffusion behaviour of Pt-modified γ-Ni and γ-Ni_3Al alloys coupled to Ni-Al-based alloys, *Metallurgical and Materials Transactions*, **36A** (2005), 1769–1775.

[108] Y. Zhang, W. Y. Lee, J. A. Haynes *et al.*, Synthesis and cyclic oxidation behaviour of a (Ni,Pt)Al coating on a desulfurised Ni-base superalloy, *Metallurgical Transactions*, **30A** (1999), 2679–2687.

[109] V. K. Tolpygo and D. R. Clarke, Surface rumpling of a (Ni, Pt)Al bond coat induced by cyclic oxidation, *Acta Materialia*, **40** (2000), 3283–3293.

[110] V. K. Tolpygo and D. R. Clarke, Morphological evolution of thermal barrier coatings induced by cyclic oxidation, *Surface and Coatings Technology*, **163–164** (2003), 81–86.

[111] M. Y. He, A. G. Evans and J. W. Hutchinson, The ratcheting of compressed thermally grown thin films on ductile substrates, *Acta Materialia*, **48** (2000), 2593–2601.

[112] M. Y. He, J. W. Hutchinson and A. G. Evans, Large deformation simulations of cyclic displacement instabilities in thermal barrier systems, *Acta Materialia*, **50** (2002), 1063–1073.

[113] M. W. Chen, M. L. Glynn, R. T. Ott, T. C. Hufnagel and K. J. Hemker, Characterisation and modeling of a martensitic transformation in a platinum modified diffusion aluminide bond coat for thermal barrier coatings, *Acta Materialia*, **51** (2003), 4279–4294.

[114] V. K. Tolpygo and D. R. Clarke, On the rumpling mechanism in nickel-aluminide coatings, part I: An experimental assessment, *Acta Materialia*, **52** (2004), 5115–5127.

[115] V. K. Tolpygo and D. R. Clarke, On the rumpling mechanism in nickel-aluminide coatings, part II: Characterisation of surface undulations and bond coat swelling, *Acta Materialia*, **52** (2004), 5129–5141.

[116] S. Darzens, D. R. Mumm, D. R. Clarke and A. G. Evans, Observations and analysis of the influence of phase transformations on the instability of the thermally grown oxide in a thermal barrier system, *Metallurgical and Materials Transactions*, **34A** (2003), 511–522.

[117] S. Darzens and A. M. Karlsson, On the microstructural development in platinum-modified nickel-aluminide bond coats, *Surface and Coatings Technology*, **177–178** (2004), 108–112.

[118] R. T. Wu, M. Osawa, Y. Koizumi, T. Yokokawa, H. Harada and R. C. Reed, Degradation mechanisms of an advanced jet engine service-retired TBC component, submitted to *Acta Materialia*.

[119] P. K. Wright and A. G. Evans, Mechanisms governing the performance of thermal barrier coatings, *Current Opinion in Solid State and Materials Science*, **4** (1999), 255–265.

[120] D. R. Mumm, A. G. Evans and I. T. Spitsberg, Characterisation of a cyclic displacement instability for a thermally grown oxide in a thermal barrier coating system, *Acta Materialia*, **49** (2001), 2329–2340.

[121] U. Schulz, M. Menzebach, C. Leyens and Y. Q. Yang, Influence of substrate material on oxidation behaviour and cyclic lifetime of EB-PVD TBC systems, *Surface and Coatings Technology*, **146–147** (2001), 117–123.

[122] P. K. Wright, Influence of cyclic strain on life of a PVD TBC, *Materials Science and Engineering*, **A245** (1998), 191–200.

[123] B. A. Pint, The role of chemical composition on the oxidation performance of aluminide coatings, *Surface and Coatings Technology*, **188–189** (2004), 71–78.

[124] B. Baufeld, M. Bartsch, P. Broz and M. Schmucker, Microstructural changes as postmortem temperature indicator in Ni-Co-Cr-Al-Y oxidation protection coatings, *Materials Science and Engineering*, **384** (2004), 162–171.

[125] J. A. Nychka and D. R. Clarke, Damage quantification in TBCs by photo-stimulated luminescence spectroscopy, *Surface and Coatings Technology*, **146–147** (2001), 110–116.

[126] D. R. Mumm and A. G. Evans, On the role of imperfections in the failure of a thermal barrier coating made by electron beam deposition, *Acta Materialia*, **48** (2000), 1815–1827.

[127] M. Gell, K. Vaidyanathan, B. Barber, J. Cheng and E. Jordan, Mechanism of spallation in platinum aluminide/electron beam physical vapor-deposited thermal barrier coatings, *Metallurgical and Materials Transactions*, **30A** (1999), 427–435.

[128] A. Rabiei and A. G. Evans, Failure mechanisms associated with the thermally grown oxide in plasma-sprayed thermal barrier coatings, *Acta Materialia*, **48** (2000), 3963–3976.

[129] A. G. Evans, M. Y. He and J. W. Hutchinson, Mechanics-based scaling laws for the durability of thermal barrier coatings, *Progress in Materials Science*, **46** (2001), 249–271.

[130]   J. S. Wang and A. G. Evans, Measurement and analysis of buckling and buckle propagation in compressed oxide layers on superalloy substrates, *Acta Materialia*, **46** (1998), 4993–5005.

[131]   A. G. Evans, J. W. Hutchinson and M. Y. He, Micromechanics model for the detachment of residually compressed brittle films and coatings, *Acta Materialia*, **47** (1999), 1513–1522.

[132]   M. Gell, E. Jordan, K. Vaidyanathan *et al.*, Bond strength, bond stress and spallation mechanisms of thermal barrier coatings, *Surface and Coatings Technology*, **120–121** (1999), 53–60.

[133]   X. Chen, R. Wang, N. Yao, A. G. Evans, J. W. Hutchinson and R.W. Bruce, Foreign object damage in a thermal barrier system: mechanisms and simulations, *Materials Science and Engineering*, **A352** (2003), 221–231.

[134]   W. S. Walston, Coating and surface technologies for turbine aerofoils, in K. A. Green, T. M. Pollock, H. Harada *et al.*, *Superalloys 2004* (Warrendale, PA: The Minerals, Metals and Materials Society (TMS), 2004), pp. 579–588.

[135]   D. R. Mumm, M. Watanabe, A. G. Evans and J. A. Pfaendtner, The influence of test method on failure mechanisms and durability of a thermal barrier system, *Acta Materialia*, **52** (2004), 1123–1131.

[136]   R. A. Miller, Oxidation-based model for thermal barrier coating life, *Journal of the American Ceramic Society*, **67** (1984), 517–521.

[137]   S. R. Choi, J. W. Hutchinson and A. G. Evans, Delamination of multilayer thermal barrier coatings, *Mechanics of Materials*, **31** (1999), 431–447.

[138]   M. Y. He, D. R. Mumm and A. G. Evans, Criteria for the delamination of thermal barrier coatings: with application to thermal gradients, *Surface and Coatings Technology*, **185** (2004), 184–193.

[139]   M. R. Begley, D. R. Mumm, A. G. Evans and J. W. Hutchinson, Analysis of a wedge impression test for measuring the interface toughness between films/coatings and ductile substrates, *Acta Materialia*, **48** (2000), 3211–3220.

[140]   A. Vasinonta and J. L. Beuth, Measurement of interfacial toughness in thermal barrier coating systems by indentation, *Engineering Fracture Mechanics*, **68** (2001), 843–860.

[141]   C. A. Johnson, J. A. Rund, R. Bruce and D. Wortman, Relationships between residual stress, microstructure and mechanical properties of electron beam-physical vapor deposition thermal barrier coatings, *Surface and Coatings Technology*, **108–109** (1998), 80–85.

[142]   J. W. Holmes and F. A. McLintock, The chemical and mechanical processes of thermal fatigue degradation of an aluminide coating, *Metallurgical Transactions*, **21A** (1990), 1209–1222.

[143]   R. C. Pennefather and D. H. Boone, Mechanical degradation of coating systems in high-temperature cyclic oxidation, *Surface and Coatings Technology*, **76** (1995), 47–52.

[144]   Y. H. Zhang, P. J. Withers, M. D. Fox and D. M. Knowles, Damage mechanisms of coated systems under thermomechanical fatigue, *Materials Science and Technology*, **15** (1999), 1031–1036.

[145] V. K. Tolpygo, D. R. Clarke and K. S. Murphy, Oxidation-induced failure of EB-PVD thermal barrier coatings, *Surface and Coatings Technology*, **146–147** (2001), 124–131.

[146] K. Vaidyanathan, M. Gell and E. Jordan, Mechanisms of spallation of electron beam physical vapour deposited thermal barrier coatings with and without platinum aluminide bond coat ridges, *Surface and Coatings Technology*, **133–134** (2000), 28–34.

[147] D. S. Rickerby, S. R. Bell and R. G. Wing, Method of applying a thermal barrier coating to a superalloy article and a thermal barrier coating. US Patent 5,667,663 (1997).

[148] Y. Zhang, J. A. Haynes, W. Y. Lee *et al.*, Effects of Pt incorporation on the isothermal behaviour of chemical vapour deposition aluminide coatings, *Metallurgical and Materials Transactions*, **32A** (2001), 1727–1741.

[149] M. Watanabe, D. R. Mumm, S. Chiras and A. G. Evans, Measurement of the residual stress in a Pt-aluminide bond coat, *Scripta Materialia*, **46** (2002), 67–70.

6 Summary and future trends

The previous chapters in this book confirm that, since the 1950s when the age of jet-powered civil aviation began, the superalloys have underpinned the improvements in performance of the modern gas turbine engine [1]. Today's designs are appreciably superior to the Ghost engine used for the first commercial flight of the Comet I; see Figure 6.1. The modern civil turbofans which power large two-engined aircraft such as the Boeing 777 at a take-off thrust approaching 100 000 lb – for example, Rolls-Royce's Trent 800 or General Electric's GE90 – have a fuel economy which has improved by a factor of about 2; the engine weight, when normalised against the thrust developed, is lower by about a factor of 4. Whilst a number of factors have contributed to this technological success, the development of the superalloys and insertion of components fabricated from them into the very hottest parts of the turbine have been absolutely critical to it. As the compositions of the alloys have evolved, the critical properties in creep and fatigue have improved markedly – and this has allowed the turbine entry temperature to be increased beyond 1700 K; see Figure 1.5. The development of new processes has played a key role too, particularly vacuum melting, directional solidification/investment casting and powder metallurgy techniques.

It is instructive to consider the technological, economic and societal pressures which have provided the incentive for these developments. In the post-war period to the late 1970s when the Cold War was at its height, most cutting-edge superalloy development was carried out in support of engine designs for military aircraft – notably in the USA, Europe and the former USSR. During this time, new superalloy and coatings technology was adopted first in fighter aircraft – with military organisations paying for it via government subsidies. With the demise of the Iron Curtain, the situation altered in a subtle way: the civil rather than the military sector set the pace, with new materials being inserted first into new civil aeroengines. In the 1990s, the need for cost reduction was apparent as the civil aviation sector became more competitive; the aeroengine business responded with emphasis being placed on 'faster and cheaper' derivative designs, on account of the variety of airframes becoming available and the speed at which these were being designed and marketed. At this time, increased use was made of computer-based design for alloy development and component optimisation via processing modelling. Around the turn of the century, substantial growth in the industrial gas turbine (IGT) business was observed – due to the 'dash-for-gas', notably in Europe and an increased thirst for electricity in the developing world, particularly in China and India. For example, in 1999 sales of IGTs increased from \$8.5 bn to \$13.0 bn, and the total share of the gas turbine market increased from 30% to 38% compared with the previous year;

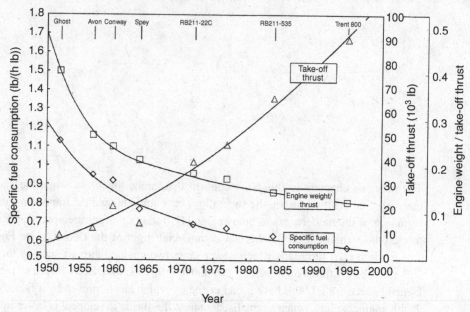

Fig. 6.1. Trends in the performance characteristics (specific fuel consumption, take-off thrust and normalised engine weight) for Rolls-Royce's large civil aeroengines.

by contrast, the aeroengine market share declined from 65% to 59%, despite an absolute increase in sales from $18.0 bn to $20.0 bn [2]. Although the downturn in the first few years of the new millennium and other unforeseen circumstances meant that some of this optimism was misplaced in the short term, most projections indicate that by 2025 the market for IGTs will outstrip that for aeroengines, which is itself projected to double by that time. In part this is due to an increased reliance on combined-cycle power plant, which requires IGTs to be used in conjunction with steam-powered turbines. For intermediate-sized plant in the range 25 to 75 MW, there is a growing market for aeroderivative IGTs – particularly for the oil and gas business in places such as Azerbaijan, Brazil, China, Kazakhstan and West Africa – which make use of materials technologies designed originally for aeroengine applications; it is anticipated that these will constitute about 10% of the total IGT market by 2025. For both aeroengines and IGTs, an increasingly important driver will be environmental regulations, introduced as the world's supply of oil and gas becomes tighter and concerns grow over the effects of CO_2 emissions [3].

So what of the future? Whatever the balance of power between the aeroengine and industrial gas turbine sectors, it seems certain that further developments in superalloy metallurgy and processing will be required over the coming decades. In the aeroengine sector, for example, it is projected that over the next 20 years a market exists for about 90 000 new civil engines, worth an estimated $550 bn; much of this requirement is driven by the rapid growth anticipated in Asia, the replacement of about 10 000 ageing aircraft, a booming business jet market and the trend for low-cost airlines. The required improvements in performance, for example, for the Trent 1000 and GEnx for the Boeing 787 Dreamliner, are ambitious; new grades of superalloy and supporting technologies are required to bring these new designs

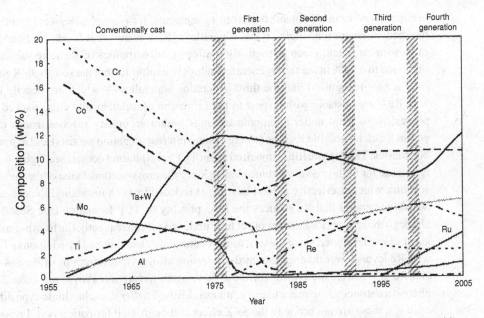

Fig. 6.2. Evolution of the compositions of the single-crystal superalloys for high-pressure turbine blading, to 2005.

to fruition. Similarly, as the operating conditions of land-based turbines become more demanding, superalloys will be used increasingly for hot section components such as turbine blading and buckets.

6.1 Trends in superalloys for turbine blade applications

Consider the evolution of single-crystal turbine blade compositions for the high-pressure turbine blading of the aeroengines. Figure 6.2 provides an approximate overview – note some subjectivity is inherent in it since, at any given time, different organisations will have been working on different alloys to exploit patent positions. The reasoning behind the changes can be rationalised given the findings of Chapter 3. Since creep resistance has been imparted by raising the γ' content of the alloys to 75%, the concentration of the γ-forming element, Cr, has dropped markedly; the Al content has been increased to achieve this, it being a strong γ' former. These modifications have, in turn, promoted changes in the oxidation behaviour – in the uncoated condition, most modern alloys favour Al_2O_3 rather than Cr_2O_3 scales; thus no great resistance to hot corrosion is displayed by them and appropriate coating technologies need to be adopted where necessary, for example, in marine environments. In many alloys, titanium is now present at very low levels since it exacerbates the rate of oxidation and has a negative effect on the retention of thermal barrier coatings (TBCs). No strong consensus has been reached on the role of Co, and consequently wide variations in its concentration are found. In the early first-generation single-crystal alloys, Ta and W additions were included to promote creep strengthening, but their use later declined as other more potent strengthening elements, for example, Re, were identified. In the second-generation single-crystal superalloys, Re was added at levels of about 3 wt%, in view of its potent creep strengthening effect. Concentrations of Re were subsequently increased to 6 wt% in the third-generation alloys to exploit further the so-called 'Re-effect'.

It is now recognised that the third-generation superalloys – which are heavily alloyed with Re – are unstable with respect to the formation of topologically close-packed (TCP) phases; these form under aluminide coatings within secondary reaction zones (SRZs), within Re-rich dendritic regions (where casting microsegregation persists) and at low angle boundaries. The rupture life is impaired when TCP precipitation occurs; see Figure 6.3. This is an issue for high-pressure turbine blading since the cross-sectional area of the thin-walled sections, which is effective at load-bearing, is reduced by such instability. Recent research has demonstrated that Ru reduces the susceptibility to TCP formation (see Section 3.2), although the reasons for this are not yet fully understood; consequently, the fourth-generation superalloys developed since the turn of the century are characterised by additions of Ru but with Re levels lower than for the third-generation alloys. Concentrations of Re at 4.5 wt% and Ru at 4.0 wt% are typical for these new fourth-generation alloys; see Table 3.5. An alternative strategy is to make use of platinum-diffused rather than aluminide-type diffusion coatings; these are not prone to the SRZ effect, although TCP formation is still possible in the bulk. At the time of writing, these fourth-generation single-crystal superalloys have yet to find application in gas turbine engines, although their potential benefit is acknowledged. The current reluctance to insert them into the latest designs is due to their cost and density; see Figure 6.4. A strong correlation exists between creep rupture life and cost, largely on account of the significant fractions of Re and Ru in the latest alloys; it can be seen that the very best creep resistance is displayed by alloys with raw material costs of about five times that of the first-generation alloys. A further difficulty is that the latest generation superalloys have densities approaching, or even exceeding, 9 g/cm^3 – partly because they are still alloyed heavily with Re, albeit at lower levels than for the third-generation superalloys. The need to limit the density is acute since the loading at the blade root scales linearly with it and because stresses imparted on the disc rim are then enhanced. Any fourth- or fifth-generation single-crystal superalloy that is to find application in the gas turbine engine will need to balance properly the very evident trade-off between creep performance, density and cost.

Traditionally, superalloys for turbine blade applications have been designed with creep, fatigue and oxidation resistance in mind. Whilst these are still very important, a new requirement of coating compatibility is now emerging, since the turbine entry temperature (TET) of the modern aeroengine is outpacing the temperature capability of the single-crystal superalloys, for which the melting temperature of nickel represents an absolute ceiling; see Figure 1.5. Hence, the use of thermal barrier coatings for protecting high-pressure blading is no longer merely advisable but absolutely critical, and consequently an increasing number of rows of aerofoils are employing cooling air; use of TBCs is also spreading to non-rotating parts such as nozzle guide vanes. This explains the great emphasis being placed on newer, lower-conductivity TBCs, ones which display resistance to sintering, and on the commercial aspects of TBC development including reduced manufacturing costs. Hence a new role for the superalloy blading relates to its requirement to support the TBC throughout service without spallation occurring; this, in turn, is focussing attention on the interfacial effects

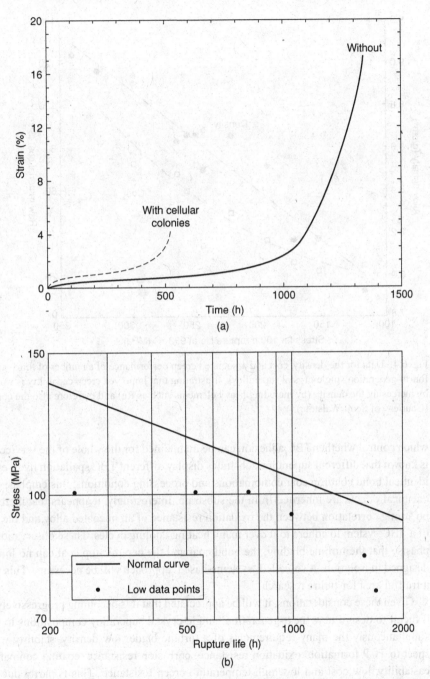

Fig. 6.3. Data confirming the debit in creep rupture life when third- and fourth-generation superalloys are tested with cellular colonies containing TCP phases. (a) Creep curves showing the detrimental effect of the presence of cellular colonies (1093 °C, 103 MPa); (b) creep rupture data confirming the reduced lifetime as a function of stress applied (1093 °C). (Courtesy of Scott Walston.)

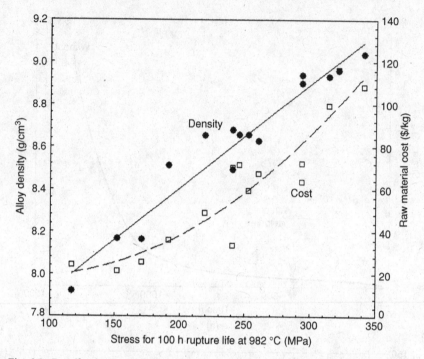

Fig. 6.4. Data for the density, cost and associated creep performance of a number of third- and fourth-generation single-crystal superalloys, illustrating that improved creep capability is won only by increasing the density (by including heavy elements such as Re) and therefore also the cost. (Courtesy of Scott Walston.)

which control whether TBC adhesion can be maintained for the whole of the service life. It is known that different superalloy substrates display different TBC spallation lives, even for identical bond coat/top coat combinations and processing conditions; this emphasises that chemical effects are inherited from the substrate. Interestingly, it appears that there exists no strong correlation between the oxidation resistance of an uncoated alloy and the ability of a TBC system to adhere to it over many heating/cooling cycles. These observations emphasise that the turbine blading, the bond coat and the ceramic top coat can no longer be designed in isolation. A so-called 'systems-based' approach will be necessary. This will be a fruitful area for future research.

Given these considerations, it will be appreciated that it is becoming progressively more difficult to design new 'next-generation' single-crystal superalloy compositions to satisfy simultaneously the many requirements of a turbine blade: low density, stability with respect to TCP formation, oxidation resistance, corrosion resistance, coating compatibility, castability, low cost and low/high temperature creep resistance. This is partly due to the 'extra' requirements of cost, density and coating compatibility, which have been become more apparent in recent years, but which traditionally were not emphasised strongly. Also of relevance is that any one alloying element is unlikely to improve all properties simultaneously, for example, Ru improves stability but increases cost, Re improves high-temperature strength but increases density, Cr improves corrosion resistance but degrades stability, etc.

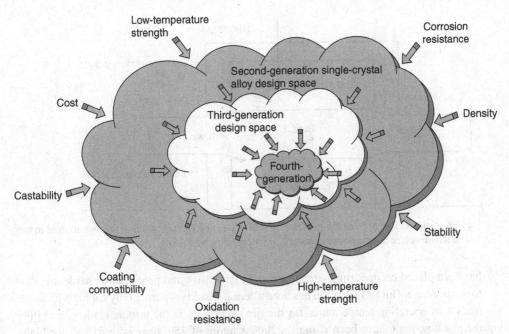

Fig. 6.5. Schematic illustration of the restricted size of the 'design space' available after the composition of each new generation of single-crystal superalloy has been chosen. (Courtesy of Scott Walston.)

Walston has illustrated this difficulty in a pictorial sense (see Figure 6.5), pointing out that the 'design space' available for next-generation single-crystal superalloys is becoming increasingly 'crowded'. From the turbine designers' perspective, this means that decisions need to be made concerning the relative importance of the requirements to be exhibited. It means also that it is unlikely that a 'one-size-fits-all' composition will be identified which is suitable for all turbine engines, so that there is scope for compositions 'tailored' for different applications. For example, the metallurgist designing a superalloy composition for an IGT is likely to place greater emphasis on corrosion resistance, cost and castability due to the large size of the blading and the operating environment [3]; retention of material properties, and thus microstructural stability, are required since there is a desire to extend component lives to 50 000 h to minimise life-cycle costs. Similarly, for aeroengine applications the oxidation resistance of the bare alloy is assuming less importance due to the increased use of diffusion-type aluminide bond coats on which thermal barrier coatings are placed; nevertheless, oxidation resistance of the superalloy substrate will remain critical when the TBC system cannot be afforded as is still the case for many turbines.

6.2 Trends in superalloys and processes for turbine disc applications

Traditionally, superalloy turbine discs were designed to resist general yielding and low-cycle-fatigue cracking at sites of stress concentration near the bore; hence, great emphasis

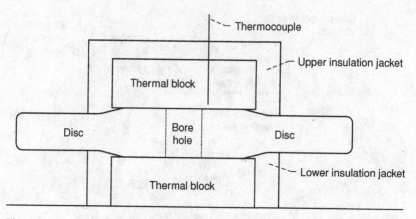

Fig. 6.6. Schematic representation of the disc, heat sinks and thermocouple arrangement used in the dual microstructure heat treatment process, as patented by NASA [4].

has been placed on imparting strengthening by grain size and precipitation hardening due to the γ' phase. But this situation has been altering steadily – particularly for high-pressure discs – as operating temperatures for the rim sections near the turbine blading, and thus the gas flow path, have been rising; in 2000 a figure of 750 °C or even 815 °C for high-performance military applications was possible for some advanced aeroengine designs. Consequently, the need to impart resistance to high-temperature 'dwell' crack propagation and creep is assuming greater importance. A corollary is that the requirements placed on the various processing steps required for component fabrication are becoming more stringent. Thirty years ago, the sole responsibility of the turbine disc supplier was to shape a superalloy billet – using a series of open/closed-die forming and machining operations – into a form suitable for heat treatment and ultrasonic inspection. Slowly, the requirements expanded to include demonstration of minimum property levels in both bore and rim regions. Nowadays, original equipment manufacturers (OEMs) are insisting on greater microstructural and property differentiation *within* any given component. Methods for this are now available; for example, the dual microstructure heat treatment technology (DMHT) technology designed and patented by NASA [4], in which both sub- and super-solvus heat treatments are carried out on bore and rim, respectively, *on a single component*. This is achieved by enhancing the thermal gradients present during heat treatment using heat sinks (typically solid metal cylinders) which chill the central portion of the disc as it resides in the furnace above the γ'-solvus; see Figure 6.6. This concept has been demonstrated for the state-of-the-art ME3 turbine disc superalloy (see Figure 6.7), with tensile and creep data which are superior in bore and rim regions, respectively, with experiments on a generic disc shape demonstrating the advantages and reliability of the dual grain structure at the component level [4]. Other propriety (and subtly different) methods are available to fabricate such 'functionally graded' discs, for example, General Electric's dual heat treatment process, which was discussed in Chapter 4. At the time of writing, dual microstructure discs have yet to be employed in the gas turbine engine, but this situation is likely to change in the near future.

A consideration of powder metallurgy (P/M) superalloys – first introduced for high-pressure turbine discs in military engines, but now being employed increasingly in civil

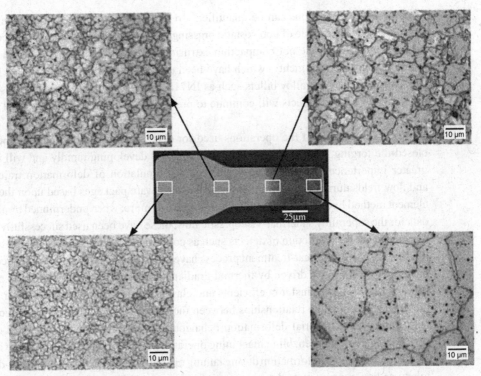

Fig. 6.7. Distribution of grain sizes in a ME3 turbine disc subjected to the dual microstructure heat treatment process [4].

aeroengines as stronger and more heavily alloyed variants are required – sheds further insight into the trends occurring. First-generation P/M superalloys (such as Rene 95) were designed with volume fractions of the gamma-prime (γ') phase which were high at around ~55% (Figure 4.17) and fine grain size of ~10 μm, to achieve maximum tensile strength. Consequently, these were usually sub-solvus heat-treated. Second-generation P/M superalloys (such as Rene 88DT) were developed with lower γ' volume fractions and were heat-treated to larger grain sizes (~50 μm) to improve resistance to fatigue crack propagation; this was to ensure adequate life consistent with the damage-tolerance lifing methods. The third-generation alloys which are now being designed (such as ME3 and LSHR) have high γ' volume fractions as for the first-generation ones; where the DMHT process is not employed, grain sizes of ~50 μm are being sought so that dwell-crack growth and creep rates are minimised. The need for creep resistance in the rim regions has ramifications for the design of new powder disc compositions. Carbon and boron levels are being increased to improve creep behaviour and to promote precipitation of carbides and borides which provide pinning sites – to ensure that γ-grain growth does not occur beyond the solvus temperature in an abnormal fashion. As the use of P/M superalloys becomes more commonplace, the turbine industry will become increasingly dependent upon probabilistic methods of lifing. Consequently, better screening methods will be required for the rapid assessment of the quality of superalloy powder prior to its consolidation, so that distributions of non-metallic inclusions can be quantified. To reduce the costs of powder material, increasing use will be made of hot isostatic pressing/conventional press conversion (cogging) practices rather than the hot compaction/extrusion routes – requiring specialised tooling and high tonnage equipment – which have been used traditionally for P/M billets [5]. For cast-and-wrought superalloy billets such as IN718 and Waspaloy, the possible presence of random melt-related defects will continue to prevent these being used to their full fatigue potential.

Process modelling of the operations used for disc production – particularly open- and closed-die forging, heat treatment and machining – is developing rapidly and will assume greater importance in the coming years. For the simulation of deformation trajectories and flow fields during forging operations, robust software packages based upon the finite-element method have been available for two decades or more; when underpinned by accurate data for the superalloy's thermal-viscoplastic flow, these have been used successfully for die design and for the prediction of defects such as cold shuts. More recently, thermal–elastic–plastic analyses of the heat-treatment process have been used increasingly for the prediction of residual stress fields driven by thermal gradients inherited from the quenching process [6]. Datasets of heat transfer coefficients and elastic/plastic constitutive behaviour are required. These allow the relationships between the heat transfer coefficient, part geometry, temperature field, material deformation behaviour and residual stress field to be quantified. Such modelling is enabling machining operations to be optimised to prevent excessive metal movement and deformation during lathing operations, which, if exceeding pre-defined tolerance levels, necessitate the scrapping of components. State-of-the-art process models make predictions of microstructure evolution, and this topic is currently an active area for research and development. There is evidence that the use of these process modelling tools is becoming increasingly necessary for quality assurance purposes. However, a further advantage is that the physical insights gained from it are stimulating the re-engineering of these processes such that component quality can be enhanced. Consider the advanced heat-treatment methods which are now in active development [7]. Traditionally, turbine discs have been quenched in oil so that the heat transfer coefficient around the periphery of the part is then uniform; consequently, thermal gradients are set up which cause the formation of a residual stress field. Whilst this cannot be entirely eliminated with the re-engineered process, its magnitude can be reduced by cooling using computer-controlled high-velocity gas jets directed at the disc, such that the heat transfer coefficient varies around its periphery in an optimal way; see Figure 6.8(a). Typically, this involves enhancing the heat transport coefficient at the bore region of the part on account of its greater thermal mass; the heat transfer coefficients required for this are deduced from the modelling. The advantage is that heat is then removed more uniformily than for traditional oil quenching, so that the thermal gradients responsible for the residual stress field are reduced; (see Figures 6.8(b) and (c). A typical arrangement for such optimised heat treatment has as many as 12 different gas-flow zones which can be controlled as a function of quenching time and part temperature [7]. Such technology will be used increasingly, probably in conjunction with dual microstructure heat-treatment furnaces in the next-generation state-of-the-art heat-treatment facilities.

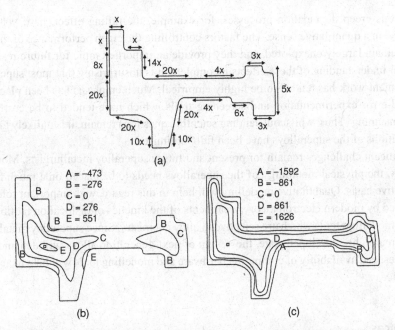

Fig. 6.8. Re-engineered process for the heat treatment of a turbine disc forging. (a) Variation of
initial heat transfer coefficients optimised via process modelling. (b) Corresponding computed
residual stress field and (c) computed residual stress field for the conventional oil-quench method
[7]. Stress contours in (b) and (c) are given in units of megapascals.

6.3   Concluding remarks

Since the middle of the twentieth century, scientists and engineers have worked to develop
the nickel-based superalloys and the processing methods required to put them to use in the
gas turbine engine. Considerable progress has been made; this has been underpinned by a
growing appreciation of the physical and chemical factors which govern their performance.
As described in this book, the design of the alloys, the process development and the component engineering have not occurred independently – progress on these topics has tended to
occur synergistically. At this stage of their development, the superalloys should rightly be
regarded as one of the most complex of all the materials engineered by man. They represent
a remarkable success story.

Nevertheless, there are reasons to believe that this class of material is not yet fully mature.
Critical questions remain about the metallurgical factors which govern performance – for
example, the creep strengthening provided by rhenium, the effect of ruthenium in imparting
alloy stability and the role of reactive elements such as yttrium in improving oxidation
performance. No well-developed theories exist for these effects. Furthermore, an adequate
explanation has yet to be offered for the chemical effects which determine compatibility
of the superalloys with thermal barrier coating systems. Bond coat chemistries have yet
to be optimised. For the single-crystal superalloys, the effects of alloying additions on

the various creep degradation processes, for example, the rafting effect, have yet to be described in a quantitative sense. The factors controlling the creep performance of the disc alloys remain largely unexplored, and they provide an important topic for future research. With our understanding of these effects incomplete, it is unsurprising that most superalloy development work has tended to be highly empirical. Much emphasis has been placed on trial-and-error experimentation and screening trials, which have tended to be costly and time-consuming. Thus, whilst outstanding scientific questions remain, it is unlikely that the compositions of the superalloys have been fully optimised.

Significant challenges remain for present and future superalloy metallurgists. Most importantly, the physical metallurgy of the superalloys needs to be placed on a much firmer quantitative basis. Quantitative modelling will help in this regard, with important roles being played by modern electron theory, treatments of the kinetics of dislocation motion and process modelling tools which are sensitive to alloy chemistry. Since superalloy metallurgy is now in a highly developed state, the design of next-generation alloys will demand this. The widespead availability of computing hardware and modelling tools will help to achieve this goal.

References

[1]   R. Schafrik and R. Sprague, The saga of gas turbine materials, Parts I, II, III *Advanced Materials and Processes*, **162**, 3, 4, 5 (2004), pp. 33–36, 27–30, 29–33.

[2]   B. B. Seth, Superalloys: the utility gas turbine perspective, in T. M. Pollock, R. D. Kissinger, R. R. Bowman, *et al.*, eds, *Superalloys 2000* (Warrendale, PA: The Minerals, Metals and Materials Society (TMS), 2000), pp. 3–16.

[3]   F. Carchedi, Industrial gas turbines: challenges for the future, in A. Strang *et al.*, eds, *Parsons 2003: Engineering Issues in Turbine Machinery, Power Plant and Renewables* (London: Maney, for the Institute of Materials, Minerals and Mining, 2003), pp. 3–23.

[4]   J. Gayda and T. Gabb, The effect of dual microstructure heat treatment on an advanced nickel-base disk alloy', in K. A. Green, T. M. Pollock, H. Harada *et al.*, eds, *Superalloys 2004* (Warrendale, PA: The Minerals, Metals and Materials Society (TMS), 2004), pp. 323–329.

[5]   A. Banik, K. A. Green, M. C. Hardy, D. P. Mourer and T. Reay, Low cost powder metal turbine components, in K. A. Green, T. M. Pollock, H. Harada *et al.*, eds, *Superalloys 2004* (Warrendale, PA: The Minerals, Metals and Materials Society (TMS), 2004), pp. 571–576.

[6]   D. U. Furrer, G. Groppi and G. Bunge, Forging the future: parts I and II, *Advanced Materials and Processes*, **163**, 5, 6 (2005), pp. 35–37, 43–45.

[7]   D. U. Furrer, R. Shankar and C. White, Optimising the heat treatment of Ni-based superalloy turbine discs, *Journal of Metals*, **55** (2003), 32–34.

Index